Lecture Not
Mathematics

Edited by A. Dold and B. Eckmann

1397

P. R. Turner (Ed.)

Numerical Analysis and Parallel Processing

Lectures given at The
Lancaster Numerical Analysis Summer School 1987

Springer-Verlag
Berlin Heidelberg New York London Paris Tokyo Hong Kong

Editor

Peter R. Turner
Mathematics Department, US Naval Academy
Annapolis, MD 21402, USA

Mathematics Subject Classification (1980): 65FXX, 65NXX, 65-06

ISBN 3-540-51645-X Springer-Verlag Berlin Heidelberg New York
ISBN 0-387-51645-X Springer-Verlag New York Berlin Heidelberg

Printed in Germany

Printing and binding: Druckhaus Beltz, Hemsbach/Bergstr.
2146/3140-543210 – Printed on acid-free paper

PREFACE

THE SERC NUMERICAL ANALYSIS SUMMER SCHOOL

University of Lancaster

12 – 31 July, 1987

The essential aims of this third Numerical Analysis Summer School were much the same as those of the first two of these meetings in 1981 and 1984. The proceedings of the earlier events are published as LNM 965 and 1129. Each week of the meeting was largely self-contained although there was an underlying theme of the effect of parallel processing on numerical analysis running through the whole three week period. During each week there was opportunity for intensive study which could broaden participants' research interests or deepen their understanding of topics of which they already had some knowledge. There was also the opportunity for continuing individual research in the stimulating environment created by the presence of several experts of international renown.

This volume contains lecture notes for most of the major courses of lectures presented at the meeting. During each of the first and third weeks there was a ten lecture course and a related shorter course while the second week contained three related courses of various lengths.

Presented here then is an account of a highly successful meeting in what is becoming a regular series of SERC Numerical Analysis Summer Schools.

Acknowledgements

The most important acknowledgement is to the Science and Engineering Research Council who once again provided very generous sponsorship for the meeting. Their contribution covered all the organisational and running costs of the meeting as well as the full expenses of the speakers and the accommodation and subsistence expenses of up to twenty participants each week. A contribution towards the expenses of the meeting was also received from the British Gas Corporation.

I also wish to thank the SERC Mathematics Secretariate for their help in the planning of the meeting and especially Professor John Whiteman who acted as their assessor and as my advisor throughout. I should also like to acknowledge the help and encouragement of many of my colleagues in the University of Lancaster and especially Mrs Marion Garner who handled nearly all the secretarial work involved both in the planning and running of the meeting. As usual her help was invaluable.

Peter R Turner

Department of Mathematics,
University of Lancaster,
Lancaster LA1 4YL
England.

Mathematics Department,
United States Naval Academy,
Annapolis,
MD 21402, USA.

CONTENTS

PARALLEL COMPUTATION AND OPTIMISATION
A series of five lectures (July 1987)

Laurence C. W. Dixon
Numerical Optimisation Centre
The Hatfield Polytechnic

1. Introduction

The material presented in this paper was prepared as a series of five lectures to be presented at the SERC Summer School in Numerical Analysis held at the University of Lancaster in July 1987.

The intention in the series of lectures was to indicate the variety of parallel processing architectures then available and to discuss how the design of software for solving optimisation problems is affected by the various architectures.

The first lecture therefore reviewed briefly the architectures then available, particular attention being paid to those on which implementations of optimisation algorithms had been attempted.

The second and third lectures discussed the design of optimisation codes on parallel processing systems. One result of this discussion is the conclusion that if optimisation codes are to be efficiently implemented on parallel systems, then sets of linear equations must be efficiently implemented as part of the code.

The fourth lecture therefore discussed the solution of sets of linear equations on parallel processing systems. This mainly concentrated on the solution of dense sets of equations as the concurrent lectures by Alan George on the solution of sparse systems formed the main topic of that week.

The fifth lecture described the results we have achieved solving structured optimisation problems on the DAP. The lecture series concluded with a brief description of some of our experiences solving sparse sets of equations. This last topic is however omitted from this paper.

In presenting the series of lectures in this paper I wish to acknowledge the help and assistance of many colleagues at the Numerical Optimisation Centre, of our research students and of their sponsors, without their support the research described could not have taken place.

2. Parallel Computers and Parallel Computing

The last few years have seen the introduction of many different designs of parallel computers. This process has been encouraged by three similar but distinct considerations. The first need that has led to the development of parallel computers was the desire to be able to solve problems that require more data storage and/or take too long to solve on a single mainframe conventional sequential computer. The type of problem that still exceeds the capabilities of such computers is the time varying three dimensional solution of flow problems. To solve such problems we must either develop a faster sequential machine or develop a parallel machine. The choice is usually determined by the relative cost and until recently sequential machines have become faster at a quicker rate than parallel machines have been developed so that at any time the largest fastest machines were sequential rather than parallel. This situation has changed for the first time recently and since (1985) parallel machines have been faster.

The second need that has led to the development of a very different type of parallel computer is the requirement to be able to solve more and more sophisticated on-line problems in real time. Here the availability of cheap microprocessors has led designers to wish to use more sophisticated algorithms in the control of systems thus improving their performance. The need to use more sophisticated algorithms has led to the need for including more computational power in on-line systems. This need can again be met by the use of a more powerful sequential chip or alternatively by the use of a number of less powerful chips in parallel. The rate of increase in power of cheap chips has to date been so great that the easier option of using a more powerful sequential chip has usually been chosen. It is, however, true that many on-line problems involve the continual computation of a number of separate but interactive tasks, that are artificially inter- weaved on a sequential chip by time slot sharing. Such problems, containing truly parallel tasks, are the third driving force behind the need for parallel computing.

These three problems are very different and obviously need different parallel hardware, however, the principles that are used to categorise the different classes of parallel computer are not based on size and cost but rather on the different levels and types of parallelism. Strictly each type of parallelism could be used in the design of hardware of all three sizes of computers though these have not all been implemented. In the next section the basic principles of each way of introducing parallelism will be described.

2.2 Parallel Computers

2.2.1 General Comments

One of the earliest classifications of parallel computers was given by Flynn (1972), who divided parallel systems in Single Instruction Multiple Data (SIMD) machines and Multiple Instruction Multiple Data (MIMD) machines.

SIMD machines use the principle that it is relatively easier to instruct a large number of units to do the same thing at the same time than it is to instruct the same number of units to carry out different individual tasks at the same time. Whilst it is easy to instruct thousands of units to simultaneously do the same task either in the army or the factory it is normally necessary in either to have a tree structure of command where one manager only instructs between five to fifteen subordinates himself if they are expected to act individually. Similarly, SIMD machines with over 1000 processors already exist but more versatile MIMD machines usually contain less than 10 processors.

As well as distinguishing between these two categories it is also necessary to distinguish carefully the level of tasks that are undertaken in parallel and the granularity (computational length) of those tasks that are being undertaken in parallel.

Other important considerations are the need to transfer data to and from the parallel processors. This can lead to conflict in accessing memory that can degrade performance. A similar problem is the frequent need to synchronise the arithmetic being performed on the processors to ensure that if processor A needs to use a new value of data being calculated by processor B then processor A does not do that calculation before processor B completes its task. This is known as the synchronisation problem and it can lead to very inefficient performance if it is not carefully considered in the design of codes. The synchronisation problem can usually be avoided on SIMD machines but is so important on MIMD machines that the possibility of using asynchronous algorithms has seriously been proposed by Baudet (1978).

2.2.2 Pipeline Machines

One of the most successful ways of speeding up arithmetic by introducing parallelism is known as the pipeline. Pipelines are designed to speed up the calculation of vector loops typified by

$$c_i = a_i + b_i \qquad i = 1 \ldots n$$

or $c_i = a_i * b_i \qquad i = 1 \ldots n.$

Let us consider the summation of two numbers

a = 1, 123, 765

b = 2, 410, 123.

We start by adding the units, then the tens, then the hundreds, etc. so when performing in decimal arithmetic on this example 7 virtually independent additions are undertaken that are only linked by the need to be able to carry forward the overflow from one operation to the next. This implies that one cannot be started until the previous one is complete. So on a decimal machine if the time for one such operation was t, then the time for performing $a_i + b_i$ would be 7t and on a sequential machine the time for the loop would approximate 7nt. Now suppose we were to put P = 7 such processors in a line and let the first processor handle the units, the second processor the tens, the third the hundreds, etc. Then the first processor can start on the second sum while the second processor is still working on the first so that all the loop will be completed in (n + 6)t units. The speed up is therefore (n+6)/7n, notice that as n gets large this is roughly 1/7 i.e. proportional to 1/P but that there is a start up time that can make the gain negligible for small values of n.

Obviously the operation of any pipeline machine is more sophisticated than this, to increase efficiency pipelines can be chained allowing the arithmetic in a second loop to be started before the arithmetic in the first is complete. Another complication but this time one that normally has an adverse effect is that due to memory access and data structure considerations there is often an upper limit N on the length of loop n that can be pipelined.

If we let I(n/N) be the first integer greater than n/N where n is now the number of data in the desired loop then this degrades the time for our example to approximately I(n/N)(N+6)t units.

The three most commonly used parallel computers the CRAY 1, CYBER 205 and the Floating Point Systems APB series all utilise the pipeline principle.

In these machines the level of the parallelism is very low being within the basic arithmetic operations. This means that sequential high level codes could in principle be directly transferred to pipeline machines, and indeed this is frequently done, but for full benefit to be obtained sequential codes frequently have to be re-organised to introduce more N loops that can be pipelined and chained.

This principle is undoubtedly the most widely used parallel method for solving large problems. While it could, in principle, also be used for real time computing, no machine based on it smaller than the Floating Point system series is known to the author.

2.2.3 Arrays of Processors

The second most common type of SIMD system consists of arrays of processors; typically a large number of identical processors are arranged in a 2 dimensional grid with processors placed at the nodes of equally spaced orthogonal lines. The processors are then connected by fast data links along these orthogonal lines and also usually by fast links connecting the ends of these lines. If we introduce axes parallel to these lines it is natural to refer to a processor at position $x_1 = i$, $x_2 = j$ as processor $P_{i,j}$ then if we wish to do typical two dimensional matrix algebra

$$C(i,j) = A(i,j) + B(i,j)$$

for all values of i,j, then all these operations can be performed in parallel by assigning the calculation of C(i,,j) to P(i,j). The level of parallelism is therefore higher than in the pipeline as it is arithmetic operations that are being performed in parallel, rather than each operation being split into parallel subtasks. For efficiency it is desirable to have a long sequence of such matrix operations before having to transfer data to the host to undertake scalar arithmetic.

Typically if we are considering a square matrix i = 1 ... n j = 1 ... n where $n^2 > P$ the number of processors then the time for performing such a sequence varies as shown in Figure 1.

Figure 1

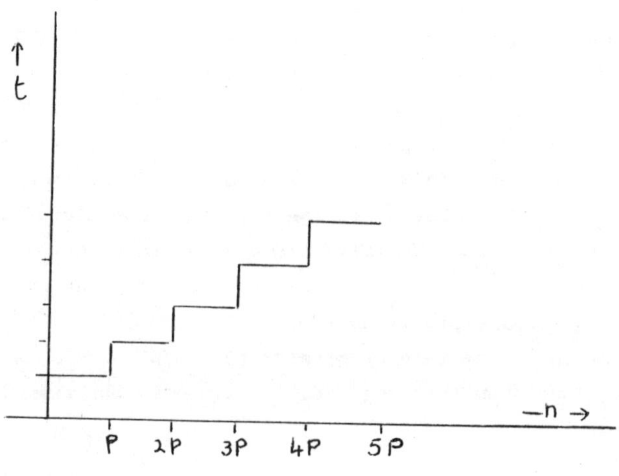

The two most well-known examples of systems based on arrays of processors are the ILLIAC IV and the ICL-DAP

2.2.4 Data Flow

When considering SIMD machines we considered one example of a small grain and one of an intermediate grain system. The data flow principle is again based on dividing arithmetic into parallel tasks at the basic arithmetic operation. A data flow machine will contain a number of processors and a free processor commences the next arithmetic operation for which all the data is available. Consider the statement

$$y = 2x^3 + 3x^2 + 4x + 5.$$

On a sequential machine this involves 5 multiplications and 3 additions i.e. 8 sequential steps, however, if we allow parallel operations and 3 processors we can construct an operation graph in which there are only 4 sequential steps. This would therefore only take half the time and of course if we had defined other tasks then some processors could already have started them.

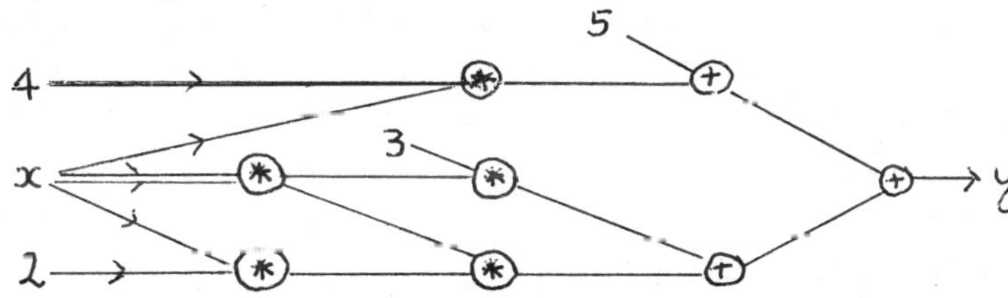

Figure 2

There is as yet no machine available commercially based on the dataflow principle, but considerable research into such machines is currently being undertaken. Such machines should be very powerful. I have seen such graphs describing the solution of a set of sparse 10 x 10 equations in under 20 sequential steps.

2.2.5 Parallel Networks

The most common way of achieving parallel MIMD computing is to connect separate processors in a network and allow them to co-operate on a task.

The pioneer work in this area took place at Carnegie-Mellon with the design of the CM* and Ccmp machines (7). This was followed by a large number of small systems containing 4 or 5 processors closely linked together. A typical example is the NEPTUNE system built by David Evans' team at Loughborough University, Barlow (2). This consisted of 4 Texas Instruments microprocessors. Experience showed that with such a system of 4 processors it was relatively easy to code least squares optimisation algorithms and global optimisation algorithms that ran more than 3 times faster than on a single such processor, Patel (1983). Simulation studies showed, however, that with this particular structure, data transfer and data access conflicts increased rapidly with P, the number of processors, and that there would be virtually no benefit in combining more than 10 processors in that mode, McKeown (1980).

In 1985 the most powerful MIMD machine in use was the CRAY XMP which links 2 CRAY 1's together. Cray have recently announced the more powerful CRAY 2 which will contain at least 4 CRAY 1's linked. These numbers again emphasise the small value of P envisaged in most MIMD systems.

Other geometries are being investigated, for instance, at Madison University a ring of Vax's is being built which should be a very powerful combination, whilst IBM and Floating Point Systems have just (1985) announced the setting up of 4 Parallel Computing Research Centres equipped with an IBM mainframe with 10 Floating Point Systems pipeline machines connected to it. This system therefore combines the MIMD mode with the pipeline principle at the lower level. It should be a very powerful research instrument.

The pace at which parallel computers are being developed can be judged by contrasting the above description which was reasonably accurate in 1985 with the situation at the UNICOM seminar [3] held in London in December 1986.

At that Seminar the producers of small parallel computers were allowed to display their current machines and twelve elected so to do. These machines ranged in size and cost from relatively cheap small transputer systems through more sophisticated MEIKO transputer systems which are simple MIMD systems with between 4 and 40 transputers; to the new ADM version of the DAP (1032 SIMD faster processors) and the rival Intel hypercube with its different network structure. Also on display were the Sequent parallel computer that contains between 2 and 30 National Semi conductor 32032 microprocessors and is capable of running both ADA concurrent tasking packages and by simple code modification of running loops in Fortran in parallel. The Alliant machine containing more powerful processors was also on display. It is very difficult to keep up with hardware developments in this area and the software that is available to run on them. In consequence any brief description of this type now soon is out of date.

In this lecture series I will not be considering pipeline machines or algorithms; but will discuss our experience on MIMD and SIMD systems. In 1984 Dew contrasted the different implications of pipeline and other parallel systems in the following words "The need to perform ever more complex scientific computations is clearly illustrated by the success of the new generation of vector processors like CRAY. The speed-up achieved by these has in the main been brought about by architectural improvements (e.g. pipelines) and the development of compilers to map sequential programs onto vector architecture. Although new algorithms are being developed to exploit pipelines, in general their introduction has had <u>little effect</u> on underlying numerical algorithms. This is not the case for <u>arrays of processors</u> like the DAP, where new approaches are required. Then, the problem, architecture and algorithm must be more closely related."

It is this challenge to design parallel software that mimics the parallelism in both the problem and the hardware that is the exciting aspect of parallel processing.

<u>References to Section 2</u>

1. Barlow, R H et al, A guide to using the Neptune processing system, Loughborough University of Technology.

2. Baudet, G, The design and analysis of algorithms for asynchronous multi-processors, PhD Dissertation, Carnegie Mellon University, 1978.

3. Proceedings of Major Advances in Parallel Processing, Unicom Seminar, 9-11 December 1986.

4. Flynn, M, IEEE Transactions on Computers, Vol C21, No 9 Sept 1972, pp 948-960.

5. McKeown, J J, (1980), Simulation of a parallel global optimisation algorithm. NOC TR 109, March 1980.

6. Patel, K D, (1983), Implementation of a parallel (SIMD) modified Newton Algorithm on the ICL DAP, NOC TR 131, 1983.

7. Wulf, W A & Bill, C G, C.MAP - a multi mini processor, AFIPS proc 1972, FJCC Vol 41, AFIPS press, pp 765-727.

3. Solving Optimisation Problems on Parallel Processing Systems

In this lecture we will be concerned with the optimisation problem

$$\text{Min} \quad F(x) \qquad x \in R^n.$$

Occasionally we will assume simple upper and lower bounds of the form $l_i \leq x_i \leq u_i$ exist.

There is of course a complete theory for the convergence of iterative algorithms which generate a sequence of estimates

$$x^{(k+1)} = x^{(k)} + \alpha p^{(k)}$$

and many efficient codes exist for the solution of such problems on a sequential machine. These have successfully solved many optimisation problems.

The question therefore arises as to why then we should be interested in introducing the parallel processing concept into numerical optimisation.

The main reasons that influenced us were:

(1) that we knew of industrial problems that took an embarassingly long time on a sequential machine and we knew too that

(2) industry only poses problems that it thinks might be soluble.

By introducing the parallel processing concept into numerical optimisation we hoped to be able to extend the range of soluble problems. We identified four different situations where we felt that the solution of optimisation problems would most benefit from the availability of parallel processing machines. These were:-

1) Small Dimensional Expensive Problems

These are typified by industrial problems which frequently have a small dimension $n < 100$ but where the time required to compute the function and gradient values at $x^{(k)}$ can be considerable and where this dominates the computation within the algorithm.

2) Large Dimensional Problems

There are many large dimensional problems $n > 2000$ where the combined processing time and storage requirements cause difficulties.

3) On-Line Optimisation

There are many on line optimisation problems, for instance the optimisation of car fuel consumption which cannot be easily solved using existing sequential optimisation codes on the type of processors that could be easily installed within a car but which might be solved on more than one such processor.

4) Multi Extremal (Global) Optimisation

Problems in which the objective function has many local minima and where the real problem is to identify the best of these, still present many difficulties because the available sequential codes are weak and expensive in computer time.

In all four of these areas the availability of parallel processors promised significant improvements and in each that promise has been achieved. In this lecture we will mainly be concerned with the first class of problem, while the second class will be discussed mainly in the final lecture. The other two will not be discussed in any depth in this series.

3.2 Optimisation Problems and Algorithms

It is usual to find on analysing the solution of most industrial optimisation problems that at least 95% of the computer time is spent in evaluating the values of the objective function F(x) at $x^{(k)}$ and only 5% of the time within the optimisation code.

It is therefore natural to concentrate on the speed up of the calculation of the engineering model F(x) rather than the code.

This can effectively be done in three distinct ways.

Approach A

The calculation of each objective function value F(x) is divided into P parallel tasks. This approach leaves the responsibility for the efficient use of parallelism in the hands of the user which is undesirable.

Approach B

The algorithm is modified so that it can accept P values of F(x) computed simultaneously. This places the responsibility for the use of parallelism on the algorithm designer.

Let us first consider two examples of Approach A.

Approach A Example 1 A H O Brown (1976) [1]

Suppose we wish to optimise the design of an aero engine so that the fuel used to cross the Atlantic is a minimum. Then each design $x^{(k)}$ implies a calculation of the performance of the engine in a number of states, e.g. take off, climb, subsonic cruise, transonic flight, supersonic flight, de-acceleration, descent, hold over final airport, final descent and landing, a total of eleven effectively separate tasks. These could be performed in parallel but unless considerable care is taken to divide the computation into roughly P equal tasks many of the processors will be idle for most of the time and hence the processor use will be inefficient. Such a calculation also requires an MIMD machine as each subprocess is different.

Approach A Example 2

A simpler example would be the least squares minimisation problem

$$\text{Min } F = \sum_{i=1}^{mP} s_i^2(x)$$

where $s_i(x)$ is the difference between an experimental data value y_i and the model value at say t_i

i.e. $s_i(x) = y_i - m(x, t_i)$.

Now if we divide the mP data points in P equal sets each parallel processor does the same SIMD task. This has been implemented on both the MIMD Neptune system [McKeown (2)] and on the ICL-DAP with the expected speed up. [Sargon, Chong & Smith (3)].

Before considering approach B let us now consider the expected speed up.

3.3 Performance Evaluation

Measuring the performance of a parallel system must be dependent on the type of parallel system used.

The Neptune system at Loughborough consisted of P = 4 small microprocessors. The only logical comparison is with itself i.e. use of P processors or 1 processor.

In contrast the DAP consisted of 4096 processors and purported to compete with a mainframe. We therefore compared it with the DEC 1091 which was The Hatfield Polytechnic mainframe at that time.

3.3.1 MIMD systems

Two concepts are usually used for measuring the performance of a parallel system, they are "speed up" and efficiency.

Let $\tau(p)$ be the processing time using P identical processors. Then the "speed up S" is defined as

$$S = \frac{\tau(1)}{\tau(P)}$$

and the efficiency

$$E = \frac{S}{P} \leq 1$$

Ideally we might expect the speed up ratio S to be P and hence the efficiency E = 1. In general, however, some degradation must be expected. The main factors that contribute to this degradation are:

a) at the system level

(1) the actual processing speeds of the processors differ

(2) input/output interrupts

(3) memory contention

(4) data transfer time between processors

b) at algorithmic level

(1) synchronisation losses, if P tasks are to be performed all usually wait for the slowest

(2) critical section losses, if the code requires all processors to access, say, global memory at once.

3.3.2 SIMD system

For the ICL-DAP we measured the speed up by

$$S = \frac{\text{processing time on Dec 1091}}{\text{processing time on ICL-DAP}}$$

for a number of standard test functions Patel (1983) reported S = 20. This was on a simple problem where 4096 function evaluations were performed on both machines. This emphasises the slow nature of the individual processors on the DAP which individually must be approximately 200 times slower than the Dec 1091. Later tests gave a value of S = 60 indicating that such comparisons cannot be expected to remain constant with time.

These ratios must be borne in mind in considering the later comparisons.

3.4 Optimisation Algorithms Approach B

For unconstrained optimisation problems it was generally accepted c.1983 that there were 4 broad categories of algorithms

(1) Modified Newton Methods $2 \leq n \leq 5$

(2) Variable Metric Methods $5 \leq n \leq 120$

(3) Conjugate Gradient Methods $n \geq 60$

(4) Conjugate Direction Methods $n \leq 30$

where the choice of algorithm was mainly determined by the dimension of the problem n.

The question that naturally arises is that if we have P processors how would the choice be effected.

3.4.1 Newton Raphson code

The modified Newton Method we chose to investigate was due to Mifflin [5].

Mifflin's Method

In Mifflin's Method at iteration k we have

Step 1 Calculate $F(x^{(k)})$ at $\frac{1}{2}n(n-1)$ places to estimate $g = \nabla F$ and $G = \nabla^2 F$ by differences. If $P \geq \frac{1}{2}n(n-1)$ these can be done in parallel.

Step 2 Estimate

$$g_i = (F(x + h_{ai}) - F(x - h_{ai}))/2h$$

$$G_{ij} = (F(x + h_{ai} + h_{aj}) - F(x + h_{ai}) + F(x + h_{aj}) - F(x))/h^2.$$

Again these are parallel computations and if $n \leq 64$ the matrix G_{ij} can be stored in a single field on the DAP.

Step 3 If $\max |g_i| \leq \varepsilon$ stop.

Step 4 Solve $Gd = -g$.

[On the DAP; DAP library routine FØ4GJNLE64 was used].

Step 5 Find α so $F(x^{k+1}) < F(x^k)$

where $x^{k+1} = x^k + \alpha d$.

The line search is essentially sequential. To evaluate 4096 points along the line would be unhelpful, so a 4D search was introduced

$$x^{k+1} = x^k + \sum_{i=1}^{4} \alpha_i p_i$$

Step 6 Return to step 1.

This code was implemented on the ICL-DAP by Patel [4] who reported the following comparison.

	N-R	V-M	C-G	DAP
Quadratic	15.00	1.8	1.2	1.2
	16.60	2.1	1.1	1.2
Extended Rosenbrock	145	103	8.4	4.9
	141	80	11.4	28.2
Extended Powell	112	53	5.9	2.5
	134	41	11.1	1.9
Trigonometric	F	66	37.6	13.9
	2604	286	78.7	8.6
Extended Box	4181	1161	137.8	107.9
	7195	3194	354.0	323.9

A comment on these results seems necessary. The problems were all in 64 dimensions and were all extended versions of standard problems. In 64 dimensions we would expect V-M and C-G codes to outperform N-R sequentially and they do. However on these extended standard problems the Hessian only has 4 distinct eigenvalues (except for the Trigonometric problem) and this further favours the C-G code. On speed up time the DAP implementation is usually more than 20 times faster than N-R but is no real improvement over the C-G code that would normally be used in this dimension.

Before leaving this algorithm I would like to comment that it is necessary to use a parallel equation solver at step 4 otherwise if 95% of the time is spent in evaluating function evaluations and 5% on overheads (mainly the equation solver) then

$$S < \frac{100}{5+95/P} < 20 \text{ all } P!$$

3.4.2 Variable Metric code

On the DAP system the variable metric code which only evaluate ∇F could do so using central differences in one step if

$$2n + 1 \leq P.$$

However the need to store an n x n Hessian matrix would seem to make such an approach undesirable. Indeed to store H in one matrix field limits $n < 64$ and then the full Newton algorithm can be implemented as we have just seen.

For other architectures i.e. for small values of P and larger store availability the possibility of updating the matrix H for P parallel steps d_j has been examined by Straeter [6] who gave the following scheme

$$r_j = H^{k-1} y_j - d_j - \sum_{i=1}^{j-1} \frac{r_i{}^t y_j}{(r_i{}^t y_i)} r_i \qquad j = 1 \ldots P$$

$$H^k = H^{k-1} - \sum_{j=1}^{P} \frac{r_j r_j{}^T}{y^{jT} r_j}$$

Here $y_j = \nabla F(x + d_j) - \nabla F(x)$.

The precise way of modifying the line search would depend upon the value of P.

3.4.3 The conjugate gradient code

On the DAP system the conjugate gradient code which also only requires ∇F can be implemented using central differences in one step if

$$2n + 1 \leq P.$$

Again as it only needs to store a few n vectors this can be implemented on the DAP easily for values of $n \simeq 4096$. It is therefore ideally suited to solve problems rather larger than those we usually consider.

This observation led to the study on finite element optimisation that will be the subject of section 6. The results will be discussed there.

3.4.4 Summary

Straeters variable metric method is intended for the range $P < n$; the conjugate gradient method is ideal if $P \sim n+1$ and the Newton Method if $P \sim (n^2 + 5n + 2)/2$. In a recent paper Byrd, Schnabel and Shultz (1987) discuss a modified variable metric method for the range $n+1 < P < (n^2 + 5n + 2)/2$. Codes have therefore been suggested for a wide range of p/n.

3.5. The Truncated Newton Algorithm

At this point in our research the direction we were following was influenced by the announcement of the Truncated Newton Algorithm by Dembo and Steihaug [7].

This was quickly implemented at Hatfield by Price [8]. It consists of essentially the following steps. It is still iterative in that $x^{(k+1)} = x^{(k)} + \alpha d$ but it differs dramatically in the way d is calculated.

In the Newton Method d is obtained by solving the set of equations

$$Gd = - g.$$

In the truncated Newton method this set of equations is approximately solved by a few iterations of the conjugate direction method for linear equations.

In particular we may note that if u_i is the prediction of d after i steps of the conjugate direction method then

(1) $||u_i|| > ||u_{i-1}||$

(2) If $Q(u) = \frac{1}{2}u^TGu + g^Tu$ then $Q(u_{i+1}) < Q(u_i)$.

So each prediction is larger than the previous one and reduces the quadratic approximation Q of F.

In the truncated Newton method there is a trust region of diameter D and the conjugate direction method is stopped if either

(1) $||u_{i+1}|| > D$

and indeed the step in the conjugate direction p_i is reduced to $\alpha_i p_i$ where

$$||u_i + \alpha_i p_i|| = D,$$

or (2) $$\frac{r_{i+1}{}^T r_{i+1}}{g_k{}^T g_k} < \text{Min}\left(\frac{0.1}{k^2}, g_k{}^T g_k\right).$$

The first prevents the step getting too large whilst the second prevents too accurate a solution of the set of equations far from the optimum.

In the Hatfield implementation we included a conical trust region to ensure Wolfe's condition [9] for effective descent is also satisfied namely

(3) $(u_{i+1}{}^{T}g_k) < -0.1\ ||u_{i+1}||\ ||g_k||$.

Again if necessary the step $u_{i-1} + \alpha p_i$ is chosen so that this is an equality.

The code is therefore

Step (1) Select $x^{(0)}$, ε, k_{max}, P_o: $k = 1$

(2) Calculate $F(x^{(k)})$, $g = \nabla F(x^{(k)})$

(3) Stop if $g^Tg < \varepsilon$ $k > k_{max}$

(4) Use the conjugate direction code

(4.1) Set $u_i = 0$, $r_1 = p_1 = -g$, $i = 1$

(4.2) Calculate $\alpha_i = r_i{}^{T}r_i/p_i{}^{T}Gp_i$

(4.3) $u_{i+1} = u_i + \alpha_i p_i$

(4.4) $r_{i+1} = r_i + \alpha_i Gp_i$

(4.5) Stop if any of the 3 truncating conditions is satisfied and go to (5)

(4.6) $\beta = r_{i+1}{}^{T}\ r_{i+1}/r_i{}^{T}r_i$

(4.7) $p_{i+1} = +r_{i+1} - \beta p_i$

(4.8) Goto (4.2) next i

(5) Evaluate $F(x + u)$

(6) Fit a parabola to $F(x)$, g^Tu, $F(x + u)$
if the predicted step is outside $0.8 < \alpha p < 1.2$ evaluate
$F(x + \alpha pu)$

(7) Test better of $F(x + u)$, $F(x + \alpha pu)$ against Wolfe's tests II and III, Dixon [9] and if necessary do an Armijo line search [10].

Let α^* be an accepted value of α.

(8) Put $x^{k+1} = x^k + \alpha^* u$
if $\alpha^* \geq 1$ put $D_{k+1} = 2\ D_k$
if $\alpha^* < 1$ put $D_{k+1} = 1/3\ D_k$
return to (2) next k.

In this code the term Gp_i occurs at steps 4.2 and 4.4. Dembo proposed calculating this approximately by differences

$$Gp_i = \{g(x + \sigma p_i) - g(x)\}/\sigma.$$

This saves the storage of G at the expense of gradient calls. The additional gradients do not have to be stored.

The tests shown in Table 1 demonstrate that this method outperforms the N-R(EO4KDF), V-M(OPVM) and C-G(OPCG) codes on all the test problems with $n > 4$.

Table 1

CODE	PROBLEM	N	NO.F	NO.G	EFE'S	C.P.U.
TRUNEW	1	4	103	229	1019	0.48
OPCG	1	4	384	15	1004	0.80
OPVM	1	4	128	77	436	0.90
E04KDF	1	4	54	202	862	0.42
TRUNEW	2	4	39	82	367	0.28
OPCG	2	4	96	40	256	0.45
OPVM	2	4	9	31	173	0.46
E04KDF	2	4	18	90	378	0.31
TRUNEW	2	60	33	90	5433	1.27
OPCG	2	60	117	48	2997	1.87
OPVM	2	60	341	194	11981	59.85
E04KDF	2	60	42	1302	78162	15.31
TRUNEW	2	80	31	95	7631	1.81
OPCG	2	80	216	93	7656	3.79
OPVM	2	80	439	251	20519	134.71
E04KDF	2	80	43	1643	131483	32.84
TRUNEW	3	10	41	59	631	0.34
OPCG	3	10	99	33	429	0.54
OPVM	3	10	133	83	963	2.00
E04KDF	3	10	41	181	1851	0.53
TRUNEW	3	20	49	71	1469	0.56
OPCG	3	20	139	40	939	1.00
OPVM	3	20	187	113	2447	6.27
E04KDF	3	20	52	312	6292	1.21
TRUNEW	4	80	21	148	11861	2.16
OPCG	4	80	186	87	7146	3.04
OPVM	4	80	201	102	8361	31.14
E04KDF	4	80	28	828	66168	13.54
TRUNEW	2	2000	41	109	218041	24.98
OPCG	2	2000	125	49	98125	50.13
TRUNEW	4	2000	29	851	1702029	206.42
OPCG	4	2000	1064	520	1041064	393.47

Naturally this result greatly influenced our program.

At about the same time that we became aware of the Truncated Newton method, we also became aware of Rall's automatic differentiation approach. This is ideal for use on certain parallel processing architectures and saves the man months often needed in obtaining and coding accurate differentiation formulae. This is our third method for efficiently using parallel processing in obtaining F and its derivatives.

3.6 Method C: Automatic Differentiation

Rall's automatic differentiation method [11] can be used to calculate the value of a function, its gradient and hessian at a known point $\underline{x} \in R^n$. He represents each variable x_i as a triplet $X = (x_i . x_i', x_i")$ where $x_i \in R$, $x_i' \in$

R^n, and x_i'' is a symmetric real (n x n) matrix. If we let T^n be the set of these triplets, then all the standard arithmetic operations can be defined in T^n.

If U = (u, u′, u″) and V = (v, v′, v″) then

(1.1) $U + V = (u + v,\ u' + v',\ u'' + v'')$

(1.2) $U - V = (u - v,\ u' - v',\ u'' - v'')$

(1.3) $U * V = (uv,\ uv' + vu',\ uv'' + u'v'^T + v'u'^T + vu'')$

(1.4) $U / V = (u/v,\ (vu' - uv')/v^2,\ (v^2u'' - v(v'u'^T) + 2uv'v'^T - uvv'')/v^3)$

v.ne.0.

To represent an independent variable in T^n space Rall lets $x_i \Rightarrow (x_i, e_i, 0)$ where e_i is the ith column of the identity matrix, and 0 the zero matrix. Similarly constants are represented by $C \Rightarrow (c, \underline{0}, 0)$. The calculation of the value, gradient vector and hessian matrix of a rational function can be done simply in FORTRAN by defining subroutines which implement (1.1) - (1.4) and calling these in the function subroutine. Instead of using a separate triplet for constants he defines rules for all the arithmetic operations when mixing constants and elements of T^n; namely

(1.5) $C + U = U + C = (c + u,\ u',\ u'')$

(1.6) $C - U = (c - u,\ - u',\ -u'')$

(1.7) $U - C = (u - c,\ u',\ u'')$

(1.8) $C * U = U.C = (c.u,\ c.u',\ cu'')$

(1.9) $C / U = \left(c/u,\ \dfrac{- cu'}{u^2},\ \dfrac{2cu'u'^T - cuu''}{u^3}\right)$ (u.ne.0)

(1.10) $U / C = (u/c,\ u'/c,\ u''/c)$ (c.ne.0)

<u>EXAMPLE</u>

If we take the two dimensional Rosenbrock function

(1.11) $F(\underline{X}) = 100\ (x_2 - x_1^2)^2 + (1 - x_1)^2$

then the standard starting point (- 1.2, 1.0) becomes in triplet form:

$X_1 = (- 1.2,\ (1, 0)^T,\ [0])$

$X_2 = (1.0,\ (0, 1)^T,\ [0])$.

Then considering the function evaluation in T^n we carry out the following steps: (ST and ST1 are workspace triplets)

Step 1	$ST = X_1 * X_1$	using 1.3
Step 2	$ST = (X_2 - St)$	using 1.2
Step 3	$ST = ST * ST$	using 1.3
Step 4	$ST = 100 * ST$	using 1.8
Step 5	$ST2 = 1 - X_1$	using 1.6
Step 6	$ST2 = ST2 * ST2$	using 1.3
Step 7	$ST = ST + ST2$	using 1.1.

ST now contains the triplet:

$$\left(24.2,\ \begin{pmatrix} - 215.6 \\ - 88.0 \end{pmatrix},\ \begin{pmatrix} 1330.0 & 480.0 \\ 480.0 & 200.0 \end{pmatrix}\right) = (F(\underline{X}),\ \nabla F(\underline{X}),\ \nabla^2 F(\underline{X}))$$

which is what would be obtained by differentiating the function analytically and evaluating $(F(\underline{X}), \nabla F(\underline{X}), \nabla^2 F(\underline{X}))$ at the point (- 1.2, 1.0). In general if g:R =>

R is a twice differentiable function then it can be extended to the mapping $g:T^n \Rightarrow T^n$ by use of the chain rule: $g(U) = g(u, u', u") = (g(u), g'(u).u', g'(u).u" + g"(u).u'u'^T)$.

So for example

$\text{Sin}(U) = \text{Sin}(u, u', u") = (\text{Sin}(u), \text{Cos}(u).u', \text{Cos}(u).u" - \text{Sin}(u)u'u'^T)$.

An interesting mapping to consider is $g = e^{f(x)}$

$$\text{since } g_i' = \frac{\partial f(x)}{\partial (x_i)} e^{f(x)} \text{ and so } g' = \nabla f(x) e^{f(x)} \tag{1}$$

$$\text{and } g_i = \frac{\partial f(x)}{\partial (x_i)} \cdot \frac{\partial f(x)}{\partial (x_i)} \cdot e^{f(x)} + e^{f(x)} \frac{\partial^2 f(x)}{\partial (x_i)^2} \tag{2}$$

and so $g" = (\nabla f(x)^2 e^{f(x)} + (\nabla^2 f(x)) e^{f(x)})$.

To automatically differentiate this function, we must first automatically differentiate f(x), to get $\nabla f(x)$ and $\nabla^2 f(x)$ and then substitute these values into (1) and (2).

In Rall[6] he states that automatic differentiation is suited to languages such as ADA and PASCAL-SC which permit the introduction of datatypes and additional definitions of the standard operator symbols to manipulate such types. When implemented in FORTRAN as in this paper, the objective function must be rewritten as a sequence of calls to subroutines; the necessary ADA data types and operator extensions have been written by Mohseninia and when these are available, the preparation of code to produce the function, gradient and Hessian becomes very simple. An example is indicated below:-

<u>AN EXAMPLE OF AN ADA PROGRAM</u>

```
with auto_diff;
procedure main is
use auto_diff;
     x1 : triplet;
     x2 : triplet;
     f  : triplet;
begin
     set (u => x1, val => - 1.2, index => 1);
     set (u => x2, val =>   1.0, index => 2);
--
-- initialise the two triplets x1 and x2 with the
-- starting solutions of (-1.2, 1.0)
-- Now we can evaluate the derivative of
-- f(x) = 100*(x2 - x1^2)^2 + (1-x1)^2
-- at the point (-1.2, 1.0)
--
     f : = 100.0*((x2 - (x1**2))**2) + (1.0 - x1)**2;
     print (f);
end main;
```

3.6.2 An Adaptation of Automatic Differentiation to Avoid Storing Any Matrices

When using the conjugate direction (or any alternative indirect method) to solve $A\underline{x} = \underline{b}$ the coefficient matrix A only occurs when multiplied by a vector $\underline{p}$, at any point in the algorithm. In an optimisation algorithm the coefficient matrix in the Newton equations is the hessian of the objective function at the current point. Therefore the hessian multiplied by a vector $\underline{p}$ is required. This can be obtained by automatic differentiation without storing the coefficient matrix. In our modification we change the third element of the triplet to a vector quantity by multiplying by a vector $\underline{p}$. So each x_i is held as $(x_i, e_i, 0.\underline{p}) = (x_i, e_i, \underline{0})$ and all the rules for the arithmetic operations are redefined appropriately. the new formulation is:

$$(2.2)\quad (u + v) = (u, u', u''\underline{p}) + (v', v''\underline{p}) = (u + v, u' + v', u''\underline{p} + v''\underline{p})$$

$$(2.3)\quad (u - v) = (u, u', u''\underline{p}) - (v, v', v''\underline{p}) = (u - v, u' - v', u''\underline{p} - v''\underline{p})$$

$$(2.4)\quad (u * v) = (u, u', u''\underline{p}) * (v, v', v''\underline{p}) = (u.v, u.v' + v.u', uv''\underline{p} + u'v'^{t}p + v'u'^{t}p + vu''\underline{p})$$

$$(2.5)\quad (u / v) = (u, u', u''\underline{p})/(v, v', v''\underline{p}) = \left(U/V, \frac{VU' - UV'}{V^2}, \frac{V^2U''\underline{P} - V((V'U'^{T}P + U'V'^{T}P) + 2UV'V'^{T}P - UVV''\underline{P})}{V^3}\right) \quad (V.ne.0)$$

It can be seen that the vector multiplied by $(\text{vector})^T$ quantities in Rall's standard automatic differentiation become $(\text{vector})(\text{vector})^T(\text{vector})$ forms, which is equivalent to a vector multiplied by the scalar arising from the inner product.

3.6.3 Sparse Automatic Differentiation

When the automatic differentiation described in the previous section was applied it was found to be inefficient for large problems when compared to Dembo and Steihaug's differencing approach [7], devised to approximate the product $H\underline{p}$. However, a further adaptation to the automatic differentiation approach can be easily implemented which takes advantage of any sparsity and structure in the hessian matrix, and thus avoids many unnecessary calculations.

In the previous approach all vectors were stored as length n, however if the operation under consideration is x_1 multiplied by x_2 within a large problem, then only the first two elements of the vectors are needed, since the other elements do not make any contribution to the gradient or hessian of $x_1 * x_2$. A similar consideration applies to most operations that occur in the calculation of a function value. Therefore by passing the correct suffices (1 and 2 in this example) through to the multiplication subroutine, the work can be substantially reduced. This approach implies that each objective function must be subdivided into a sequence of subfunction calculations. After the function, gradient and $H\underline{p}$ of each subfunction has been evaluated, they have to be added to the workspace triplet storing the function value, gradient and $H\underline{p}$ of the overall objective function. This addition will probably have to be performed in n-space, however this work is minimal since the automatic differentiation for addition simply involves (2n + 1) additions. The calculation of the triplet corresponding to each subfunction can be performed with vectors of a reduced length providing the subfunction only involves a limited number of the optimisation variables. Results using this second adaptation, reported in the next section, show it to be far more efficient than always working in full n-space.

Rall's automatic differentiation technique as described above has been merged with the truncated Newton method to calculate g and Gp. Details of the test results are given in TR 170. Summarising those for larger problems we obtained

	N=60	N=80	N=2000
TNEWAD	7.13	9.99	280.19
E04KDF	15.31	32.84	
OPVM	26.28	134.7	
OPCG	1.27	3.79	50.13

We can see that the price paid for using automatic differentiation in the Truncated Newton method is that it is no longer faster than OPCG. The time to analytically differentiate, implement and correct formula for $g = \nabla F$ is however saved.

Again we comment that Automatic Differentiation is ideal for parallel processing as the doublets and triplet elements can be computed independently on separate processors.

3.7 Conclusion

In this lecture we have first reported our results implementing the classical unconstrained optimisation algorithms on the ICL-DAP; then our results with the new Truncated Newton method and finally introduced the concept of automatic differentiation and shown that it is suitable for parallel processing.

References to Section 3

1. Brown, A H O, The development of computer optimisation procedures for use in aero engine design, In Optimisation in Action, L C W Dixon (ed.), Academic Press, 1976.

2. McKeown, J J, Experiments in implementing a nonlinear least squares algorithm on a dual processor computer, NOC TR102, The Hatfield Polytechnic, 1979.

3. Sargon, J D, Chong, Y Y and Smith, M, Nonlinear Econometric Modelling on a Parallel Processor, DAP Support Unit, Queen Mary College, London University.

4. Dixon, L C W, Patel, K D and Ducksbury, P G, Experience running optimisation algorithms on parallel processing systems, NOC TR138, 1983.

5. Mifflin, R, A superlinearly convergent algorithm for minimization without evaluating derivatives, Yale University TR65, 1973.

6. Straeter, T A, A parallel variable metric optimisation algorithm, NASA TR D 7329, (1973).

7. Dembo R, and Steihaug, T, Truncated Newton Method for Large Scale Optimisation: Mathematical Programming 26, (1983).

8. Dixon, L C W and Price, R C, Numerical Experience with the Truncated Newton code, NOC TR169, 1986 (to appear in JOTA).

9. Dixon, L C W, Nonlinear Optimisation: A survey of the state of the art, In D Evans, Software for Numerical Mathematics and its Application, Academic Press, 1973.

10. Armijo, L, Pacific J. Maths. 16 p 1-16.

11. Rall, L B, Automatic Differentiation Techniques and Applications, Lecture Notes in Computer Science, No. 120, Springer Verlag, 1981.

12. Dixon, L C W and Price, R C, The Truncated Newton Method for sparse unconstrained optimisation using automatic differentiation, NOC TR170, 1986. (to appear in JOTA).

13. Mohseninia, M, The use of the extended operator set of ADA with automatic differentiation and the Truncated Newton method, NOC TR175, The Hatfield Polytechnic, 1986.

14. Byrd, R H, Schnabel, R B and Shultz, G A, "Using parallel function evaluations to improve Hessian approximation for unconstrained optimization," Technical Report CS-CU-361-87, University of Colorado, Boulder, 1987.

4. Solution of Sets of Linear Equations

4.1 Introduction

As mentioned earlier it is quite usual to find on analysing the solution of most optimisation problems that at least 95% of the computer time is spent in evaluating the values of the objective function and its gradient at x^k.

It was therefore natural to first discuss the way parallel processing could be used to reduce this 95% of the time. With the advent of parallel automatic differentiation or by the use of subfunction calculation we will assume this can be done efficiently.

The speed up factor is therefore 1/0.95/p+0.05 which tends to 20 as P goes to infinity. It is therefore essential that the optimisation code in which the 5% of the time is spent also uses the parallel hardware efficiently. As most of this time is usually spent solving a set of linear equations, in this lecture we will consider the solution of small dense sets of equations on parallel systems. The problem of solving a set of equations

$$Ax = b$$

is one of the most frequently occurring problems in numerical computation. Because of the frequency with which it occurs it has been intensely researched and there are very many variants of most algorithms designed to solve it. It is probably however true that on a single processor machine most numerical analysts would prefer to use one of four standard methods if, as will be assumed in this lecture, A is dense.

The four broad classes of method we will consider are:

1. Gaussian Elimination
2. QU Decomposition
3. Iteration
4. Conjugate Direction Techniques.

Within the first category we will consider four variations:

1.1 Choleski Decomposition

1.2 Total Pivoting

1.3 Partial Pivoting

1.4 Gauss-Jordan.

Similarly within the second category we have:

2.1 Householder's Method

2.2 Givens' Method

2.3 The Gramm Schmidt Method

and in the third category:

3.1 S.O.R.

3.2 Gauss-Seidel Method

3.3 Jacobi Method.

In each of the above three classes of method, I have attempted to list the variations in the popular order for a sequential machine, i.e. on a sequential machine I suspect most people would use Choleski decomposition if the matrix is symmetric and positive definite. Whilst on a nonsymmetric or nonpositive matrix most people prefer to use a partial pivoting code, the case for total pivoting on really badly conditioned problems is accepted but disliked. The Gauss-Jordan variant of either is rejected as more expensive.

The questions we shall consider in this lecture are which algorithm we should use in R^N with p processors and secondly how does that answer vary with n and p and the relative compute and data transfer times of the system.

4.2 Gaussian Elimination Methods

Most Gaussian codes consist of three parts

(1) triangularisation

(2) pivoting

(3) back substitution

in which the triangulation and pivoting interact and are both completed before the back substitution starts.

4.2.1 Triangularisation

Without pivoting triangularisation takes the simple form

FOR k = 1 TO N - 1

FOR i = k + 1 TO N

FOR j = k + 1 TO N F1

$a_{ij} = a_{ij} - a_{ik}a_{kj}/a_{kk}$

NEXT j, i, k.

Most early numerical analysis texts commented that it is preferable to rewrite this as:

FOR k = 1 TO N - 1

FOR i = k + 1 TO N

$c = a_{ik}/a_{kk}$ F2

FOR j = k + 1 TO N

```
    a_ij = a_ij - c a_kj
NEXT j, i, k.
```

This change was recommended because on most sequential machines

t{1 DIV + (N - k) mults} < t{(n - k)Divs}.

Parkinson observed that this is no longer true on either a pipeline or a DAP processor. On a pipeline

t{1 Div + pipeline mult} > t{chained mult/divisions}

whilst on the DAP the original code is 3 simple matrix operations, a property lost in the modified form.

For those using Fortran it is often further recommended that for data access reasons it is preferable to do

```
FOR k = 1 TO N - 1
FOR i = k + 1 TO N
c_i = a_ik/a_kk
NEXT i
FOR j = k + 1 TO N
FOR i = k + 1 TO N
a_ij = a_ij - c_i a_kj
NEXT i, j, k.
```

While these changes may seem trivial such considerations can significantly alter the efficiency of codes and their effect in a particular architecture should be borne in mind.

4.2.2 The partial pivoting operation

In partial pivoting the largest element of n - k elements in the pivot row is determined between the k and i loops. On a sequential machine this implies n - k comparisons. On a parallel machine the number of comparisons needed to compare R numbers on P machines depends on R and P.

For instance if P = 4, R = 16 it takes 5 steps

Step 1 $C(N_1; N_2)$, $C(N_3, N_4)$, $C(N_5, N_6)$, $C(N_7, N_8)$

2 $C(M_{1,2}, M_{3,4})$, $C(N_9, N_{10})$, $C(M_{5,6}, N_{11})$, $C(N_{7,8}, N_{12})$

3 $C(M_{1,2,3,4}, M_{9,10})$, $C(N_{13}, N_{14})$, $C(M_{5,6,11}, N_{15})$, $C(M_{7,8,12}, N_{16})$

4 $C(M_{1,2,,4,9,10}, M_{13,14})$, - , $C(M_{5,6,11,15}, M_{7,8,12,16})$, -

5 $C(M_{1,2,3,4,9,10,13,14})$, $M(_{5,6,11,15,7,8,12,14})$, - , - , - .

The case usually quoted assumes p = R/2 when the comparison requires $\log_2 R$ steps if $R = 2^k$.

4.2.3 The partial pivoting algorithms

On a sequential machine the algorithm requires $\frac{1}{3}n^3 + O(n^2)$ multiplications and divisions, and $O(n^2)$ comparisons. The times for the comparisons are usually ignored.

On a parallel machine with $P = (n - 1)^2$ processors Sameh and Kuck (1977) showed the most active processor only needed 3(n - 1) multiplications/ divisions, but O(n log n) comparisons. For large n pivot comparisons dominate! Lord,

Kowalik & Kumar (1983) who tested many algorithms on the HEP (P = 8) computer for which the assumption $P = (n - 1)^2$ was inappropriate, re-examined the algorithm assuming $P = [n/2]$, they showed that the algorithm then requires $n^2 - 1$ multiplications and divisions on the critical path.

Their Speed up $= \frac{1}{3}n^3/n^2 - 1 = \frac{1}{3}n$ Efficiency $= \frac{1}{3}n/\frac{n}{2} = \frac{2}{3}$.

The Sameh/Kuck and Lord/Kowalick & Kumar methods are contrasted below for n = 4.

<u>Sameh/Kuck P = 9</u>

1. Select Pivot and form $r_k = 1/_{akk}$
2. Calculate $c_i = a_{ik}r_k$ (PAR)
3. $a_{ij}^{(2)} = a_{ij}^{(1)} - c_i a_{kj}^{(1)}$ (PAR)
4. Select pivot and form $r_k = 1/a_{kk}$
5. Calculate $c_i = a_{ik}^{(2)}\ r_k$ (PAR)
6. $a_{ij}^{(3)} = a_{ij}^{(2)} - c_i a_{kj}^{(2)}$ (PAR)
7. Select pivot and form $r_k = 1/a_{kk}$
8. Calculate $c_i = a_{ik}^{(3)}\ r_k$
9. $a_{ij}^{(4)} = a_{ij}^{(4)} - c_i(a_{kj}^{(4)}$

<u>Note</u>. L/K/K could be improved at the beginning but this would alter their calculation

<u>Lord Kowalik & Kumar P =2</u>

1.	The same	
2.	$c_2 = a_{2k}r_k$	
3.	$c_3 = a_{3k}r_k$	
4.	$c_4 = a_{4k}r_k$	
5.	$a_{22}^{(2)}$	$a_{23}^{(2)}$
6.	$a_{32}^{(2)}$	$a_{33}^{(2)}$
7.	$a_{42}^{(2)}$	$a_{43}^{(2)}$
8.	Pivot and r_k	$a_{24}^{(2)}$
9.	c_3	$a_{34}^{(2)}$
10.	c_4	$a_{44}^{(3)}$
11.	$a_{33}^{(3)}$	$a_{34}^{(3)}$
12.	$a_{43}^{(3)}$	$a_{44}^{(3)}$
13.	Pivot and r_k	
14.	c_4	
15.	$a_{44}^{(4)}$.	

Amusing variants can be constructed, for instance P = 3

1.	Select pivot and form $r_k = 1/a_{kk}$		
2.	c_2	c_3	c_4
3.	$a_{22}^{(2)}$	$a_{23}^{(2)}$	$a_{24}^{(2)}$
4.	$a_{32}^{(2)}$	$a_{33}^{(2)}$	$a_{34}^{(2)}$
5.	$a_{42}^{(2)}$	$a_{43}^{(2)}$	$a_{44}^{(2)}$
6.	Select pivot and form $r_k = 1/a_{kk}$		
7.	c_3	c_4	
8.	$a_{33}^{(3)}$	$a_{34}^{(3)}$	
9.	$a_{43}^{(3)}$	$a_{44}^{(3)}$	
10.	Select pivot and form $r_k = 1/a_{kk}$		
11.	c_4		
12.	$a_{44}^{(4)}$.		

Comparing the four codes we have

P = 1	C = 23	C*P = 23
P = 2	C = 15	C*P = 30
P = 3	C = 12	C*P = 36
P = 9	C = 9	C*P = 81

so our speed up is always obtained at a processor * time cost.

No attention has been given so far to data transfer times i.e. the time necessary to transfer the data to the processor undertaking the compute. We will return to this point in section 4.5.

4.2.4 Total pivoting

In total pivoting at iteration k we need to compare $(n - k)^2$ numbers. This would dominate the cost even more for large n and small P; but is ideal on the ICL-DAP where there is a special fast $\mathrm{Max}(A_{ij})$ instruction for determining the maximum element of a matrix.

4.2.5 The Gauss Jordan method

Gaussian elimination is usually completed by back substitution.

Kuck (1977) has shown that this can be done in 3(n - 1) steps on n - 1 processors. However it should never be done on such a system since the unused processors are available precisely when they are needed to do the Gauss Jordan calculation. If we insert n additional rows to update RHS we now only need one additional step to complete the solution.

For data transfer considerations it is better to use n or (n + 1) processors than
n - 1 as then computations involving one row or column can be done on a particular processor. If this is done then it is efficient to update the pivot row in parallel at each step by making the pivot element unity.

Gauss Jordan $n = 4$ $P = 4$

1. Select pivot and form $r_k = 1/a_{kk}$
2. c_1 c_2 c_3 c_4
3. $a_{12}^{(2)}$ $a_{22}^{(2)}$ $a_{32}^{(2)}$ $a_{42}^{(2)}$
4. $a_{13}^{(2)}$ $a_{23}^{(2)}$ $a_{33}^{(2)}$ $a_{43}^{(2)}$
5. $a_{14}^{(2)}$ $a_{24}^{(2)}$ $a_{34}^{(2)}$ $a_{44}^{(2)}$
6. $b_1^{(2)}$ $b_2^{(2)}$ $b_3^{(2)}$ $b_4^{(2)}$
7. Select pivot and form $r_k = 1/a_{kk}$
8. c_1 c_2 c_3 c_4
9. $a_{13}^{(3)}$ $a_{23}^{(3)}$ $a_{33}^{(3)}$ $a_{43}^{(3)}$
10. $a_{14}^{(3)}$ $a_{24}^{(3)}$ $a_{34}^{(3)}$ $a_{44}^{(3)}$
11. $b_1^{(3)}$ $b_2^{(3)}$ $b_3^{(3)}$ $b_4^{(3)}$
12. Select pivot and form $r_k = 1/a_{kk}$
13. c_1 c_2 c_3 c_4
14. $a_{14}^{(4)}$ $a_{24}^{(4)}$ $a_{34}^{(4)}$ $a_{44}^{(4)}$
15. $b_1^{(4)}$ $b_2^{(4)}$ $b_3^{(4)}$ $b_4^{(4)}$
16. Form $r_k = 1/a_{kk}$
17. c_1 c_2 c_3 c_4
18. $b_1^{(5)} = x_1$ $b_2^{(5)} = x_2$ $b_3^{(5)} = x_3$ $b_4^{(5)} = x_4$

It is probably worth repeating that on most parallel machines it will be more efficient to omit the separate calculation of r_k and compute c_i by division rather than multiplication. If the pivot calculation is ignored the number of steps is

$$\sum_{k=1}^{n+1} k - 1 = \tfrac{1}{2}(n + 1)(n + 2) - 1.$$

4.3 Orthogonal QU Decomposition Techniques

The parallel analysis of both Householder's and Givens' method was given by Sameh and Kuck (1977). They showed that given $P = O(n^2)$ processors Householder's method required $O(n \log n)$ operations but Givens' method only required $O(n)$. This contrasts with the sequential situations where Householder's method is the more efficient. This comparison was tested and confirmed by Sorenson (1985).

The Sequential Givens routine consists of

1. FOR Q = 1 TO N - 1
2. FOR P = Q + 1 TO N

   $S = a_{QQ}/\sqrt{a_{PQ}^2 + a_{QQ}^2} \qquad c = a_{PQ}/\sqrt{a_{PQ}^2 + a_{QQ}^2}$
3. FOR J = 1 TO N

   $TEMP = C\ a_{PJ} + S\ a_{QJ}$

   $a_{QJ} = - S/\ a_{PJ} + C\ a_{QJ}$

   $a_{PJ} = TEMP$
4. Next j, P, Q.

Note that the J loop is an ideal pipeline calculation. Typically operation 3 is considered 1 task.

The essential feature of Givens' method is that a pair PQ only alters rows P and Q so a number can be performed in parallel.

The earliest discussion of a parallel Givens code was that due to Sameh and Kuck (77) who reduced A_{PQ} to 0 in the following order for an n x n matrix

$$\begin{bmatrix} * \\ 1 & * \\ 2 & 3 & * \\ 3 & 4 & 5 & * \\ 4 & 5 & 6 & 7 & * \\ 5 & 6 & 7 & 8 & 9 & * \\ 6 & 7 & 8 & 9 & 10 & 11 & * \\ 7 & 8 & 9 & 10 & 11 & 12 & 13 & * \end{bmatrix} \qquad \begin{bmatrix} * \\ 3 & * \\ 2 & 5 & * \\ 2 & 4 & 7 & * \\ 1 & 3 & 6 & 8 & * \\ 1 & 3 & 5 & 7 & 9 & * \\ 1 & 2 & 4 & 6 & 8 & 10 & * \\ 1 & 2 & 3 & 5 & 7 & 9 & 11 & * \end{bmatrix}$$

Sameh & Kuck (77)   Modi & Clarke (84)

Their code performed 13 parallel Givens steps on 4 processors. In general (2n - 3) parallel Givens steps on $\frac{n}{2}$ processors. As each parallel Givens step involves 4n + 4 operations the speed up is dominated by $4n^3/3(2n - 3)(4n + 4)$; so $S = \frac{1}{6}n$ and $E = \frac{1}{3}$. Many attempts at improving the Sameh & Kuck method have been proposed but Cosnard (1985) has shown that the "Greedy" algorithm, Modi and Clarke (1984), is optimal. This is also shown above, and only requires 11 parallel Givens steps on 4 processors. Cosnard states that the efficiency of this method is asymptotically $\frac{1}{2}$ but if we utilise parallel pipelines when performing the inner loop the algorithm is much more powerful than this would indicate.

Kowalik, Kumar and Kangria (1983) noted that if 2(k + 1) processors are used to perform the inner loop it only requires 4 steps. Again analysing the nonoptimal sort they concluded that with $P = \frac{3}{4}n^2 + \frac{5}{2}n$ they could obtain a speed up $S = n^2/6 + O(n)$, $C = 8n - 12$. Adding the 3n - 3 operations for parallel back substitution they state the parallel operation count as 11n - 5. They also analysed the case with $P = [(n-1)/2]$ and claimed $C = 6n^2 + 8n - 25$ giving $S = \frac{2}{9}n$ and $E = \frac{4}{9}\frac{n}{n-1}$. These times were verified on their test runs on the HEP.

4.4.6. Iterative Methods

4.4.1 Jacobi's method

To solve $Ax = b$ first scale so $A_{ii} = 1$ then iterate
$x^{(k+1)} = (I - A)x^{(k)} + b$.

This method is convergent if $||I - A|| < 1$ but very slow when $||I - A|| > .9$.

The method is however ideal for parallel processing as we can do all the rows in parallel if $P \geq n$ and P rows in parallel if $P < n$. If each processor is a pipeline it is even better, as each row update is an ideal pipeline operation.

It is therefore possible to speed up the very poor method significantly.

4.4.2 Gauss-Seidel method

The ordinary Gauss-Seidel algorithm

$$x_i^{(k+1)} = x_i^{(k)} - \sum_{j=1}^{i-1} A_{ij}x_j^{(k+1)} - \sum_{j=i}^{n} A_{ij}x_j^{(k)} + b_i$$

is not suitable for parallel computing due to the $x_j^{(k+1)}$ term on RHS.

This can however be adapted for a block Gauss-Seidel approach with $P < n$. If q s.t. $Pq < i < P(q+1)$

$$x_i^{(k+1)} = x_i^{(k)} - \sum_{j=1}^{Pq} A_{ij}x_j^{(k+1)} - \sum_{j=Pq+1}^{n} A_{ij}x_j^{(k)} + b_i$$

allows P values of x to be evaluated in parallel.

4.4.3 Assynchronous Jacobi or Gauss Seidel Method

As an alternative to the Block Gauss Seidel method Baudet (1978)[17] proposed the assynchronous algorithm where

$$x_i^{(k+1)} = x_i^{(k)} - \sum_{j=1}^{n} A_{ij}x_j^{L} + b_i$$

is used where x_j^L is the latest value in the store when it is accessed. Allowing $P = n$ processors, 1 for each i, and slight variations of performance x_j^L could be either $x_j^{(k)}$ or $x_j^{(k+1)}$. His analysis indicated there would be no loss of performance.

4.5 Conjugate Gradient Method

The sequential code consists of the following steps:

1. Choose $x^{(0)}$ (usually 0); ε; $r_0 = b - Ax_0$, $d_0 = r_0$ while $||r|| > \varepsilon$ do 2,3,4.
2. $\underline{x}_i^+ = \underline{x}_i + \alpha\, \underline{d}_i$ $\qquad \alpha = r^T d/d^T Ad$
3. $\underline{r}_i^+ = \underline{b}_i - A\underline{x}^+$
4. $d^+ = r^+ + \beta d$ $\qquad \beta = r^{+T}r^+/r^T r$.

Steps 2,3,4 are ideal for a parallel processor.

With $P = n$, the calculation of Ax and Ad requires n parallel steps
$r^T d$, $d^T Ad$, $r^T r$ requires log n parallel steps.

So one iteration costs $2n + 3 \log n + 4$ steps.
If $p = n^2$ we can do all the multiplications in Ax and Ad in parallel so one iteration is only $3 \log n + 6$ steps.

4.6 Preconditioning

In practice on a sequential machine both Jacobi and Conjugate Gradients are preconditioned before use. In the Jacobi method this involves the use of either a matrix C or a matrix M where the equation becomes

$$CAx = Cb \quad \text{or} \quad M^{-1}Ax = M^{-1}b.$$

For efficiency we require C to approximate A^{-1}, or M to approximate A; if we are using M we need $Mz = w$ to be much easier to solve than $Ax = b$; while if we are using C we need C to have a nice sparsity pattern.

The preconditioned Jacobi method is then either

$$x^{k+1} = (I - CA)x^k + Cb \quad \text{or} \quad x^{k+1} = x^k + z$$
$$\text{where} \quad Mz = b - Ax^k.$$

In the conjugate gradient method we must maintain symmetry so we use

$$CAC^Ty = Cb \quad \text{where} \quad x = C^Ty$$

and we require CAC^T to be wellconditioned and C to have a nice sparsity pattern.

Whilst the method could be programmed in terms of y it is more usual to substitute into x space then the initial step

$$d_1 = \Delta x_1 = C^TC(Ax - b)$$
$$r_1 = Ax_1 - b; \quad z_1 = C(Ax_1 - b)$$

and at subsequent iterations

$$x^+ = x + \alpha d$$
$$r^+ = r + \alpha Ad$$
$$z^+ = z + \alpha CAd$$
$$d^+ = Cz^+ + \beta d$$

where $\alpha = - r^Td/d^TAd$ and $\beta = z^{+T}z^+/z^Tz$.
The usual choices of conditioning matrix on sequential machines are the Neumann or incomplete Choleski.

The Neumann conditions used the principle if $G = (I - A)$ then $(I - G)^{-1} = I + G + G^2 + G^3 \ldots$ and truncate this series for C.

On sequential machines these are not as fast as the incomplete Choleski factorisation, where a Choleski factorisation is done without fill in to preserve sparsity. On pipelines and parallel computers the presence of these matrices puts up the CPU time, as the algorithms are no longer readily adapted for parallel processing.

A typical set of results is that given by Blumenfield (1983)

Preconditioner	No. of Iterations	Pipeline Time
I	120	584
$I + G$	65	445
$I + G + G^2$	70	606
$I + G + G^2 + G^3$	45	493
Incomplete Choleski	37	1293

The problem of preconditioners for parallel and pipeline systems needs further research.

4.7 Conclusions

In this lecture I have tried to illustrate some of the considerations that must be taken into account when selecting an algorithm for use on a parallel processor architecture. I hope to have shown convincingly that the assumption that any sequential algorithm can be selected without careful consideration is false.

Of the Gaussian type methods the Gauss-Jordan version is recommended. For QR decomposition the Greedy parallel Givens method is the obvious choice. For iteration methods the Block Gauss Seidel is effective and simple to program as is the conjugate gradient method. Great care needs to be taken when introducing preconditioning matrices into parallel computation.

References

1. Sameh, A H and Kuck, D J, Parallel Direct Linear Equation Solvers - a survey, In M Feilmer (ed.) Parallel Mathematics, 1977.

2. Lord, R E, Kowalik, J S and Kumar, S P, Solving Linear Algebraic Equations on an MIMD Computer, J. of ACM, Jan 1983.

3. Kuck, D J, A survey of parallel machine organization and programming, Computing Surveys 9, 29-59, 1977.

4. Sameh, A H and Kuck, D J, On Stable Linear System Solvers, J.A.C.M. 25, 31-91, 1978.

5. Sorenson, D, Lecture presented at XIV Mathematical Programming Symposium, Boston, 1985.

6. Cosnard,M and Robert, Y, Complexity of the Parallel QR Decomposition of a Rectangular Matrix In Feilmeir, Joubart and Schendel (eds.) Parallel Computing 1985, Elsevier Science.

7. Modi, J J and Clarke, M R B, An alternative Givens ordering, Numer. Math. 43, 83-90, 1984.

8. Kowalik, J S, Kumar S P and Kangria, E R, An implementation of the fast Givens transformation on an MIMD computer, Applications Mathematicae, Polish Academy of Sciences, 1983.

9. Baudet, G M, Asynchronous iterative methods for multiprocessors, JACM 2, 226-244, 1978.

10. Blumenfield, M, Preconditioning Conjugate Gradient Methods on Vector Computers In Feilmeier, Joubert and Schendel (eds.) Parallel Computer 83, North Holland, 1983.

5. Finite Element Optimisation of the ICL DAP

Many optimisation problems can be written in the partially separable form

$$F(x) = \sum_i F_i(x) \qquad x \in R^n$$

where $F_i(x)$ ony depends on a small number of the components of x.

This is especially true of all optimisation problems that originate from finite element analysis. If, for instance, we wished to solve a set of nonlinear partial differential equations such as the Navier-Stokes equations

$$e_1 = u\frac{\partial u}{\partial x_1} + v\frac{\partial u}{\partial x_2} + \frac{\partial p}{\partial x_1} - \mu\left(\frac{\partial^2 u}{\partial x_2^2} + \frac{\partial^2 u}{\partial x_2^2}\right) = 0$$

$$e_2 = u\frac{\partial v}{\partial x_1} + v\frac{\partial v}{\partial x_2} + \frac{\partial p}{\partial x_1} - \mu\left(\frac{\partial^2 v}{\partial x_2^2} + \frac{\partial^2 v}{\partial x_2^2}\right) = 0$$

$$e_3 = \frac{\partial u}{\partial x_1} + \frac{\partial v}{\partial x_2} = 0$$

then these can always be converted into an optimisation problem in one of two ways.

If the equations are self-adjoint then using the calculus of variations a variational function I exists

$$I = \int_n f \, dA$$

such that the Euler Lagrange equations are simply the partial differential equations.

A well known example of this is the fact that

$$\text{Min } I = \int \left(\frac{\partial u}{\partial x}\right)^2 + \left(\frac{\partial u}{\partial u}\right)^2 dA$$

corresponds to the Laplace equation

$$\frac{\partial^2 u}{\partial x^2} + \frac{\partial^2 u}{\partial y^2} = 0.$$

Appropriate boundary values must of course be imposed.

For non self-adjoint systems a least squares approach may be used as any set of equations

$$e_i(u) = 0$$

can be solved by minimising

$$\text{Min } I = \int e_i^2(u) \, dA.$$

In practice a preconditioning operator J(x) is introduced and the minimisation becomes

$$\text{Min } I = \int e^T J^T J e \, dA$$

which also has a global minimum I = 0 at the solution of the partial differential equations if J is nonsingular.

To minimise such a 2 dimensional integral on the ICL-DAP we first imbed the area of interest into a rectangle, the rectangle is then divided into a regular mesh of up to 64 x 64 elements. Each element is then associated with a processor of the ICL-DAP maintaining the 2 dimensional geometric connections. If an element is outside the area of interest the corresponding processor is turned off.

Shape functions $\phi(x)$ are now introduced into each element and the unknowns u(x) approximated in the simplest case by

$$u_A(x) = \sum_i u_i \phi_i(x)$$

where u_i are the values of u at the nodes of the element and $\phi_i(x)$ is in this case bilinear.

The approximation $u_A(x)$ is then substituted into the integral

$$I = \int F(u)\, dA$$

which is then evaluated element by element

$$I = \sum_{el} I_{el} = \sum_{el} \int_{el} F(u)\, dA.$$

The values of I_{el} are evaluated in parallel on their respective processors. We note that I_{el} only depends on the values u_i at the nodes of that element and if these u_i are stored on the appropriate processors all information needed to evaluate I_{el} is available by fast data transfer from nearest neighbour processors. The gradient

$$\nabla I = \sum_{el} \nabla I_{el}$$

and as I_{el} only depends on neighbouring values of u_i ∇I_{el} has essentially only 4 N_{Node} nonzero elements where N_{Node} is the number of unknowns at each mode. We may therefore store ∇I_{el} as a $4N_{el}$ vector and distribute N_{Node} components to each processor holding values of u_i associated with the element.

In the conjugate gradient algorithm the direction of search p_i will be distributed identically to u_i and the direction is simply formed by summing the contribution from 4 elements on each processor.

The Truncated Newton method with Gp estimated by difference can and has been implemented in an identical way.

When storage permits the Truncated Newton methods is much faster if $\nabla^2 I$ is calculated and stored

$$\nabla^2 I = \sum_{el} \nabla^2 I_{el}$$

and $\nabla^2 I_{el}$ contains at most 16 x N^2_{Node} nonzeros. Ducksbury showed that when these can be stored on each processor the resultant algorithm is much faster than the Gp difference approach.

The observant reader will have noticed that the structure used above is precisely that which would have been used by structured automatic differentiation which would naturally evaluate I_{el}, ∇I_{el}, $\nabla^2 I_{el}$ in terms of the local variables as no others would appear. For efficiency it would be necessary to ensure that all the elements were evaluated before the summation commenced as otherwise the automatic differentiation code might cease to only use local variables.

It is interesting to note that as $\nabla^2 I_{el}$ only depends on 4 N_{Node} elements, it could be calculated using 4 N_{Node} evaluations of ∇I_{el} by using the SR1 variable metric formula. The 4 N_{Node} evaluations of all the ∇I_{el} is not identical to 4 N_{Node} evaluations of ∇I (it is less!) but the computation is essentially the same. This is therefore an alternative very efficient way of computing $\nabla^2 I$ for this type of problem.

It is possible with this type of approach to achieve efficiencies considerably greater than 1 because it is no longer necessary even to number the variables or to assemble the matrix or to perform most of the data transfer needed on a sequential machine. Essentially the code uses the two dimensional geometry of the processor grid to avoid all such overheads.

Ducksbury reports speed ups of 100 in CPU time on a 64 x 64 solutions of Laplace's problem; and on a 128 x 128 problem a speed up of 1000 using the achieved DAP time and a log linearly extrapolated DEC 1091 time. The DEC 1091 could not be used with on-line store at this dimension and the log linear extrapolation is therefore more than fair. Given our earlier comments on the relative speed of the machines this is impressive.

Ducksbury's results regularly confirmed the superiority of the Truncated Newton code over the conjugate gradient code on the DAP.

However the most impressive such result is due to Dolan when solving the Navier-Stokes flow past a circular cylinder. The finite element used was a complete cubic formulation of the stream function ψ; the element contained 13 unknowns, the values of ψ, u, v, p at the nodes and ψ at the centroid.

After 900 gradient calls, the conjugate gradient method had a value of $I = 6.61 \text{ D } 10^{-2}$ and $|| \nabla I ||^2 = 1.19 \text{ D } 10^{+3}$ whilst the Truncated Newton code using Gp by differences had $I = 1.83 \text{ D } 10^{-5}$ and $|| \nabla I ||^2 = 2.34 \text{ D } 10^{-4}$. This is probably one of the most dramatic improvements so far reported.

It should be stressed that all the Truncated Newton methods discussed in this paper have solved the sparse symmetric set of equations

$$\nabla^2 I d = - \nabla I$$

by the conjugate direction method. Ducksbury used a diagonal preconditioner, but even so it is only fair to admit that problems are known where due to the instability of the conjugate direction code, the direction d is not particularly good.

Essentially the inclusion of either a more stable sparse equation solver or a preconditioner that could be distributed over the processors of a DAP array seem desirable. Both are current topics of research.

References to Section 5

1. Dolan, P S, Viscous incompressible flow solutions via divergence free least squares finite element optimisation, PhD thesis, The Hatfield Polytechnic, 1987.

2. Ducksbury, P G, An investigation of the relative merits of optimisation algorithms on the ICL-DAP, PhD thesis, The Hatfield Polytechnic, 1985, published as "Parallel Array Processing", Ellis Horwood, 1986.

Solution of Sparse Systems of Equations on Multiprocessor Architectures *

Alan George †
Departments of Computer Science and Mathematics
The University of Tennessee
Knoxville, Tennessee 37996-1301

Abstract

The primary objective of these lecture notes is to introduce the reader to some basic techniques for solving large sparse systems of equations, along with some currently active areas of research on the subject. In addition, introductory material is included on the basic problems and techniques associated with performing numerical computations on multiprocessors, using the problem of solving large sparse systems as an application example.

*This work was supported in part by Canadian Natural Sciences and Engineering Research Council grant A8111, by the Applied Mathematical Sciences Research Program, Office of Energy Research, U.S. Department of Energy under contract DE-AC05-84OR21400 with Martin Marietta Energy Systems Inc., by the U.S. Air Force Office of Scientific Research under contract AFOSR-ISSA-87-00013, and by the Science Alliance, a state supported program at the University of Tennessee.

†Also affiliated with the Mathematical Sciences Section, Oak Ridge National Laboratory, through the UT/ORNL Distinguished Scientist program.

1 Introduction

These notes are intended to give the reader an understanding of some of the basic techniques involved in solving large sparse systems of equations, along with some introductory material on parallel processors and how they may be used to solve such systems. The material is intended to be self-contained in the sense that little background in either sparse matrix computation or parallel computation is assumed. However, the reader should be familiar with the basic techniques and results of numerical linear algebra, such as those found in the book by Forsythe and Moler [14], and with standard matrix notation and manipulation. Only direct methods based on Gaussian elimination are considered, and no attempt has been made to be complete in the coverage of even this restricted class of methods for solving sparse systems.

These notes borrow heavily from a number of sources. The basic material on sparse matrix computation dealing with positive definite systems follows the development in Chapters 1 and 5 of the book by George and Liu [22], and the material dealing with Gaussian elimination with partial pivoting draws heavily from articles [21,28,29,30] and [32]. The material on parallel architectures and algorithms is derived from the work contained in articles [17,18,15,19,20,26] and [27]. The author is grateful to his collaborators Drs. Micheal Heath and Esmond Ng of the Oak Ridge National Laboratory for their reading of the manuscript, and numerous helpful suggestions.

2 Basic Material on Sparse Matrix Computation

2.1 Sparsity versus Stability

Suppose the given $n \times n$ system of equations to be solved is

$$Ax = b,$$

where A is nonsingular and sparse, and n is assumed to be large.

If A has no special properties, it is well-known that applying Gaussian elimination to A may fail. A zero element might turn up in the pivot position. For the moment, let us ignore this difficulty, and assume that the application of Gaussian elimination to A yields the triangular factorization

$$A = LU,$$

where L is unit lower triangular and U is upper triangular. The solution x can then be obtained by solving the two triangular systems

$$Ly = b,$$

and

$$Ux = y.$$

An important fact about this process in the context of A being sparse is that the matrix normally suffers *fill* during the factorization. That is, L and U usually have nonzeros in positions which are zero in the corresponding positions in A.

A second important fact is that permuting the rows and columns of A can have an enormous effect on the amount of fill that occurs during the factorization. A classical example is the so-called "arrow" matrix, shown on the left in Figure 1, where $\times$ denotes a nonzero element in the matrix. The matrix suffers complete fill when Gaussian elimination is applied. The structures of its factors are displayed on the right in Figure 1. On the other hand, a simple reordering of the rows and columns of the matrix yields the structure shown on the left in Figure 2, which suffers no fill at all. This is an extreme example, but it illustrates the point. Most sparse matrices do suffer some fill, but the amount depends very strongly on the ordering of their rows and columns. Of course we have ignored the fact that we might not have been allowed to reorder the matrix in this manner because of numerical stability considerations.

$$\begin{pmatrix} \times & \times & \times & \times & \times & \times \\ \times & \times & & & & \\ \times & & \times & & & \\ \times & & & \times & & \\ \times & & & & \times & \\ \times & & & & & \times \end{pmatrix} = \begin{pmatrix} \times & & & & & \\ \times & \times & & & & \\ \times & \times & \times & & & \\ \times & \times & \times & \times & & \\ \times & \times & \times & \times & \times & \\ \times & \times & \times & \times & \times & \times \end{pmatrix} \begin{pmatrix} \times & \times & \times & \times & \times & \times \\ & \times & \times & \times & \times & \times \\ & & \times & \times & \times & \times \\ & & & \times & \times & \times \\ & & & & \times & \times \\ & & & & & \times \end{pmatrix}$$

Figure 1: Nonzero pattern of A and its triangular factors.

$$\begin{pmatrix} \times & & & & & \times \\ & \times & & & & \times \\ & & \times & & & \times \\ & & & \times & & \times \\ & & & & \times & \times \\ \times & \times & \times & \times & \times & \times \end{pmatrix} = \begin{pmatrix} \times & & & & & \\ & \times & & & & \\ & & \times & & & \\ & & & \times & & \\ & & & & \times & \\ \times & \times & \times & \times & \times & \times \end{pmatrix} \begin{pmatrix} \times & & & & & \times \\ & \times & & & & \times \\ & & \times & & & \times \\ & & & \times & & \times \\ & & & & \times & \times \\ & & & & & \times \end{pmatrix}$$

Figure 2: Nonzero pattern of reordered A from Figure 1 and its triangular factors.

Note that here and at all times we make the usual *no-cancellation assumption.* That is, whenever two nonzero quantities are added or subtracted, the result is nonzero. This means that in the analysis we ignore any zeros which might be created through exact cancellation. Such cancellation rarely occurs, and any such prediction would be difficult in general, particularly in floating point arithmetic which is subject to rounding errors.

Now let us return to the issue of whether A has a triangular factorization, and even if it does, whether it can be computed in a numerically stable manner. It is well-known that for a general indefinite matrix A, some form of *pivoting* (row and/or column interchanges) is necessary to ensure numerical stability. Thus given A, one normally obtains a factorization of P_rA or P_rAP_c, where P_r and P_c are $n \times n$ permutation matrices. When A is sparse, these permutations are determined *during the factorization* by a combination of (usually competing) numerical stability and sparsity requirements. Different matrices, even though they may have the same nonzero pattern, will normally yield different P_r and P_c, and therefore have factors with different sparsity patterns. In other words, it is in general not possible to predict where fill will occur for general sparse matrices before the numerical computation begins. Thus, if we wish to store only the nonzeros during the computation, we are obliged to use some form of *dynamic* storage scheme that allocates storage for fill as the computation proceeds. This is the approach adopted by Duff in his code MA28 [6], Gilbert and Peierls [34], Sherman [48], and Zlatev et al. [53]. The approach we describe in these notes is somewhat different: in exchange for storing *some* zeros, we are able to use a *static* data structure during the numerical computation. Whether this approach is superior or inferior to other schemes depends on the type of matrix problem being solved. Some experimental comparisons bearing on this point are contained in [28].

2.2 Positive Definite Systems

In this section we consider the case when A is symmetric and positive definite. Many linear equation problems arising in science and engineering have this property, and the problem of solving such systems is both different from and easier than that for general sparse A.

Cholesky's method is a symmetric variant of Gaussian elimination tailored to positive definite matrices. Applying Cholesky's method to A yields the triangular factorization

$$A = LL^T,$$

where L is lower triangular with positive diagonal elements.

One of the key advantages of these matrices is that Gaussian elimination applied to a symmetric positive definite matrix does not require interchanges (pivoting) to maintain numerical stability. Since PAP^T is also symmetric and positive definite for any permutation matrix P, this means we can choose to reorder A symmetrically to reduce fill *without regard to numerical stability* and *before the actual numerical factorization begins.*

These options, which are not normally available to us when A is a general indefinite matrix, have enormous practical implications. Since the ordering can be determined before the factorization begins, the locations of the fill suffered during the factorization can also be determined. Thus, the data structure used to store L can be constructed prior to the actual numerical factorization, and spaces for fill components can be reserved. The computation then proceeds with the storage structure remaining *static* (unaltered). Thus, the three problems of i) finding a suitable ordering, ii) setting up the appropriate storage scheme, and iii) the actual numerical computation, can be isolated as separate objects of study, as well as separate computer software modules.

Thus, solving large sparse positive definite systems $Ax = b$ typically involves four distinct steps:

1. (Ordering) Find a good ordering P for A; that is, determine a permutation matrix P so that PAP^T has a sparse Cholesky factor L.

2. (Symbolic factorization) Determine the structure of L, and set up a data structure for this factor.

3. (Numerical factorization) Place the elements of A into the data structure, and then compute L.

4. (Triangular solution) Using the computed L, solve the triangular systems $Ly = Pb, L^T z = y$, and then set $x = P^T z$.

This independence of tasks has a number of distinct advantages. It encourages software modularity, and, in particular, allows us to tailor storage methods to the given task at hand. For example, the use of lists to store matrix subscripts may be quite appropriate for an implementation of an ordering algorithm, but decidedly inappropriate in connection with actually storing the matrix or its factors. In the same vein, knowing that we can use a storage scheme in a static manner during the factorization sometimes allows us to select a method which is very efficient in terms of storage requirements, but would be a disaster in terms of bookkeeping overhead if it had to be altered during the factorization. Finally, in many engineering design applications, numerous *different* positive definite matrix problems having the *same* structure must be solved. Obviously, the ordering and storage scheme set-up need to be performed only once, so it is desirable to have these tasks isolated from the actual numerical computation.

In numerous practical situations matrix problems arise which are unsymmetric but have symmetric structure, and for which it can be shown that pivoting for numerical stability is not required when Gaussian elimination is applied. Almost all the ideas and algorithms described here extend immediately to this class of problems.

In these notes we do not deal with the ordering problem, although the efficiency of algorithms for steps 2, 3 and 4 depends critically on finding a good ordering. Currently, the most successful general-purpose algorithm for this task is the *minimum degree algorithm*. A good description of it, along with a highly successful implementation, can be found in [41]. A historical perspective on its development is described in [24].

2.2.1 Solving Triangular Systems

Before we consider the problem of computing the factorization of matrices, let us consider the simpler problem of solving triangular systems. To be specific, we will consider the system

$$Tx = b$$

where T is nonsingular and lower triangular. The analogous results for upper triangular matrices will be obvious.

There are two common ways of solving the system, which differ only in the order in which the operations are performed. The first one, which we will call the *scalar-product form*, is given below.

$$\begin{array}{l} \textbf{for } i = 1,2,\ldots,n \textbf{ do} \\ \quad x_i \leftarrow \left(b_i - \sum_{k=1}^{i-1} t_{ik}x_k\right)/t_{ii} \end{array}$$

The second method uses the matrix components of T column by column, and we will refer to it as the *vector-sum form*.

$$\begin{array}{l} \textbf{for } i = 1,2,\ldots,n \textbf{ do} \\ \quad x_i \leftarrow b_i/t_{ii} \\ \quad \begin{pmatrix} b_{i+1} \\ \vdots \\ b_n \end{pmatrix} \leftarrow \begin{pmatrix} b_{i+1} \\ \vdots \\ b_n \end{pmatrix} - x_i \begin{pmatrix} t_{i+i,1} \\ \vdots \\ t_{ni} \end{pmatrix} \end{array}$$

Note that this scheme lends itself to exploiting sparsity in the solution x. If b_i turns out to be zero at the beginning of the i-th step, x_i is zero and the entire step can be skipped. The former solution method accesses the components of the lower triangular matrix row by row and therefore lends itself to row-wise storage schemes. If the matrix is stored column by column, the latter method is more appropriate. Note that this column-oriented method is often used to solve the *upper* triangular system

$$L^T x = y,$$

where L is a lower triangular matrix stored using a row-wise scheme.

2.2.2 Cholesky Factorization: $A = LL^T$

Let the symmetric positive definite matrix A be partitioned as shown below, where d_1 is a scalar and v_1 is the remaining part of the first column. The first step of the Cholesky factorization algorithm can then be described in matrix terms as follows.

$$\begin{aligned} A &= A_0 = E_0 = \begin{pmatrix} d_1 & v_1^T \\ v_1 & \bar{E}_1 \end{pmatrix} \\ &= \begin{pmatrix} \sqrt{d_1} & 0 \\ v_1/\sqrt{d_1} & I \end{pmatrix} \begin{pmatrix} 1 & 0 \\ 0 & \bar{E}_1 - v_1 v_1^T/d_1 \end{pmatrix} \begin{pmatrix} \sqrt{d_1} & v_1^T/\sqrt{d_1} \\ 0 & I \end{pmatrix} \\ &= L_1 \begin{pmatrix} 1 & 0 \\ 0 & E_1 \end{pmatrix} L_1^T \\ &= L_1 A_1 L_1^T. \end{aligned}$$

The matrix E_1 can then be similarly partitioned, and the process repeated, yielding $A_1 = L_2 A_2 L_2^T$ after the second step, as shown below.

$$\begin{aligned} A_1 &= \begin{pmatrix} 1 & 0 \\ 0 & E_1 \end{pmatrix} = \begin{pmatrix} 1 & 0 & 0 \\ 0 & d_2 & v_2^T \\ 0 & v_2 & \bar{E}_2 \end{pmatrix} \\ &= L_2 A_2 L_2^T. \end{aligned}$$

After n steps of the algorithm, we have

$$A = L_1 L_2 \cdots L_n L_n^T \cdots L_2^T L_1^T,$$

where it can easily be shown that

$$L = L_1 + L_2 + \cdots + L_n - (n-1)I_n.$$

Thus the ith column of L is precisely the ith column of L_i.

In this scheme, the columns of L are computed one by one. At the same time, each step involves the modification of the submatrix $\bar{E}_i$ by the outer product $v_i v_i^T / d_i$ to give E_i, which is simply the submatrix remaining to be factored. This formulation of the algorithm is referred to as the *outer product form* or *submatrix-Cholesky*.

We now consider an alternative formulation of the factorization process, often referred to as the *bordering method* or *row-Cholesky*. Suppose the matrix A is partitioned as

$$A = \begin{pmatrix} M & u \\ u^T & s \end{pmatrix},$$

where the symmetric factorization $L_M L_M^T$ of the $(n-1) \times (n-1)$ leading principal submatrix M has already been obtained. Then the factorization of A is given by

$$A = \begin{pmatrix} L_M & 0 \\ w^T & t \end{pmatrix} \begin{pmatrix} L_M^T & w \\ 0 & t \end{pmatrix},$$

where

$$w = L_M^{-1} u$$

and

$$t = \left(s - w^T w\right)^{1/2}.$$

Note that the factorization $L_M L_M^T$ of the submatrix M is also obtained by the bordering technique. Thus, the scheme can be described as follows.

for $i = 1, 2, \ldots, n$ **do**

Solve $\begin{pmatrix} l_{11} & & O \\ \vdots & \ddots & \\ l_{i,i-1} & \cdots & l_{i-1,i-1} \end{pmatrix} \begin{pmatrix} l_{i1} \\ \vdots \\ l_{i,i-1} \end{pmatrix} = \begin{pmatrix} a_{i1} \\ \vdots \\ a_{i,i-1} \end{pmatrix}$

$l_{ii} \leftarrow \left(a_{ii} - \sum_{k=1}^{i-1} l_{ik}^2\right)^{1/2}$

In this scheme, the rows of L are computed one at a time; the part of the matrix remaining to be factored is not accessed until the corresponding part of L is to be computed.

Note that the solution of a triangular system is involved here, and in the previous section we saw two different forms for triangular solution. Thus, depending upon which algorithm is used, we will have either the *scalar-product row-Cholesky* algorithm or the *vector-sum row-Cholesky* algorithm.

The final scheme we consider for computing the components of L will be referred to as *column-Cholesky*. It can be described as follows.

$$
\begin{aligned}
&\textbf{for } j = 1, 2, \ldots, n \textbf{ do} \\
&\quad l_{jj} \leftarrow \left(a_{ii} - \textstyle\sum_{k=1}^{j-1} l_{jk}^2\right)^{1/2} \\
&\quad \textbf{for } i = j+1, j+2, \ldots, n \textbf{ do} \\
&\quad\quad l_{ij} \leftarrow \left(a_{ij} - \textstyle\sum_{k=1}^{j-1} l_{ik} l_{jk}\right) / l_{jj}
\end{aligned}
$$

These formulae can be derived directly by equating the elements of A to the corresponding elements of the product LL^T.

Like the outer product (or submatrix) version of the algorithm, the columns of L are computed one by one, but the part of the matrix remaining to be factored is not accessed during the scheme.

Note that this scheme need not compute inner products. It can instead be formulated as shown below, where all the inner products can be regarded as being computed "in parallel".

$$
\begin{aligned}
&\textbf{for } j = 1, 2, \ldots, n \textbf{ do} \\
&\quad \textbf{for } k = 1, 2, \ldots, j-1 \textbf{ do} \\
&\quad\quad \begin{pmatrix} a_{jj} \\ \vdots \\ a_{nj} \end{pmatrix} \leftarrow \begin{pmatrix} a_{jj} \\ \vdots \\ a_{nj} \end{pmatrix} - l_{jk} \begin{pmatrix} l_{jk} \\ \vdots \\ l_{nk} \end{pmatrix} \\
&\quad l_{jj} \leftarrow \sqrt{a_{jj}} \\
&\quad \textbf{for } k = j+1, j+2, \ldots, n \textbf{ do} \\
&\quad\quad l_{kj} \leftarrow a_{kj} / a_{jj}
\end{aligned}
$$

Thus, we again have two forms of the algorithm, depending upon whether computation is done using scalar products or vector sums. We will refer to these as the *scalar-product column-Cholesky* and *vector-sum column-Cholesky* algorithms, respectively.

2.2.3 A Storage Scheme for L

The computer storage used for a sparse matrix involves two components: *primary storage* and *overhead storage*. Primary storage refers to that which is used for the actual numerical values, while overhead storage refers to that which is used to represent the *structure* of the matrix. That is, the latter information provides the locations in the matrix where the numbers contained in the primary storage belong, along with a mechanism for accessing those numbers. Ideally, the overhead storage should be small, and the accessing mechanism should be rapid and convenient to use. The scheme discussed here is column-oriented and is due to Sherman [49]. We begin with the so-called *uncompressed subscript* version.

The scheme has a main storage array LNZ which contains all the nonzero entries in the lower triangular factor. A storage location is provided for each nonzero in the factor. The nonzeros in L, excluding the diagonal, are stored column after column in LNZ. An accompanying vector $NZSUB$ is provided, which gives the row subscripts of the nonzeros. In addition, an index vector $XLNZ$ is used to point to the start of nonzeros in each column in LNZ (or equivalently $NZSUB$). The diagonal entries are stored separately in the vector $DIAG$.

An example of this storage scheme is given by the matrix in Figure 3 and the arrays in Figure 4.

$$A = \begin{pmatrix} l_{11} & & & & & & \\ l_{21} & l_{22} & & & & & \\ & & l_{13} & & & & \\ l_{41} & l_{42} & & l_{44} & & & \\ & & l_{53} & l_{54} & l_{55} & & \\ & & l_{63} & & l_{65} & l_{66} & \\ & & & & l_{75} & l_{76} & l_{77} \end{pmatrix}$$

Figure 3: Structure of the matrix L.

$DIAG$	l_{11}	l_{22}	l_{33}	l_{44}	l_{55}	l_{66}	l_{77}		
LNZ	l_{21}	l_{41}	l_{42}	l_{53}	l_{63}	l_{54}	l_{65}	l_{75}	l_{76}
$XLNZ$	1	3	4	6	7	9	10		
$NZSUB$	2	4	4	5	6	5	6	7	7

Figure 4: Uncompressed subscript representation of L in Figure 3.

Although this scheme is not particularly well suited for random access of nonzero entries, it lends itself quite readily to sparse factorization and solution.

We now consider an enhancement of this basic scheme, which is motivated by the fact that very often successive columns of L have large numbers of nonzeros in common positions. More specifically, apart from the diagonal position of column i, column i and column $(i+1)$ often have nonzeros in exactly the same rows. This is exploited by storing only one copy of their common subscripts.

In exchange for the compression, we need to have an auxiliary index vector $XNZSUB$ which points to the start of row subscripts in $NZSUB$ for each column. The compressed scheme for the example in Figure 3 is given in Figure 5.

In the compressed scheme, the primary storage remains the same, but the overhead storage is increased. The example given is too small to bring out the significance

$DIAG$	l_{11}	l_{22}	l_{33}	l_{44}	l_{55}	l_{66}	l_{77}		
LNZ	l_{21}	l_{41}	l_{42}	l_{53}	l_{63}	l_{54}	l_{65}	l_{75}	l_{76}
$XLNZ$	1	3	4	6	7	9	10		
$NZSUB$	2	4	5	6	5	6	7		
$XNZSUB$	1	2	3	5	6	7	8		

Figure 5: Compressed subscript representation of L in Figure 3.

of the compressed scheme. However, for problems of practical size, the subscript storage is often reduced by at least fifty percent, compared to the uncompressed scheme, with only a very slight increase in access time.

2.2.4 Symbolic Factorization

As its name implies, *symbolic factorization* is the process of simulating the numerical factorization of a given matrix A in order to obtain the nonzero structure of its factor L.

A naive way to determine the structure of L is to carry out the Cholesky factorization symbolically on the structure of A. The execution time of this straightforward algorithm is $O(\rho)$, where ρ is the number of floating-point operations performed in the numerical factorization. However, we can do much better than this, by looking more carefully at the structure of L. We begin with a fundamental result due to Parter [45].

Lemma 1 *Let $i > j$. Then $L_{ij} \neq 0$ if and only if at least one of the following conditions holds:*

a) $A_{ij} \neq 0$

b) *For some $k < j, L_{ik} \neq 0$ and $L_{jk} \neq 0$.*

In words, the lemma above states that L_{ij} is nonzero as a result of one *or both* of the following: a) the corresponding element of A is already nonzero, or b) some step of the factorization previous to step j caused fill to occur in position (i, j).

Here and elsewhere it is convenient to be able to discuss the structure of matrices and vectors. In order to do so we introduce the function Ω. For a vector v, $\Omega(v)$ is the set of subscripts corresponding to nonzeros in v, and for the matrix M, $\Omega(M)$ is the set of subscript pairs corresponding to nonzero elements of M. The notation $M^{\triangleleft}$ refers to the matrix obtained from M by setting all of its elements below the main diagonal to zero, and $M^{\triangleright}$ denotes the matrix obtained by setting all the elements above the main diagonal to zero. The notation M_{*i} and M_{i*} denotes the ith column and row respectively of the matrix M. Thus, for example, $\Omega(M^{\triangleright}_{*i}) - \{i\}$ is the set of row subscripts of nonzeros in column i of M below the diagonal.

Let m_k denote the row index of the first off-diagonal nonzero in column k of L, if $|L_{*k}| > 1$. If column k of L does not have any off-diagonal nonzeros, we set $m_k = k$.

Lemma 2

$$\{\Omega(L_{*k}) - \{k\}\} \subset \Omega(L_{*m_k})$$

Proof: We assume that $k < m_k$, since the result is trivial otherwise. For $i \in \{\Omega(L_{*k}) - \{k\}\}$, we have $k < m_k \leq i$. For $i = m_k$, there is nothing to prove, since $L_{m_k,m_k} \neq 0$. Otherwise, by Lemma 1, $L_{ik} \neq 0$ and $L_{m_k,k} \neq 0$ implies that $L_{i,m_k} \neq 0$. □

Thus, for $i \in \{\Omega(L_{*k}) - \{k\}\}$ and $i > m_k$, it is redundant to consider column k when determining the structure of column i, since the relevant structure can be found in column m_k.

This result enables us to show that $\Omega(L_{*i})$ can be characterized as follows [22].

Lemma 3

$$\Omega(L_{*i}) = \Omega(A^{\triangleright}_{*i}) \cup (\cup_{k<i}\{\Omega(L_{*k}) \mid m_k = i\}) - \{1, 2, \cdots, i-1\}.$$

Proof: Consider any k for which $i \in \{\Omega(L_{*k}) - \{k\}\}$. Using the notation $m(k)$ to mean m_k, $m^2(k)$ to mean m_{m_k}, and so on, we can generate an increasing sequence of subscripts that is bounded above by i:

$$k < m(k) < m^2(k) < m^3(k) < \cdots < i.$$

Thus, there exist an integer p such that $m^{p+1}(k) = i$. Let $\mathcal{T} = \{1, 2, \ldots, i-1\}$. Then by Lemma 2 we have

$$\{\Omega(L_{*k}) - \mathcal{T}\} \subset \{\Omega(L_{*,m(k)}) - \mathcal{T}\} \subset \cdots \subset \{\Omega(L_{*,m^p(k)}) - \mathcal{T}\}$$

The result then follows. □

In words, the lemma says that the structure of column i of L is given by the structure of column i of A (excluding the portion above the diagonal), together with the structures of those columns of L whose first off-diagonal nonzeros are in row i. Consider the example in Figure 6, where $\times$ denotes a nonzero and $+$ denotes a fill due to factorization. The structure of column 4 of L is given by the union of the structure of column 4 of A and the structures of columns 2 and 3 of L.

$$A = \begin{pmatrix} \times & & & & \times & & & \times \\ & \times & & \times & \times & & & \\ & & \times & \times & & & & \times \\ & \times & \times & \times & & & \times & \\ \times & \times & & & \times & \times & & \\ & & & & \times & \times & & \times \\ & & & \times & & & \times & \\ \times & & \times & & & \times & & \times \end{pmatrix}, \quad L = \begin{pmatrix} \times & & & & & & & \\ & \times & & & & & & \\ & & \times & & & & & \\ & \times & \times & \times & & & & \\ \times & \times & & + & \times & & & \\ & & & & \times & \times & & \\ & & & \times & + & + & \times & \\ \times & & \times & + & + & \times & + & \times \end{pmatrix}$$

Figure 6: Structure of the matrices A and L.

We can now formulate a symbolic factorization algorithm using the characterization of L provided by Lemma 3 above. The algorithm is given in Figure 7. At the end of the symbolic factorization algorithm, the structure of column k of L, $\Omega(L_{*k})$, is given by the elements of $\mathcal{L}_k$. The set $\mathcal{S}_k$ is used to record the columns of L whose

for $k = 1, 2, \ldots, n$ **do**
 $\mathcal{S}_k \leftarrow \phi$
for $k = 1, 2, \ldots, n$ **do**
 $\mathcal{L}_k \leftarrow \Omega(A^{\triangleright}_{*k})$
 for $i \in \mathcal{S}_k$ **do**
  $\mathcal{L}_k \leftarrow \mathcal{L}_k \cup \mathcal{L}_i - \{i\}$
 determine m_k
 if $m_k > k$ **then**
  $\mathcal{S}_{m_k} \leftarrow \mathcal{S}_{m_k} \cup \{k\}$

Figure 7: Symbolic Factorization Algorithm for Symmetric Matrices.

structures will affect that of column k of L. It is not hard to see that the complexity of the algorithm is $O(|L|)$. This is certainly more efficient than the straightforward approach mentioned earlier.

Note that when $\mathcal{S}_k$ contains only one column, say i, and $\Omega(A^{\triangleright}_{*k}) \subset \Omega(L_{*i})$, then clearly $\Omega(L_{*k}) = \Omega(L_{*i}) - \{k\}$. This can be used to speed up the symbolic factorization algorithm. In fact, the conditions are often satisfied when k is large, since column k of L tends to become dense as k increases.

To conclude our discussion in this section, let us describe the data structure used in the implementation of the algorithm on serial machines. The set $\mathcal{S}_k$ can be represented using a linked list. When a new column, say i, has to be added to $\mathcal{S}_k$, i is inserted at the beginning of the linked list. Moreover, because of the way the set $\mathcal{S}_k$ is defined, $\mathcal{S}_k \cap \mathcal{S}_l = \phi$ for $k \neq l$, and the total number of elements in all these sets can never exceed n. Hence the linked lists for the sets $\mathcal{S}_k$ can be stored in a one-dimensional integer array of size n. The set $\mathcal{L}_k$ can be represented using an ordered linked list. Since the columns of L are computed sequentially, only one such linked list is present at any given time. Thus, the linked lists for the $\mathcal{L}_k$'s can be implemented using only one integer array of size n as well. When $\mathcal{L}_k$ is fully determined, its elements can be copied to consecutive locations in a subscript array. See [22] for more details on the implementation.

2.2.5 Numerical Factorization

In this section we describe some important implementation details associated with using the column-oriented data structure described in section 2.2.3. These techniques are used in numerous algorithms that we describe later. It is assumed that the data structure contains the elements of A, and the factorization is performed "in place". The vector-sum column-Cholesky form of the algorithm is used, because the matrix is stored by columns.

We make use of three temporary vectors $LINK, FIRST$ and $TEMP$, all of size n. Consider computing a column L_{*i} of the factor. The columns that are involved in the formation of L_{*i} are exactly those L_{*j} with $l_{ij} \neq 0$. The modification can be done one column at a time as follows:

$$\textbf{for } j \in \{\Omega(L_{i*}) - \{i\}\} \textbf{ do}$$
$$\begin{pmatrix} l_{ii} \\ \vdots \\ l_{ni} \end{pmatrix} \leftarrow \begin{pmatrix} l_{ii} \\ \vdots \\ l_{ni} \end{pmatrix} - l_{ij} \begin{pmatrix} l_{ij} \\ \vdots \\ l_{nj} \end{pmatrix}$$

The vector $LINK$ is maintained so that at step i, all the columns that affect L_{*i} are given by the list

$$LINK(i), LINK(LINK(i)), LINK(LINK(LINK(i))), \cdots$$

To minimize subscript searching, a work vector $FIRST$ is used so that $FIRST(j)$ points to the location in the storage vector LNZ where the nonzero l_{ij} resides for

$$j = LINK(i), LINK(LINK(i)), \cdots.$$

In this way, the modification of L_{*i} by L_{*j} can start at the location $FIRST(j)$ in LNZ. Note that the success of this technique depends on the fact that the elements of each column are stored in increasing order of their row subscripts. The third working vector $TEMP$ is used to accumulate the modifications to the column A_{*i}.

2.2.6 The Elimination Tree of a Cholesky Factor L

In this section we introduce the notion of the *elimination tree* corresponding to a sparse symmetric matrix A [7,8,38,43,47]. As we shall see in subsequent sections, this structure is useful in the design of data structures, and also useful in identifying and exploiting parallelism. In this section we simply define the tree and show how it can be economically represented.

Consider the structure of the Cholesky factor L of A. For each column $j \leq n$, if column j has off-diagonal nonzeros, define $parent[j]$ by

$$parent[j] = \min\{i \mid l_{ij} \neq 0, i > j\};$$

that is, $parent[j]$ is the row subscript of the first off-diagonal nonzero in column j of L. If column j has no off-diagonal nonzero, we set $parent[j] = j$. (Hence $parent[n] = n$.)

We now define an elimination tree corresponding to the structure of L. The tree has n vertices, labelled from 1 to n. For each j, if $parent[j] > j$, then vertex $parent[j]$ is the *parent* of vertex j in the elimination tree, and vertex j is one of possibly several *child* vertices of vertex $parent[j]$. We assume that the matrix A is *irreducible* , so that n is the only vertex with $parent[j] = j$ and it is the *root* of the tree. Thus, for $1 \leq j < n$, $parent[j] > j$. (If A is reducible, then the elimination tree defined above is actually a forest which consists of several trees.) There is exactly one *path* from each vertex to the root of the tree. If vertex i lies on the path from vertex j to the root, then vertex i is an *ancestor* of vertex j, and vertex j is a *descendant* of vertex i. Vertices which are the same distance from the root are said to be at the same level in the tree. By the *subtree rooted at node* i, or simply *subtree* i, we mean the subtree consisting of vertex i together with *all* of its descendants in the tree.

An example to illustrate the notion of elimination tree is provided by the structure of the Cholesky factor shown in Figure 8, with the associated elimination tree being shown in Figure 9. Elimination trees have been used either implicitly or explicitly in numerous articles dealing with sparse symmetric factorization [5,8,9,10,11,13,25,36,38,42,43,46,47]. In particular, in [39], Liu uses the elimination tree as a model to study the parallel sparse Cholesky factorization algorithm on multiprocessors.

$$L = \begin{pmatrix} \times & & & & & \\ & \times & & & & \\ \times & & \times & & & \\ & & \times & \times & & \\ \times & \times & \times & \times & \times & \\ & \times & & \times & \times & \times \end{pmatrix}$$

Figure 8: Structure of a Cholesky factor.

The elimination tree has a simple structure that can be economically represented using *parent*, as shown in Figure 10. Thus, the representation requires only a single vector of size n.

In some contexts it is necessary to be able to obtain the elimination tree prior to determining the structure of L. Fortunately, there is a very efficient algorithm that can be used to obtain the tree directly from the structure of the matrix A [38]. Its complexity can be shown to be $O(|A| \log_2 n)$ [52].

2.3 Sparse Orthogonal Factorization: $A = QR$

We saw in the previous section that there are practical reasons why it is desirable to isolate the ordering, data structure set-up, and numerical computation aspects of solving sparse systems. This appears to be impossible for general A, for which pivoting is required, since the positions of the nonzeros in the triangular factors depend not only on the structure of the matrix, but also depend on the interchanges. Since the interchanges are not known until the numerical decomposition is performed, the actual structures of these matrices are revealed only during the actual computation.

An alternative approach to Gaussian elimination for solving indefinite systems of equations is to employ orthogonal techniques to factor A. Such methods have not been popular for sparse systems because it has been believed that the factors Q and

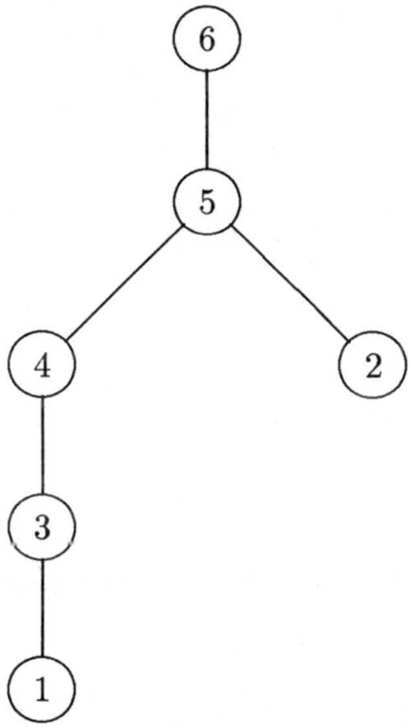

Figure 9: The elimination tree associated with the Cholesky factor in Figure 8.

j	1	2	3	4	5	6
$parent[j]$	3	5	4	5	6	6

Figure 10: Computer representation of the tree of Figure 9 .

R of A would have unacceptably large numbers of nonzeros. For certain classes of problems, this turns out not to be so, as shown in [30].

In the following several sections we assume that A has a zero-free diagonal. This assumption is convenient in some of our arguments, and in the implementation of our codes. However, the bounds we derive are true, and our algorithms can be modified to work correctly, irrespective of whether A has a zero-free diagonal. Our development follows that contained in [28].

We now consider the use of Householder transformations in the orthogonal decomposition of a matrix. Partition A into

$$A = \begin{pmatrix} d & v^T \\ u & E \end{pmatrix},$$

and consider applying one step of Householder reduction to A.

Using standard procedures[51], a scalar β and vector w can be computed to construct a Householder reflection Q_1 to annihilate the vector u in the matrix A. The matrix Q_1 has the form

$$Q_1 = I - \frac{1}{\beta}\begin{pmatrix} \beta \\ w \end{pmatrix}\begin{pmatrix} \beta \\ w \end{pmatrix}^T.$$

The vector $\binom{\beta}{w}$ is referred to as the *Householder vector.* After the transformation the matrix becomes

$$Q_1 A = \bar{A} = \begin{pmatrix} \sigma_d & v^T - y^T \\ 0 & E - wy^T/\beta \end{pmatrix} = \begin{pmatrix} \sigma_d & z^T \\ 0 & \bar{E} \end{pmatrix}, \tag{1}$$

where

$$y^T = \beta v^T + w^T E. \tag{2}$$

That is, y^T is a linear combination of v^T and some rows of E. The diagonal element σ_d is the Euclidiean length of the first column of A, with the sign chosen appropriately for numerical stability. Also note that since w is a scalar multiple of u, it follows that

$$\Omega(w) = \Omega(u). \tag{3}$$

Thus, the matrix A can be reduced to upper triangular form by a sequence of Householder transformations

$$Q_1, Q_2, \cdots, Q_{n-1}$$

defined by the corresponding Householder vectors

$$\begin{pmatrix} \beta_1 \\ w_1 \end{pmatrix}, \begin{pmatrix} \beta_2 \\ w_2 \end{pmatrix}, \cdots, \begin{pmatrix} \beta_{n-1} \\ w_{n-1} \end{pmatrix},$$

where β_i is a scalar, and w_i is a vector of size $n-i$. We are interested in the structure of this set of Householder vectors when A is sparse. For this purpose, we define an $n \times n$ lower triangular matrix H to be

$$H = \begin{pmatrix} \beta_1 & 0 & 0 & \cdots & 0 \\ & \beta_2 & 0 & \cdots & 0 \\ w_1 & & \beta_3 & \cdots & \vdots \\ & w_2 & & \ddots & \\ & & w_3 & \cdots & \beta_n \end{pmatrix}.$$

Since we want to relate the structure of the matrix H to that of the upper triangular factor matrix R, it is convenient to define H to be square, and to include the (redundant) β_n in the (n, n) position of H. Recall that the diagonal elements of A are assumed to be nonzero.

Lemma 4 $\Omega(H) \subset \Omega(R^T)$.

Proof: Since A has a zero-free diagonal, so does E, which implies that

$$\Omega(w^T) \subset \Omega(w^T E) \subset \Omega(y^T) \subset \Omega(v^T - y^T).$$

Thus the structure of the first column of H is contained in the structure of the first row of R. The result then follows since the reduction process is simply applied repeatedly to matrices of successively smaller order. □

If $A = QR$, then since $B = A^T A = R^T Q^T Q R = R^T R$, apart from possible sign differences of the rows, R^T is the Cholesky factor of B. In [16], George and Heath used this fact to set up a static data structure for R by performing a symbolic factorization of the symmetric matrix B. If A is reducible, the structure obtained may be too generous. That is, the structure of the Cholesky factor of B may contain nonzeros in positions that must necessarily be zero in R^T. However, as shown by Coleman et.al. [3], if A is irreducible, the structures of the two factors will be identical. (Both articles cited above deal with non-square matrices.)

Lemma 5 *Let $B = A^T A$ and let L_B be the Cholesky factor of B. Then $\Omega(R) \subset \Omega(L_B^T)$, and if A is irreducible, then $\Omega(R) = \Omega(L_B^T)$.* □

Lemmas 4 and 5 give us insight on how to order the rows and columns of A in order to obtain a sparse H and R. First, note that since for any permutation matrix P, $(PA)^T PA = A^T A$, it follows that permuting the *rows* of A has no effect on the structure of H and R. Second, note that permuting the *columns* of A is equivalent to a symmetric reordering of the matrix B, since $(AP)^T(AP) = P^T(A^T A)P$. Thus, an apparently reasonable strategy is to find a symmetric reordering for B that will yield a sparse R. Of course we can use well-developed existing techniques for reordering symmetric matrices, such as the minimum degree algorithm mentioned earlier. Note that this idea will not work if A happens to have a dense row, since all structural information in A will be lost in B. For more details on this ordering problem, and strategies for dealing with dense rows, see [32].

2.3.1 Data Structures

In this section we describe data structures for storing H and R. We have already examined the method we will use for storing R in connection with storing the Cholesky factor L in section 2.2.3. For completeness, however, and because the scheme is used in connection with storing an upper triangular matrix R rather than L, we will include discussion of storage for both R and H. The main new idea in this section is the way in which the elements of H are stored and accessed.

The elimination tree introduced earlier was defined in terms of a lower triangular matrix. Here we need a similar definition for upper triangular matrices. The elimination tree for the upper triangular matrix R is precisely the tree obtained using the original definition, applied to the matrix R^T. Thus, for the elimination tree of R, $parent[j]$ is the column subscript of the first off-diagonal nonzero in row j of R.

In order to explain why the data structure for H is designed as it is, we need to examine its structure more closely, and relate it to the structure of R. Let us re-examine the effect of applying the Householder transformation Q_1 to A in order to annihilate the nonzero entries below the diagonal in the first column. From the transformation process, we make the following observations, where we assume as usual that exact cancellations do not occur.

Observation 1: The first row of the factor matrix R is given by

$$R_{1*} = (\sigma_d, z^T).$$

Observation 2: The *structure* of R_{1*} is the *union* of the row structures of A_{i*} where $a_{i1} \neq 0$.

Observation 3: $R_{1,parent[1]} \neq 0$, and $R_{1,j} = 0$ for $1 < j < parent[1]$.

Observation 4: If $A_{i1} \neq 0$, then

$$(Q_1 A)_{i,parent[1]} \neq 0$$

and

$$(Q_1 A)_{i,j} = 0, \;\; 1 < j < parent[1].$$

In fact, apart from the first element of the "active" rows, the structure of $(Q_1 A)_{i*}$ is the same as that of R_{1*}.

Observation 4 implies that if $a_{i1} \neq 0$, then after the Householder transformation is applied to A the next nonzero entry in row i of $Q_1 A$ appears in column $parent[1]$. Since the algorithm is to be applied recursively to the submatrix $\bar{E} = E - wy^T/\beta$, Observation 4 can be used repeatedly and the next theorem then follows [28].

Theorem 6 *If f_i is the column subscript of the first nonzero in row A_{i*}, then the locations of the nonzeros in row H_{i*} are given by*

$$f_i, parent[f_i], parent[parent[f_i]], \cdots, i.$$

□

Theorem 6 says that the structure of each *row* of the Householder matrix H corresponds to a *chain* in the elimination tree of R. Since the structure of this tree is already given implicitly in the row structure of R (by the *parent*'s), it is sufficient to know for each row of H where the associated chain starts in the tree (that is, the f's). This can be viewed as a special case of row structures in [38], where row structures are generally subtrees of the elimination tree. As we shall see later it is useful to know the location of a given vertex in the elimination tree relative to the root. Hence we define the *level number* l_i of vertex i in the elimination tree to be the number of vertices on the chain joining the root and vertex i. The level numbers will be useful in setting up a mechanism for directly accessing the nonzeros in a row of H.

We now provide a 6×6 example in Figures 11 and 12 to illustrate these ideas. It is easy to verify that the structures of its Householder matrix H and its triangular

$$A = \begin{pmatrix} \times & & & & & \times \\ & \times & & \times & & \\ \times & & \times & & & \\ & \times & & \times & & \times \\ & & & & \times & \times \\ \times & & & & & \times \end{pmatrix}$$

Figure 11: Structure of the matrix A.

factor R are as shown in Figure 12. Here $+$ denotes fill-in. The elimination tree

$$H = \begin{pmatrix} \times & & & & & \\ & \times & & & & \\ \times & & \times & & & \\ & \times & & \times & & \\ & & & & \times & \\ \times & & + & + & & \times \end{pmatrix}, \quad R = \begin{pmatrix} \times & & + & & & \times \\ & \times & & \times & & + \\ & & \times & + & & + \\ & & & \times & & \times \\ & & & & \times & \times \\ & & & & & \times \end{pmatrix}$$

Figure 12: Structure of the matrices H and R.

associated with R is given in Figure 13.

Note that the column subscripts of the nonzeros in row 6 of H are $1, 3, 4, 6$, and $(1, 3, 4, 6)$ is a chain in the elimination tree.

We are now in a position to describe the data structure we employ to store H and R. The diagonal elements of R are stored in an array $DIAG$. The off-diagonal nonzeros in the matrix R are stored by rows in a main storage vector RNZ and an accompanying index vector $XRNZ$ is used to point to the beginning of each

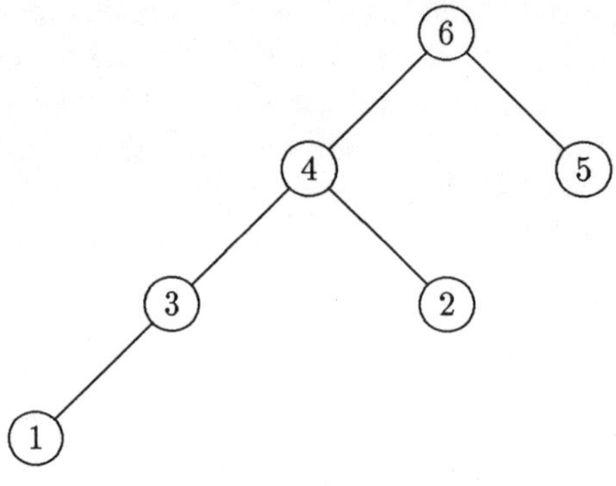

Figure 13: The elimination tree associated with R in Figure 12.

row in the storage array. The row structure of R is represented using a *compressed subscript* structure, as discussed in section 2.3.1. The compressed subscripts are stored in an integer vector $NZSUB$, and an accompanying index vector $XNZSUB$ is used to point to the beginning of the subscript sequence for each row of R. As noted in section 2.2.3, this scheme is due to Sherman [49]. Note that the values of $parent[i]$ can be retrieved quite readily from the compressed subscript structure, if the subscripts for each row are stored in ascending order. There are efficient algorithms for generating the vector $XRNZ$ and the compressed subscript structure. For example, the symbolic factorization algorithm described in the next section [32], which applies directly to A, can be used for this purpose. Alternatively, symbolic factorization algorithms for sparse symmetric positive definite matrices, such as that described in Chapter 5 of [22], can be applied to A^TA. However, experience has shown that the latter approach is usually more time consuming and requires more storage.

The discussion leading to Theorem 6 suggests a simple data structure to store the Householder vectors used in triangularizing the given sparse matrix A. The diagonal elements of the Householder matrix H are stored in a one-dimensional array $HDIAG$. The off-diagonal nonzeros in the Householder matrix H can be stored by rows in a one-dimensional array HNZ. We shall assume that the nonzeros in HNZ are stored in ascending order of their column subscripts. This turns out to be easy to arrange because we have the tree available.

We need only two additional overhead integer vectors, $HFIRST$ and $LEVEL$, where $HFIRST[i]$ keeps the column subscript of the first nonzero in row i of the matrix H (or equivalently, in row i of the original matrix A) and $LEVEL[i]$ is the level number l_i of node i in the elimination tree of the triangular factor R. Thus, the amount of overhead storage for the Householder matrix is $2n$ integer locations.

As an example, we have provided in Figure 14 the storage representation of the Householder matrix H and the upper triangular factor R given in the example in the previous section. The values of $parent[*]$ are also given for completeness; however, they are not explicitly stored in the actual storage scheme since they can be obtained directly from the row structure of R.

parent	3	4	4	6	6	6		
LEVEL	4	3	3	2	2	1		
HFIRST	1	2	1	2	5	1	3	2
HDIAG	h_{11}	h_{22}	h_{33}	h_{44}	h_{55}	h_{66}		
HNZ	h_{31}	h_{42}	h_{61}	h_{63}	h_{64}			
DIAG	r_{11}	r_{22}	r_{33}	r_{44}	r_{55}	r_{66}		
RNZ	r_{13}	r_{16}	r_{24}	r_{26}	r_{34}	r_{36}	r_{46}	r_{56}
XRNZ	1	3	5	7	8	9	9	
NZSUB	3	6	4	6				
XNZSUB	1	3	3	4	4	4	5	

Figure 14: Storage representation of H and R in Figure 12

For example, to retrieve the structure of row 6 of H, we note that $HFIRST[6] = 1$ so that the nonzero subscripts are given by

$$HFIRST[6] = 1,\ parent[1] = 3,\ parent[3] = 4,\ parent[4] = 6.$$

Sometimes it is necessary to access an individual nonzero in the Householder matrix H. An example is when we have to put the nonzeros of the original matrix A into the appropriate locations in the storage array HNZ or RNZ. If the nonzero is in the upper triangular portion of A, we can use the information in the compressed subscript structure, together with the array $XRNZ$, to put the nonzero in the appropriate location in the storage array RNZ. However, if the nonzero is in the lower triangular portion of A, it is slightly more complicated to find the correct location in the storage array HNZ where the nonzero should be, since we do not have the column subscripts explicitly.

Suppose for the moment that we have an index array $XHNZ$ where $XHNZ[i]$ points to the beginning of row i of H in the storage array HNZ. Assume again that A^TA is irreducible. This implies that the elimination tree for the structure of R has exactly one tree and n is its root. Now using the observation that the locations of the nonzeros in row i of H form a chain in the elimination tree (with starting vertex f_i and terminating vertex i), the location for nonzero $a_{ij}, i > j$, must be $HNZ[p]$, where

$$p = XHNZ[i] + (LEVEL[f_i] - LEVEL[j]).$$

Here $LEVEL[f_i] - LEVEL[j]$ is the distance between vertex j and vertex f_i (the first nonzero in row i) in the elimination tree.

Note that we do not really have to include $XHNZ$ in the storage scheme since it can be generated very easily from the information in $HFIRST$ and $LEVEL$. Let $XHNZ[1] = 1$. Recalling again the discussion that led to Theorem 6, it is easy to see that

$$XHNZ[i+1] = XHNZ[i] + (LEVEL[f_i] - LEVEL[i]),\ 1 \le i \le n.$$

When A^TA is reducible, the definition of $XHNZ[i+1]$ will be slightly different. In particular, instead of a single elimination tree, there will be a *forest* of trees, and $XHNZ$ must be evaluated using the $LEVEL$ values from the appropriate tree.

With the use of this row-oriented storage representation for the Householder matrix H, the sequence of Householder transformations can be stored in factored form. To use it effectively, we need to adapt the computation to this data structure. There is a version of the Householder transformation described in [23] which consists of only row operations.

The data structure described here is also appropriate for computing orthogonal decompositions using Givens rotations. If each plane rotation is represented by a single real number as proposed by Stewart [50], it was shown in [31] that there is always enough room in the main storage vector HNZ to store all the rotations required to reduce A to R. In this case, the storage scheme may overestimate the number of rotations actually required.

Finally, we note that the techniques described in this section for efficiently storing H and R can be easily generalized to handle orthogonal factors of rectangular matrices. For details about these extensions, see [28].

2.3.2 Symbolic Factorization

In this section we describe a symbolic factorization scheme that can be used to set up the data structures introduced in the previous section for H and R. Our development follows that described in [32].

The success of the procedure relies heavily on the efficiency of identifying the "active" rows at each step of the factorization and updating their structures. An important fact about the structure of the "pivot" row and column helps us achieve high efficiency. Specifically, it follows directly from Lemma 4 that the index of the first offdiagonal nonzero in the pivot column is always at least as large as the index of the first nonzero in the pivot row (after computing the union).

Consider the first step of the symbolic factorization, as depicted by equation (1). The structure of row j ($j > 1$) of $\bar{E}$ is either unchanged or the same as that of z^T. Thus, all we need is an indicator $mask(j)$ for each row j that tells us whether row j is involved in step 1. In subsequent steps, if $mask(j)$ is set, then the structure of row j has been used and its new structure should be obtained from (in this case) z^T.

We must also maintain information that tells us when the new structure of the modified rows will be used again and where the new structure can be found. Suppose

$$w_{\sigma_1}, w_{\sigma_2}, \cdots, w_{\sigma_p}$$

are nonzero, and

$$z_{\rho_1}, z_{\rho_2}, \cdots, z_{\rho_q}$$

are nonzero. As we noted above, $\rho_1 \leq \sigma_1$. In addition, $\bar{A}_{ij}$ must be zero for $2 \leq j < \rho_1$ and $i = \sigma_1, \sigma_2, \cdots, \sigma_p$. Thus the new structures of the modified rows will be used again (at the earliest) at step ρ_1. We should emphasize that the new structures are identical to that of z^T. Also, at step ρ_1, the active rows will include those of step 1 (whose row indices are greater than ρ_1), but the structure of these rows can be conveniently found by examining the structure of z only. To summarize, the only other information we have to record is the fact that at step ρ_1, we have to look at z to obtain the correct structure of the modified matrix. In the algorithm described below, we have used a set $\mathcal{S}_{\rho_1}$ for this purpose. After the first step, we have $\mathcal{S}_{\rho_1} = \{1\}$.

With these preliminary observations, we can now present in Figure 15 the entire symbolic factorization algorithm, which is a generalization of what we have just described. Recall that $\Omega(A_{*j})$ is the set of row subscripts of the nonzeros of column j of A, and $\Omega(A_{j*})$ is the set of column subscripts of the nonzeros of row j of A.

The algorithm makes use of the sets $\mathcal{R}_k$, which contain the structure of z_k. Note that z_k is just the part of R_{k*} to the right of the diagonal.

Initialization:

> **for** $k := 1, 2, \ldots, n$ **do**
>
> $mask[k] \leftarrow 0$
>
> $nrows[k] \leftarrow |\{m > k \mid f_m = k\}|$
>
> $\mathcal{S}_k \leftarrow \phi$

Main Loop:

> **for** $k := 1, 2, \ldots, n$ **do**
>
> $\mathcal{R}_k \leftarrow \phi$
>
>   **for** $i \in \Omega(A_{*j}^{\triangleright})$ **do**
>
>   **if** $mask[i] = 0$ **then**
>
> $\mathcal{R}_k \leftarrow \mathcal{R}_k \cup \Omega(A_{i*})$
>
> $mask[i] \leftarrow 1$
>
>   **for** $i \in \mathcal{S}_k$ **do**
>
> $\mathcal{R}_k \leftarrow \mathcal{R}_k \cup \mathcal{R}_i - \{1, \cdots, k-1\}$
>
> $\rho \leftarrow \min\{t \mid t \in \mathcal{R}_k - \{k\}\}$
>
>   **if** $nrows[k] > 1$ **then**
>
> $\mathcal{S}_\rho \leftarrow \mathcal{S}_\rho \cup \{k\}$
>
> $nrows[\rho] \leftarrow nrows[\rho] + nrows[k] - 1$

Figure 15: Symbolic Factorization Algorithm for Orthogonal Factors.

The complexity of the algorithm is linear in the number of elements in $\cup_i \mathcal{R}_i$, as the following discussion shows. Since each $i, i = 1, 2, \cdots, n$, will be included in at most one $\mathcal{S}_\rho$, $\mathcal{S}_\rho \cap \mathcal{S}_{\rho'} = \phi$, for $\rho \neq \rho'$. Thus we access the elements of $\mathcal{R}_i$ *at most once* for each i. Furthermore, we access the elements of $\Omega(A_{*j})$ and $\Omega(A_{j*})$ at most once. Hence the complexity of the algorithm is

$$O\left(\max(|A|, \sum_{k=1}^{n-1} |\mathcal{R}_k|)\right).$$

The overhead in storage is also very small. There are n elements in *mask* for marking the rows. Since at step k, k is included in at most one set $\mathcal{S}_\rho$, $(\sum_i |\mathcal{S}_i|) \leq n$.

2.4 Gaussian Elimination with Partial Pivoting: $A = LU$

2.4.1 Structural Relationships Among L, U, Q, and R

As we noted in the previous section, for general A, it appears to be impossible to predict the structures of L and U in advance of the actual numerical factorization, even though it is desirable to be able to do so. In this section we outline a recently developed approach to solving general systems which attempts to "have it both ways." The basic ingredients of the method are the use of Gaussian elimination with partial pivoting and the idea of setting up a data structure that is large enough to accomodate *all* possible fill that might occur, irrespective of the actual row interchanges that are performed. The effectiveness of the scheme depends on the way in which the columns of the matrix are ordered, but we will defer this aspect of the problem until a later section.

A standard approach for solving $Ax = b$ involves reducing A to upper triangular form using elementary row eliminations (i.e., Gaussian elimination). In order to maintain numerical stability, one may have to interchange rows at each step of the elimination process. Thus, we may express the result of the process as follows:

$$A = P_1 M_1 P_2 M_2 \cdots P_{n-1} M_{n-1} U,$$

where P_k is an $n \times n$ elementary permutation matrix corresponding to the row interchange performed at step k, M_k is an $n \times n$ unit lower triangular matrix whose k-th column contains the multipliers used at the kth step, and U is an $n \times n$ upper triangular matrix. As noted earlier, when A is sparse, fill normally occurs during the triangular decomposition, so there are usually collectively more nonzeros in $M = \sum_{k=1}^{n-1} M_k$ and U than in A.

Suppose that we could create from the *structure* of A a data structure which can accommodate all the nonzeros in M and U, *irrespective of the actual pivot sequence chosen.* Let $\bar{L}$ and $\bar{U}$ be matrices whose structures contain respectively the structures of M and U, irrespective of the pivot sequence $P_1, P_2, \cdots, P_{n-1}$. Although the data structure for these matrices would be more generous than it needs to be for any *specific* sequence, it could be set up in advance of the numerical computation, which could then be done efficiently using a static data structure. The advantages noted earlier for positive definite matrices would then be available. Our objective in this section is to demonstrate that the structures of $\bar{L}$ and $\bar{U}$ turn out to be precisely those of H and R respectively.

As we did in the previous section, partition A into

$$A = \begin{pmatrix} d & v^T \\ u & E \end{pmatrix},$$

and consider applying one step of Gaussian elimination to A with partial pivoting.

$$\begin{aligned} P_1^T A &= P_1^T \begin{pmatrix} d & v^T \\ u & E \end{pmatrix} = \begin{pmatrix} \hat{d} & \hat{v}^T \\ \hat{u} & \hat{E} \end{pmatrix} \\ &= \begin{pmatrix} 1 & 0 \\ \hat{u}/\hat{d} & I \end{pmatrix} \begin{pmatrix} \hat{d} & \hat{v}^T \\ 0 & \hat{E} - \hat{u}\hat{v}^T/\hat{d} \end{pmatrix} \\ &= \begin{pmatrix} 1 & 0 \\ l & I \end{pmatrix} \begin{pmatrix} \hat{d} & \hat{v}^T \\ 0 & A_1 \end{pmatrix}. \end{aligned}$$

Note that only the first row of A and rows j of A where $u_j \neq 0$ are involved in the first step of the elimination. Thus only the structures of these *candidate pivot rows* will be affected by the first step of Gaussian elimination with partial pivoting. Moreover, assuming structural and numerical cancellations do not occur, the new structure of each of these rows must be contained in the *union* of the structures of all the candidate pivot rows, which is given by the structure of

$$\bar{v}^T = v^T + u^T E.$$

Thus, if we let

$$\bar{A}_1 = E - u\bar{v}^T/d$$

we have

$$\begin{aligned} \Omega(l) &= \Omega(u), \\ \Omega(\hat{v}^T) &\subset \Omega(\bar{v}^T), \end{aligned}$$

and

$$\Omega(A_1) \subset \Omega(\bar{A}_1).$$

Hence we can use the structures of u, $\bar{v}$ and $\bar{A}_1$ to *bound* those of l, $\hat{v}$ and A_1 respectively, *regardless of the choice of* P_1.

Note that the structure of $\bar{A}_1$ can be obtained by applying one step of Gaussian elimination *without* row interchanges to the $n \times n$ matrix $\bar{A}$ below.

$$\bar{A} = \begin{pmatrix} d & v^T + u^T E \\ u & E \end{pmatrix}.$$

This forms the basis of the symbolic factorization algorithm developed in [32], which we have already presented in the context of setting up a data structure for H and R.

Since E has a zero-free diagonal, $\bar{A}_1 = E - u(v^T + u^T E)/d$ must also have a zero-free diagonal under the assumption that exact cancellation does not occur. Thus the argument and idea above can be applied recursively to $\bar{A}_1$, and so on, yielding a procedure for generating matrices $\bar{L}$ and $\bar{U}$ such that

$$\Omega(M) \subset \Omega(\bar{L})$$

and

$$\Omega(U) \subset \Omega(\bar{U}).$$

Now using (3), we can conclude that the first row of R has the same structure as the first row of $\bar{U}$, since

$$\Omega(v^T - y^T) = \Omega(v^T + w^T E) = \Omega(v^T + u^T E). \tag{4}$$

Referring to (1), and using (3) and (4), we have

$$\Omega(E - wy^T/\beta) = \Omega(E - u(v^T + u^T E)/d) = \Omega(\bar{A}_1).$$

Thus, we can state the following lemma, which relates the structures of M, H and $\bar{L}$, and the structures of U, R and $\bar{U}$ [30].

Lemma 7 *If A is irreducible, then*

$$\Omega(M) \subset \Omega(\bar{L}) = \Omega(H)$$

and

$$\Omega(U) \subset \Omega(\bar{U}) = \Omega(R).$$

□

2.4.2 Numerical Factorization

As mentioned in the previous section, the data structure required for $\bar{L}$ and $\bar{U}$ is identical to that for the Householder matrix H and the upper triangular factor R. The only difference is that, in the case of Gaussian elimination, we do not need the array $HDIAG$ since all the diagonal elements of M_k are equal to 1. This data structure can be set up efficiently using the symbolic factorization algorithm described in section 2.3.2.

Now let us consider the implementation of the numerical factorization in detail. In order to implement partial pivoting, we have to be able to search for the pivot and, after finding the appropriate pivot, to perform the row interchange and row modifications efficiently. Note that all the *candidate* pivots are below the diagonal and in the same column. That is, they are in $\bar{L}$.

However, since the proposed storage scheme for $\bar{L}$ is row-oriented, these candidate pivots are sprinkled in the data structure for $\bar{L}$. This raises two problems. First, how do we identify the *candidate* pivot rows at a given step, and second, how do we locate the candidate pivot in a candidate pivot row in the data structure?

The techniques we use to solve these problems are similar to those used previously by Eisenstat et. al. in [12] and others.

Consider the first step of the algorithm. For $1 \le k \le n$, define Z_k as follows.

$$Z_k = \{i \mid f_i = k\}.$$

(Recall that f_i is the column subscript of the first nonzero in row i of A.) That is, Z_k is the set of (indices of) rows such that their first nonzeros are all in column k. Clearly, at the first step of Gaussian elimination, the candidate pivot rows are given by the rows in Z_1. As shown in Theorem 6, after step 1, the next step in which these rows (except row 1) will again be used as candidate pivot rows has to be step *parent*[1]. Thus, all we need to do is to "merge" Z_1 and $Z_{parent[1]}$ after step 1; that is,

$$Z_{parent[1]} \leftarrow Z_{parent[1]} \cup Z_1 - \{1\}.$$

This is done at each step. That is, at the end of step k, we replace $Z_{parent[k]}$ by $Z_{parent[k]} \cup Z_k - \{k\}$. Thus, Z_k will contain the correct indices of candidate pivot rows at the beginning of step k. These sets $Z_1, Z_2, \cdots, Z_n$ can be maintained efficiently using linked lists. We can use two integer vectors $LNKBEG$ and $ROWLNK$ for this purpose. At any time, the elements in Z_k are given by the sequence

$$\begin{aligned} &LNKBEG[k], \\ &ROWLNK[LNKBEG[k]], \\ &ROWLNK[ROWLNK[LNKBEG[k]]], \\ &\quad\vdots \end{aligned}$$

If p is the last row in the set, then $ROWLNK[p] = 0$. The linked list of candidate pivot rows allows the search for the pivot row to be carried out efficiently. The merging of two sets can also be performed easily. We simply modify the values of $ROWLNK[p]$ and $LNKBEG[parent[k]]$. Note that at the beginning of the elimination process, $LNKBEG$ and $ROWLNK$ can be generated easily using the information in the $HFIRST$ vector.

It is also necessary to be able to efficiently locate the candidate pivot in a candidate pivot row in the row-oriented data structure. As in section 2.3.1, we shall assume that the nonzeros in a row of $\bar{L}$ are stored in the storage array in ascending order of their column subscripts. Before the elimination begins, we generate a vector $NXTNZ$ of integers, where $NXTNZ[k]$ points to the beginning of row k of $\bar{L}$. The initial content of $NXTNZ$ is identical to that of the vector $XHNZ$ described in section 2.3.1 and can be constructed from $HFIRST$ and $LEVEL$. Thus, the candidate pivots at step 1, for example, are found in $HNZ[NXTNZ[i]]$, where $i \in Z_1$. After the first step of elimination, we simply increment $NXTNZ[i]$ by 1, for $i \in Z_1$, to indicate the location of the nonzero in row i to be used *next* time. (Alternatively, one can use the index vector $XHNZ$ and the integer vectors $HFIRST$ and $LEVEL$ to *compute* the location of a given nonzero. However, experimental results have shown that this approach is much more time consuming than the one described above.)

Note that it is possible to exploit completely the sparsity of the pivot row when we perform row modifications. If we are at step k of the elimination process and if row i is a row to be modified, then we only have to modify column j of row i, where $\bar{U}_{kj}$ is *actually* nonzero. This can be achieved by maintaining a list of locations of the actual nonzeros in row k of $\bar{U}$. In our implementation, we use an integer vector $OFFSET$ for this purpose. (Its name is intended to suggest that we keep the location of each real nonzero in row k of $\bar{U}$ relative to the first predicted nonzero in that row.)

Finally, we note that the techniques described in this section for efficiently performing the LU factorization are equally applicable in computing the orthogonal factorization of A. Moreover, as noted earlier, the storage scheme can be easily generalized to handle factors of rectangular matrices, and the techniques of this section apply equally well in that context.

3 Multiprocessor Architectures

Multiprocessors fall into two general categories: *shared-memory* multiprocessors, and *local-memory* (or *distributed-memory*) multiprocessors. As their name implies, they are distinguished by whether each processor can directly access the entire memory available, or whether the memory is partitioned into portions which are private to each processor. There are some architectures that display aspects of both, but we will limit our attention to these two categories.

We also assume implicitly throughout that the number of processors p in our multiprocessor is *much less* than the order of the matrix problems we wish to solve. Thus, we are mainly interested in so-called medium-to-coarse grained parallelism, about which we will say more later.

Finally, we assume that the computers we deal with are so-called *multiple instruction stream* multiprocessors. That is, each processor is capable of executing its own program independantly of the other processors. This is in contrast to *single instruction* multiprocessors, where all processors execute the same instruction stream in lockstep.

3.1 Shared-Memory Architectures

Shared-memory multiprocessors consist of several processors all of which can access all of the available memory through some type of very fast bus. Since the bus must service memory requests serially, it could be argued that such architectures really cannot achieve speed-up at all. However, through various hardware features, such as cache memories, one can effectively serve the data needs of multiple processors. In effect, the serial execution time imposed by the bus is a "low order term" in the overall execution time. However, as more processors are added, memory conflicts tend to become significant. In other words, the basic architecture does not "scale up" very well. Typically, for this class of very successful architecture, p does not range above 25 or 30.

Programmers of such computers are typically provided with several mechanisms to allow processes to be synchronized, and to prevent processes from simultaneously updating the same location, and thus corrupting the data. Note that we use the term process, rather than processor. Programming is typically done in terms of processes, which may or may not run on individual processors. The problem of simultaneous update is present in any case.

A common device that is provided is a so-called *synchronization* or *lock* variable. If the programmer wishes to have several processes share an array y, but allow only one of them to access it at any given time, the processes can share a synchronization variable *sync*, which imposes a consecutive read-write discipline. Once it has been read from (accessed), it cannot be read by another process until the first has written back to it. Thus, "critical sections" of program can be protected by a sequence as shown in Figure 16.

$$\vdots$$

$temp \leftarrow sync$

(critical section involving y)

$sync \leftarrow temp$

$$\vdots$$

Figure 16: Protecting a critical section.

Another common device that allows the programmer to synchronize a set of processes is a so-called *barrier* function, which again makes use of a shared variable, say *count*. Typically, a master process will spawn p processes, each of which is to do a task and then wait until the other processes have completed their tasks. The master process creates a barrier with *count* having the value p. As each process completes its task, it calls the barrier function, with *count* as its argument. The process is then suspended, and the value of *count* is reduced by one. When the value of *count* is reduced to zero (all tasks have been completed), the processes are allowed to resume execution.

3.2 Local-Memory Architectures

This class of architectures avoids the memory conflict problem by providing each processor with its own private or local memory, not accessible by other processors. Since the processors can no longer communicate via the globally shared memory, they must be connected by some type of network. Information is sent from one processor to another by sending messages over the network. This presents a different set of challenges for both the hardware designer and the algorithm designer.

For the hardware architect, tradeoff must be made between the cost of equipment and the richness of the connection topology. Various topologies are used. Some of the more common ones are a broadcast bus, star, ring, two dimensional grid, and binary hypercube. The latter is probably the most common topology currently in use. These are depicted in Figures 17 and 18.

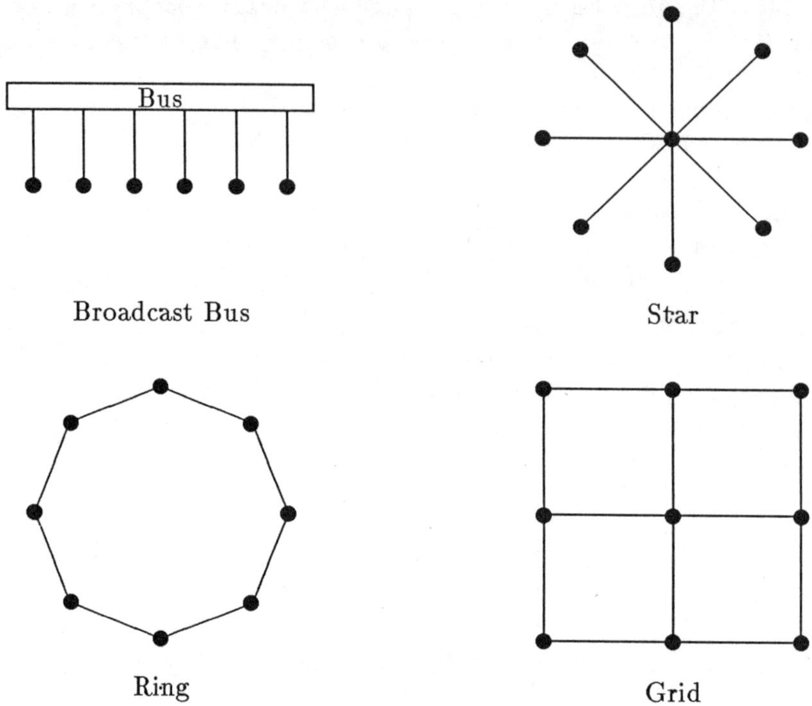

Figure 17: Bus, star, ring, and grid interconnection networks

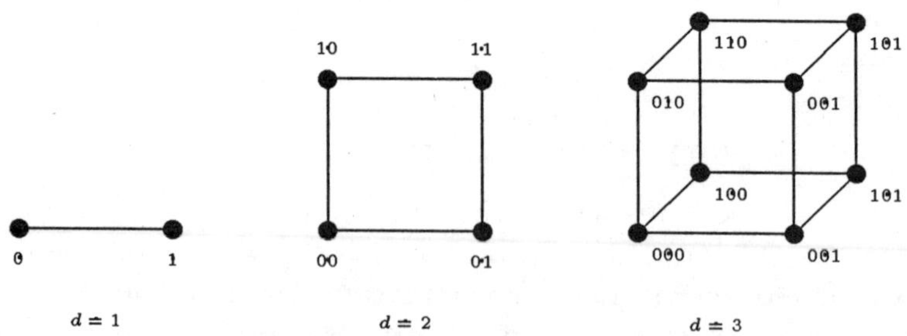

Figure 18: Binary hypercubes of dimension 1, 2 and 3.

In terms of the design of algorithms, local-memory machines impose the additional burden of requiring that the data, as well as the computation, be partitioned. In addition to identifying parallelism in the computation, and assigning computational tasks to individual processors, the data associated with the computation must be distributed among the processors, and communicated among them as necessary. The challenge is to do this in such a way that each processor has the data it needs in its local memory at the time that it needs it, and the amount of communication required among the processors during the computation is kept acceptably low.

Operating systems for distributed-memory computers normally provide two message passing primitives, which for narrative purposes we will refer to as *send* and *receive*. Execution of a *send* does not cause the sending process to wait for a reply. On the other hand, execution of a *receive* causes the process executing it to be suspended until the message is received. Messages that arrive at the destination process before the execution of the receiving *receive* are placed in a queue until needed.

The cost of communication between adjacent processors in a message-passing system can usually be modeled with reasonable accuracy by the expression

$$t = \alpha + \beta M,$$

where α is a start-up cost for any message independent of its length, β is the incremental cost per unit length, and M is the length of the message in bytes or words. The relative magnitudes of α and β are important factors in the design of algorithms. For example, if α is relatively large, then for a given communication volume, small messages will be much less efficient than large messages.

Another important hardware characteristic that influences the design of parallel algorithms is the relative cost of communication and computation. If communication is relatively slow, then coarse-grain algorithms in which a relatively large amount of computation is done between communications will be more efficient than fine-grain algorithms.

For machines currently available:

1. Start-up cost is relatively high ($\beta \ll \alpha$).

2. Communication is relatively slow compared to floating-point computation.

These characteristics may change over time, but the current situation tends to make algorithm designers favour relatively coarse-grain algorithms.

Many algorithms require some form of global communication: upon completing one of its assigned tasks, a given processor may be required to make its results available to *all* other processors. Such a communication pattern is referred to as *broadcasting*. The method used to implement broadcasting, and its resulting efficiency, depend on the details of the underlying interconnection network among the processors. In a bus-based system, for example, broadcasting can usually be accomplished as a single *send* operation since all of the processors are listening to the common bus. (On the other hand, write access to the common bus is necessarily serial; i.e., all processors can receive simultaneously, but only one processor can send at a time.) In a completely connected network (e.g., a crossbar switch), broadcasting is accomplished by $p-1$ separate *send* operations, one to each of the other processors. If some processors are not directly connected by the network, then broadcasting requires that the message be forwarded by intermediate processors in order to reach some destinations. In a ring network, for example, broadcast messages must be propagated around the ring, with a resulting delay of at least $p/2$ communication steps.

A hypercube interconnection network offers a number of possibilities for implementing broadcasting. First, the "diameter" of a hypercube network is relatively small: if $p = 2^d$, then the largest distance between any two processors is d, and thus a message never has to be forwarded more than $d-1$ times. Moreover, in many hypercubes the operating system automatically forwards messages as necessary to reach arbitrary destination nodes. Thus, one option for implementing broadcasting is simply to write the program as though the network were completely connected, with a separate *send* from the source node to each destination node.

Finally, it is easy to design a strategy for broadcasting that takes full advantage of the recursive structure and small diameter of the hypercube. This approach is based on embedding a minimal spanning tree in the hypercube network, rooted at the source node of the broadcast. The root node of the broadcast sends the message to all of its neighbors, who in turn send the message to all of their neighbors who have not already received the message, etc., until after d stages all nodes have received the message.

4 Parallel Algorithms

4.1 Speed-up, Granularity, Concurrency, Load Balance

Simply stated, our objective in designing parallel algorithms is to achieve *speed-up*. More specifically, given a p-processor computer, we would like to solve our problem in time t that is as close as possible to a factor of p less than the time t_s needed to solve the same problem on a single processor version of the machine, *using the best serial algorithm available.* Normally, in a multiprocessor, all processors in the machine have the same execution speed, and we assess the implementation of a parallel algorithm by its *speed-up* "s" and *efficiency* "e" as defined below.

$$s \equiv \frac{t_s}{t} \tag{5}$$

and

$$e \equiv \frac{s}{p}. \tag{6}$$

For most algorithms, it is not possible to achieve a speed-up of p. Usually there is a sequential order in which certain computations must be performed because of data dependencies. Nevertheless, most major computations do admit a great deal of concurrency. The objective of the parallel algorithm designer is to identify that parallelism, and to exploit it using a given architecture.

There are several important factors which determine the efficiency of an implementation of an algorithm on a *given* parallel computer. These are *concurrency*, *granularity*, *load balance*, and *communication cost.*

Problems for which a large fraction of the computation can be performed in parallel are said to display a high degree of concurrency. For example, adding two vectors of length n has a high degree of inherent parallelism. Given n processors, the addition can be done in one basic time step. On the other hand, summing the elements of a vector of length n displays much less concurrency, since it requires approximately $\log_2 n$ time steps even if $n/2$ processors are available.

Normally, computational problems to be solved on a parallel computer are divided into *tasks* which are then assigned in some way to the individual processors. These tasks may be as small as adding two numbers together, or a significant fraction of the entire computational problem. The size of the tasks, relative to the computation required for the entire problem, is referred to as the granularity of the parallel algorithm. Since tasks need not all be the same size, the term is not very precise.

Suppose we consider the problem of factoring a dense $n \times n$ matrix. With a single processor, the amount of work required is $O(n^3)$ arithmetic operations. With more processors, the total number of arithmetic operations performed remains the same, but the total execution time will be reduced as a result of sharing the work among the processors, even though some additional overhead may be introduced by necessary communication or synchronization among the tasks and processors. If the number of processors available is very large, say $O(n)$ or $O(n^2)$, then the tasks performed by each processor will be correspondingly small, perhaps just a few arithmetic operations. Some appropriate parallel algorithms for this case include systolic arrays [2] and data-flow [44], and the corresponding architectures involve very simple processors with limited communication among them and only local memory. As noted earlier, in these notes we deal only with the case $p \ll n$.

The choice of tasks will normally have a significant effect on how much concurrency can be achieved, since tasks themselves will be executed serially. Generally, the finer the granularity, the more concurrency is possible, provided that the data dependencies in the problem itself allow it.

Irrespective of the amount of concurrency that is available and allowed by the choice of tasks (granularity), if efficiency is to be achieved, that concurrency must be exploited. The work must be allocated to available processors in such a way that they are working simultaneously on the problem for as large a fraction of the time as possible. This is the load balancing problem.

Finally, tasks in a parallel algorithm normally satisfy some precedence relations, so that the results computed by one processor must be made available (i.e., communicated) to some or all of the others. For shared-memory architectures, communication is a relatively minor issue because global data access is available through the common memory (although memory contention can still be a problem). Thus, in designing algorithms for these architectures, the primary focus is on the effect of precedence constraints and task size on concurrency and load balancing. However,

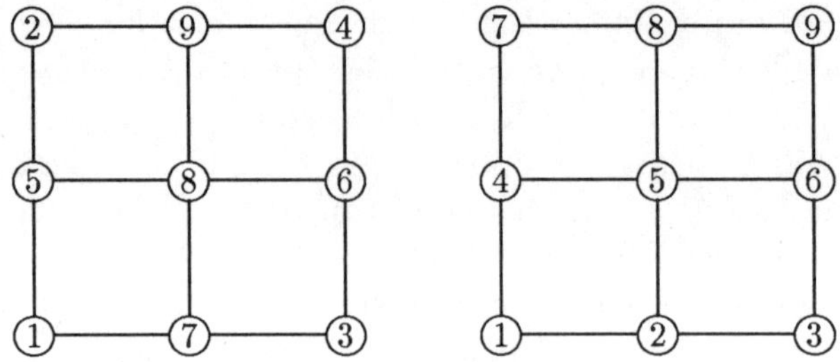

Figure 19: Two orderings of a 3 by 3 grid.

$$\begin{pmatrix} \times & & & & & & & & \\ & \times & & & & & & & \\ & & \times & & & & & & \\ & & & \times & & & & & \\ \times & \times & & & \times & & & & \\ & & \times & \times & & \times & & & \\ \times & & \times & & \times & \times & \times & & \\ \times & \times & \times & \times & \times & \times & \times & \times & \\ & \times & & \times & \times & \times & \times & \times & \times \end{pmatrix} \begin{pmatrix} \times & & & & & & & & \\ \times & \times & & & & & & & \\ & \times & \times & & & & & & \\ \times & \times & \times & \times & & & & & \\ \times & \times & \times & \times & \times & & & & \\ & \times & \times & \times & \times & \times & & & \\ & & & \times & \times & \times & \times & & \\ & & & \times & \times & \times & \times & \times & \\ & & & & \times & \times & \times & \times & \times \end{pmatrix}$$

Figure 20: Structure of the Cholesky factors for the orderings of Figure 19.

in a distributed-memory system, communication is accomplished by passing messages among processors, and thus the necessity of communication becomes a critical factor affecting efficiency.

4.2 Elimination Trees and Parallelism

In this section we explore the role that elimination trees can play in identifying and exploiting parallelism in matrix computations. We begin by considering two different orderings of the same problem, and study their corresponding elimination trees. Consider a 3 by 3 grid problem as shown in Figure 19, where the 9 vertices of the grid are numbered in some manner, and the associated symmetric positive definite matrix A has the property that $a_{ij} \neq 0$ if and only if vertex i and vertex j are associated with the same small square in the grid. Two different orderings of the grid are given in Figure 19, the associated Cholesky factors are displayed in Figure 20, and their corresponding elimination trees are shown in Figure 21.

The elimination tree on the left is typical of those generated by orderings that are good in the sense of yielding low fill and low operation counts. Its tree structure is short and wide. The fortunate fact is that such trees and their associated orderings lend themselves well to parallel computation.

Consider computing the Cholesky decomposition of A, and for definiteness, suppose we use the vector-sum column-Cholesky version of the algorithm. As we saw

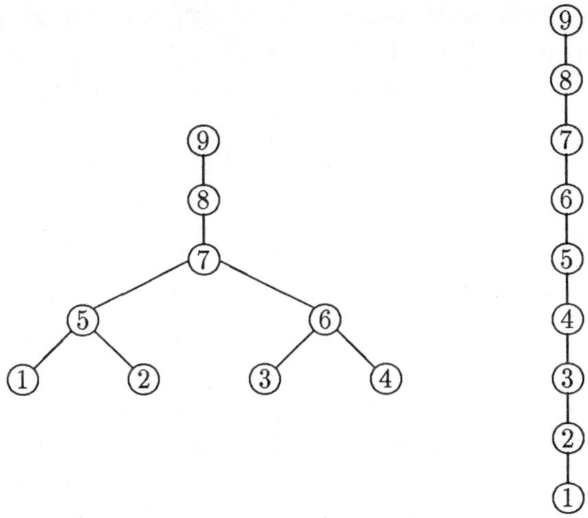

Figure 21: The elimination trees associated with the matrices in Figure 20.

earlier, when we compute the ith column, we do so by modifying it using all columns j for which $L_{ij} \neq 0$. That is, column i *depends* on these columns; it cannot be computed until the columns upon which it depends are themselves computed. Lemma 8 below provides the connection between this idea and the elimination tree.

Lemma 8 *Let $i > j$ and $L_{ij} \neq 0$. Then vertex i is an ancestor of vertex j in the elimination tree.*

Proof: If $i = parent[j]$, there is nothing to prove. Otherwise, $parent[j] < i$ and by repeated application of Lemma 1, we generate an ascending sequence of vertices (subscripts) k bounded above by i and for which $L_{ik} \neq 0$:

$$j < parent[j] < parent[parent[j]] < \cdots < i.$$

Thus, there exists an integer p such that $parent^p[j] = i$. □

An important implication of Lemma 8 is that in the elimination tree, if vertex i and vertex k belong to the same level of the tree, the column sets upon which columns i and k depend *are disjoint*.

For discussion purposes, suppose that the basic task in a parallel algorithm for Cholesky decomposition is to compute a column of L, and let $Tcol(i)$ denote the task of computing column i of L. Then for the tree on the left in Figure 21, we see that $Tcol(1)$,$Tcol(2)$,$Tcol(3)$, and $Tcol(4)$ can start immediately in parallel. Moreover, when they have completed execution, $Tcol(5)$ and $Tcol(6)$ may proceed independently.

On the other hand, the band-oriented ordering whose tree is on the right in Figure 21 is less desirable because the $Tcol(i)$ must be executed serially. Moreover, the operation counts and fill are inferior to that of the first ordering.

Thus, another way of interpreting Lemma 8 is that if vertex i and vertex j belong to the same level of the tree, tasks $Tcol(i)$ and $Tcol(j)$ can be performed independently so long as the tasks associated with their descendant vertices have all been completed. In order to gain high processor utilization, it is therefore desirable to assign vertices on the same level of the tree to different processors, if possible.

4.3 Elimination Trees and Communication

The structure of elimination trees can also provide guidance on how to allocate data to processors in order to reduce communication costs. We will again use the example from the previous section for illustration. We saw that it is desirable to allocate vertices (columns) at the same level of the tree to different processors. Within this constraint, there are several ways that such an allocation can be made. As an example, consider the elimination tree shown in Figure 22. This tree corresponds to a matrix derived from a grid like the one in Figure 19, except it is 7×7 rather than 3×3. Two different options for allocating the columns to the processors are illustrated in Figures 23 and 24. Note that there are four processors, numbered from 0 to 3.

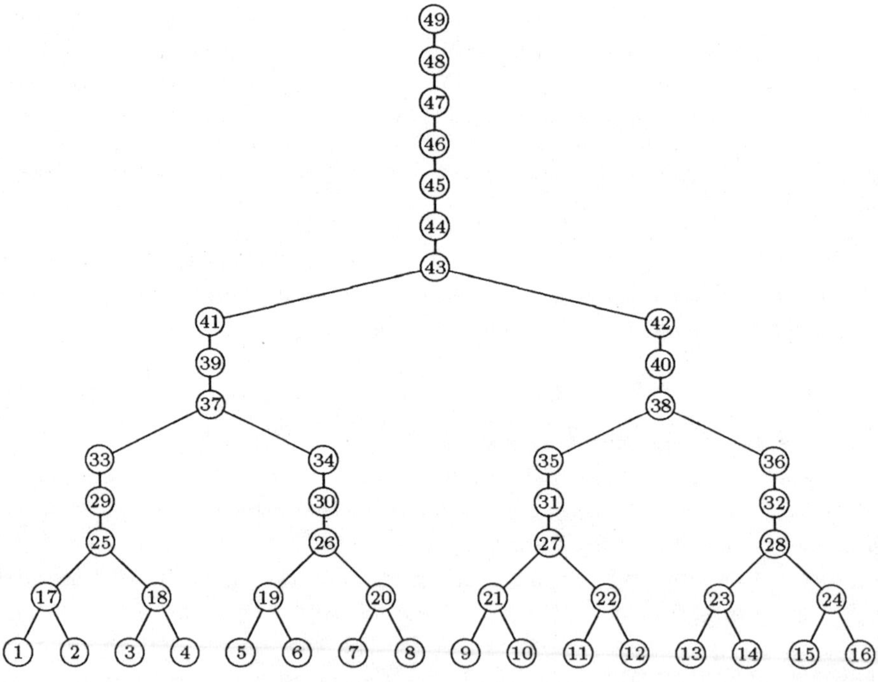

Figure 22: An example of an elimination tree.

In a local memory environment, the subtree task assignment strategy will reduce the communication requirements, since no communication at all will be required until 5 levels of the tree have been processed. After that, the amount of communication

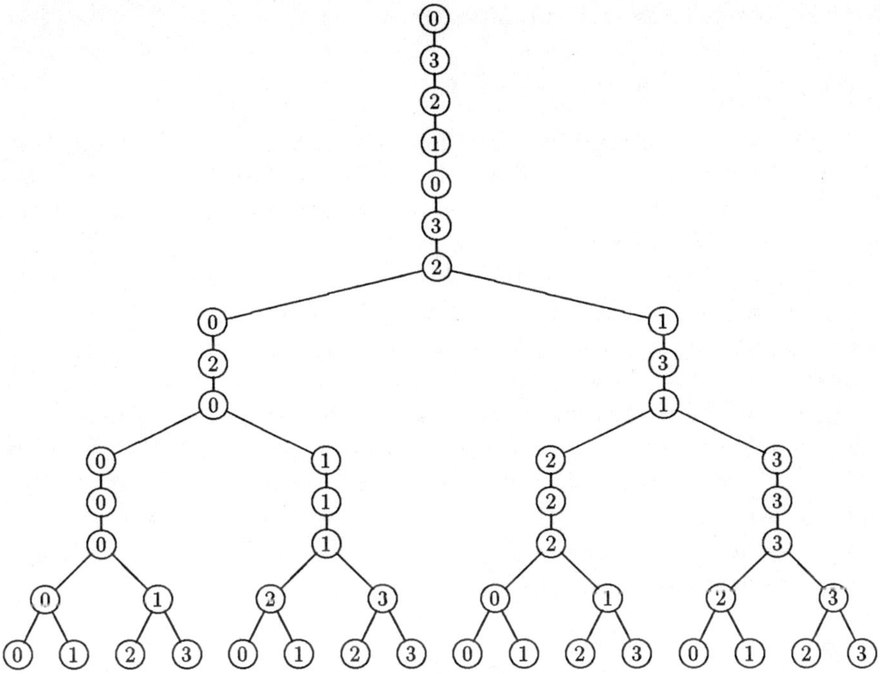

Figure 23: Bottom-up task assignment.

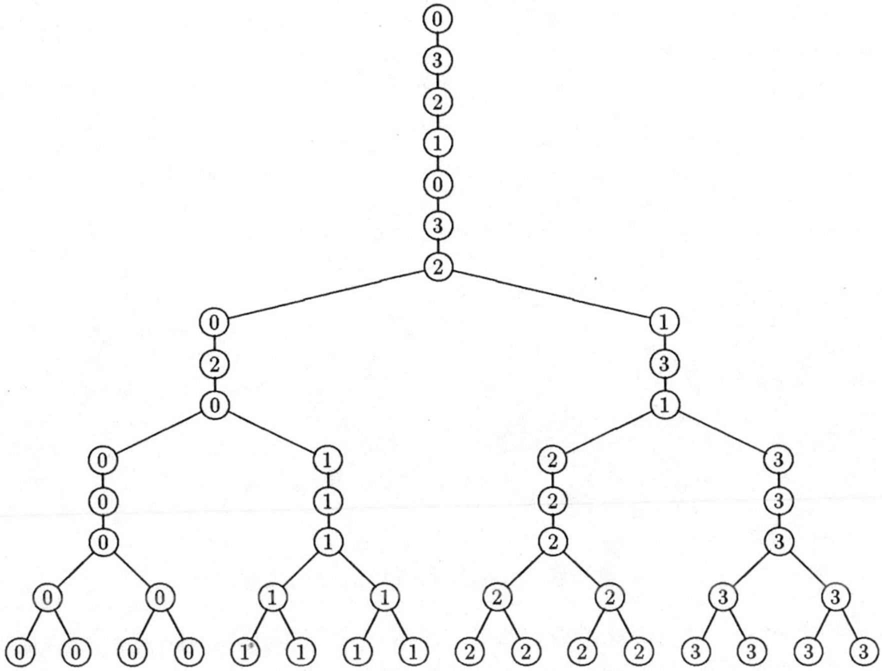

Figure 24: Subtree task assignment.

will be about the same for the two allocations.

4.4 Reordering and Balancing Elimination Trees

It is widely known in the literature (although there seems to be no published proof) that given an ordering of a sparse matrix, and therefore an elimination tree, any symmetric *reordering* of the matrix based on a relabelling of the tree that numbers each vertex ahead of its parent is *equivalent* to the original ordering in terms of fill and computation. This raises a subtle point in connection with the example in the previous section. If one examines the labelling of the tree in Figure 22, and the processor assignment in Figure 23, we see that the assignment of the columns to the processors was done in a so-called *wrap-around manner*. That is, the columns were assigned to the processors in much the same manner one would deal cards, assigning consecutive columns to consecutive processors until all processors have been allocated a column, and then wrapping back to the first processor with further columns. However, note that using wrap mapping on an *equivalent reordering* of the matrix would have produced the processor assignment given in Figure 24. This equivalent reordering is shown in Figure 25.

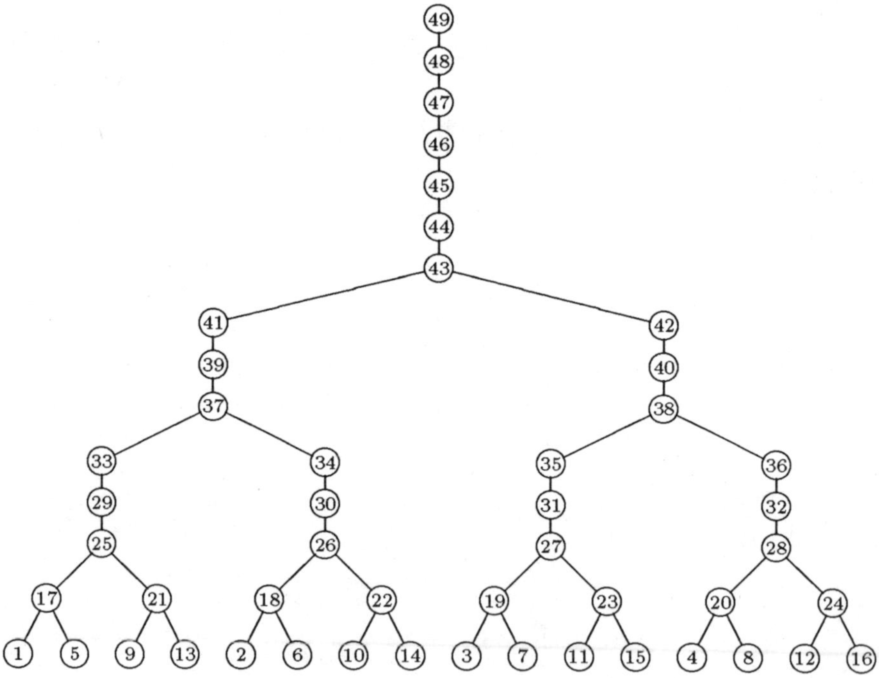

Figure 25: Reordered elimination tree.

Thus, we see that in terms of commmunication reduction, the fundamental structure is the tree, rather than the ordering of the matrix.

There is a second important aspect about elimination trees that should be noted.

Recently Liu [40] has shown how to produce equivalent reorderings of matrices which *change* the elimination tree. That is, the amount of fill and the amount of computation required for the factorization do not change, but the structure of the tree does change. It would take us too far afield here to examine his algorithm. From our previous discussions, it would seem desirable to have the best ordering we can find in terms of fill and computation, and then within the class of reorderings that preserve that level of fill and computation, choose one which produces a "short" and "wide" tree.

Liu has developed a fast algorithm [43] which will reorder a sparse matrix problem in order to either increase or decrease the height of the elimination tree, while preserving the level of fill and computation. In order to illustrate how dramatic an effect this reordering can have, we applied his algorithm (in "unbalancing mode") to the problem which produced the elimination tree shown in Figure 22. The result is shown in Figure 26. The problem of factoring the matrix ordered in this way exhibits considerably less parallellism than the same problem ordered in the original way.

Thus, if we were trying to allocate columns to processors for vector-sum row-Cholesky so as to reduce communication costs on a distributed-memory multiprocessor, a reasonable strategy to adopt is as follows:

1. Find a good sparsity-preserving ordering for A.

2. Find the elimination tree corresponding to the reordered A.

3. Reorder the problem so as to reduce the height of its elimination tree using Liu's height-reducing algorithm [43].

4. Assign the column tasks to the processors in a "bottom up" manner with respect to the tree, but as much as possible, assign subtrees of the tree to subsets of the processors, as illustrated in the example in Figure 24.

5 Algorithms for Shared-Memory Machines

5.1 The Self-Scheduling Paradigm

The computational paradigm we employ for shared-memory machines is that of a *pool of tasks* whose parallel execution is governed by a *self-scheduling* discipline. We will assume that the tasks are assigned some well ordered sequence of task numbers or task id's, in order to satisfy whatever precedence relations are imposed by the computation.

In some parallel algorithms, specific sets of tasks are mapped onto specific processors in advance of initiating the computation, and therefore effective load balancing among the processors requires that the aggregate amount of work associated with each set be reasonably uniform in size. Self-scheduling can be viewed as a technique for automatic and dynamic load balancing that does not necessarily require

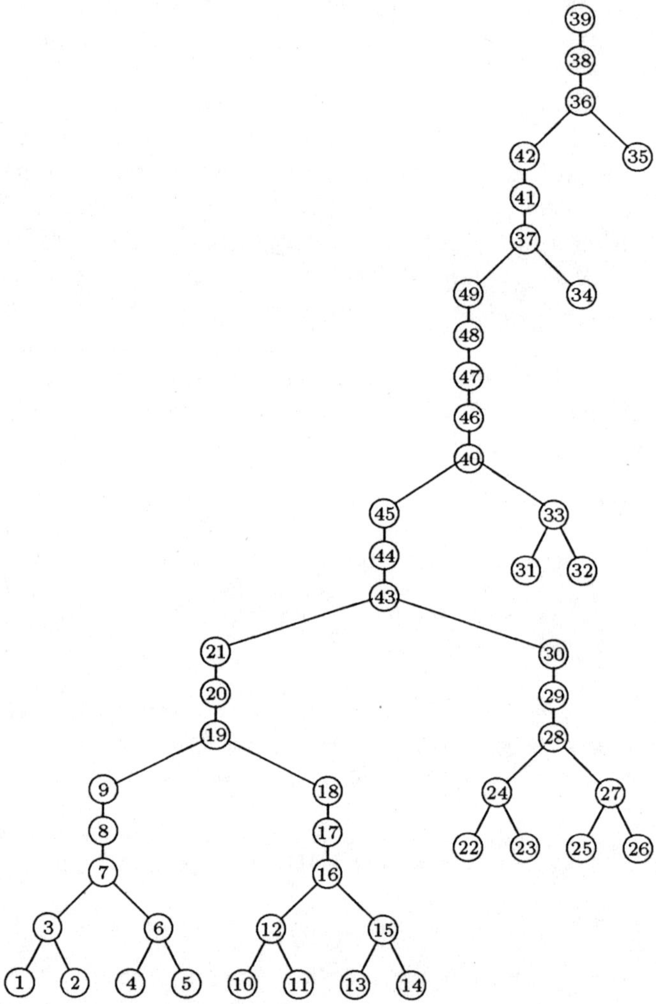

Figure 26: Unbalanced elimination tree.

uniformly sized tasks. In self-scheduling, p processes are invoked to perform a job consisting of q tasks ($p \leq q$). When a given process completes an assigned task, it checks whether any unassigned tasks remain in the pool, and if so it is assigned the next one. Thus, if a processor happens to have drawn a relatively small task, it will become free to take on yet another task from the pool sooner than a processor occupied by a larger one. In this way all of the processors tend to be kept busy even if the tasks vary in their computational difficulty. Similar advantages are gained from this approach when processors having different computational speeds are employed; i.e., faster processors tend to share a greater portion of the total work load.

As noted above, tasks are claimed by free processors in an order specified by the task numbers or id's, and we assume that these have been assigned to satisfy any precedence relations that may hold among the tasks. Between two tasks, say task a and task b, there are several possible types of precedence relations affecting their potential parallel execution. Three that are of interest to us in this context are:

Type 1: task a must finish before task b can begin

Type 2: task a must finish before task b can finish

Type 3: task a must begin before task b can begin

The first type permits only serial execution of the two tasks, but the other two types permit at least some degree of concurrent execution. Obviously the order chosen for the pool of tasks should be consistent with any such precedence relations in order to take maximum advantage of parallelism. We shall see examples of all three types of precedence relations in various forms of Cholesky factorization.

This self-scheduled pool-of-tasks approach is flexible in that it is not strongly dependent on the number of processors available, but it is best suited to large-to-medium grained parallelism, which, in the present context, means problems for which the total number of tasks q exceeds the number of processors p by a substantial margin. Since the pool of tasks must be made available to each processor, this paradigm is appropriate for shared-memory architectures, but much less so for distributed-memory machines.

The concept of self-scheduling, at least in the sense we are using it, seems to be due to the designers of the Denelcor HEP multiprocessor. It is mentioned briefly in [37] and is given as an example in the HEP Fortran manual [1].

Parallel programs on shared-memory multiprocessors typically have a structure consisting of a "master" or "mainline" program which performs initialization, such as setting up the pool of tasks, and then initiates the execution of multiple copies of a "slave" subroutine(s). The master program and the slave subroutines are normally executed by individual processors. Thus, the mainline program usually also performs its share of the tasks after initiating the execution of the slave subroutines. This latter detail will be ignored in our pseudo-code descriptions of algorithms for shared-memory machines. They will have the general form shown in Figure 27.

```
Master Process
        (initialization)
        for j = 1,2,...,n do
            schedule Task(j)
Slave Process : Task(j)
            ⋮
```

Figure 27: Structure of shared-memory pseudo-code algorithms.

5.2 Cholesky's Method - Dense Case

5.2.1 Overview

We saw in section 2.2.2 that there are numerous ways of organizing the computation of the Cholesky factorization. Each has advantages in specific circumstances (memory access patterns, vectorization, etc.). A systematic study of these variations on Gaussian elimination and their implications for particular computer architectures is given by Dongarra, Gustavson and Karp [4].

Recall from section 2.2.2 that we identified three basic versions of the algorithm:

- row-Cholesky
- column-Cholesky
- submatrix-Cholesky

We further refined the first two categories into scalar-product row-Cholesky, vector-sum row-Cholesky, scalar-product column-Cholesky, and vector-sum column-Cholesky

For the three basic forms, let $Trow(i)$ be the task to compute the ith row L_{i*} of the Cholesky factor L, $Tcol(j)$ be the task to determine the jth column L_{*j}, and $Tsub(k)$ be the task to perform the submatrix modification from the kth column of L. We shall consider the self-scheduling of these tasks for the three algorithms and compare their respective performance in a multiprocessor environment.

Thus, at a very general level, our algorithms have the following forms:

```
for i = 1,2,...,n          for j = 1,2,...,n          for k = 1,2,...,n
  schedule Trow(i)           schedule Tcol(j)           schedule Tsub(k)
end                        end                        end
```

Of course there are many other ways of setting up self-scheduling loops to perform Cholesky factorization, depending on how we split the computation into tasks. However, these are appropriate for the level of granularity we wish to exploit, and they serve to illustrate the various techniques used in later sections to compare different parallel algorithms.

5.2.2 Row-Cholesky

In order to compute the ith row of the Cholesky factor, we require access to the previous $i-1$ rows of L. Computationally, these $i-1$ rows are used to do a lower triangular solution to determine L_{ij}, for $j = 1, 2, \ldots, i-1$. Then the diagonal element L_{ii} can be obtained from these computed entries of the ith row.

Depending on how the previous $i-1$ rows are accessed, whether by row or by column, we have the scalar-product or the vector-sum version of row-Cholesky.

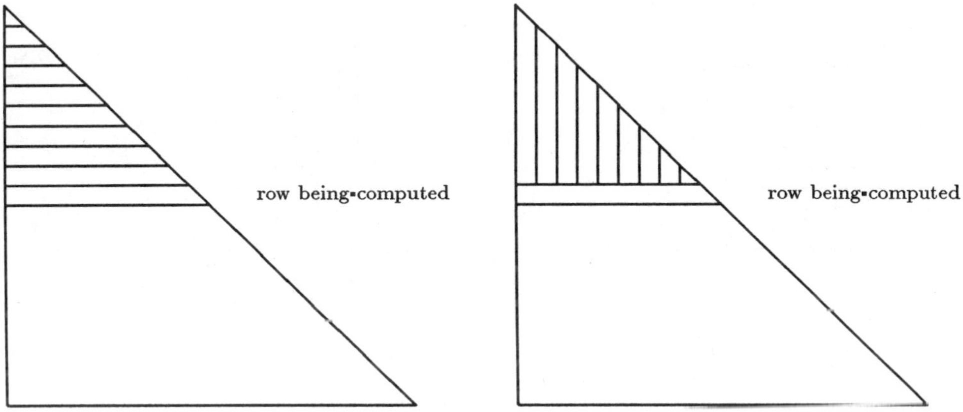

Figure 28: Two forms of row-Cholesky

The ith task $Trow(i)$ depends on results from all the previous $i-1$ tasks. Note, however, that in the scalar-product form, since the first $i-1$ rows of L are being accessed row by row in the execution of task $Trow(i)$, the part involving the first r rows can be performed once the first r tasks $Trow(1), \ldots, Trow(r)$ to compute $L_{1*}, \ldots, L_{r*}$ have been completed. In other words, although $Trow(i)$ uses results from $Trow(i-1)$, a major portion of these two tasks can be executed concurrently, except when $i \leq 2$.

Thus there is a Type 2 precedence relation among the n tasks

$$Trow(1) \rightarrow Trow(2) \rightarrow \cdots \rightarrow Trow(n)$$

since task $Trow(i)$ cannot be completed until $Trow(i-1)$ has finished, but $Trow(i)$ can begin before $Trow(i-1)$ is finished. Thus, the scheduling of these tasks on a number of parallel processors becomes potentially advantageous. The vector-sum form of row-Cholesky, on the other hand, leads to a Type 1 (i.e., serial) precedence relation among the tasks $Trow(*)$. Since we want to maximize parallelism, we shall henceforth use "row-Cholesky" to refer to the scalar-product form of row-Cholesky.

5.2.3 Column-Cholesky

Let us now consider the column-oriented version of Cholesky's method. To compute column j of the Cholesky factor, we require access to the rectangular submatrix enclosed (inclusively) by the jth row and j^{th} column of L. This rectangular submatrix can be accessed either by row or by column, so that we have the scalar-product or vector-sum versions of column-Cholesky. These two forms are illustrated in Figure 29.

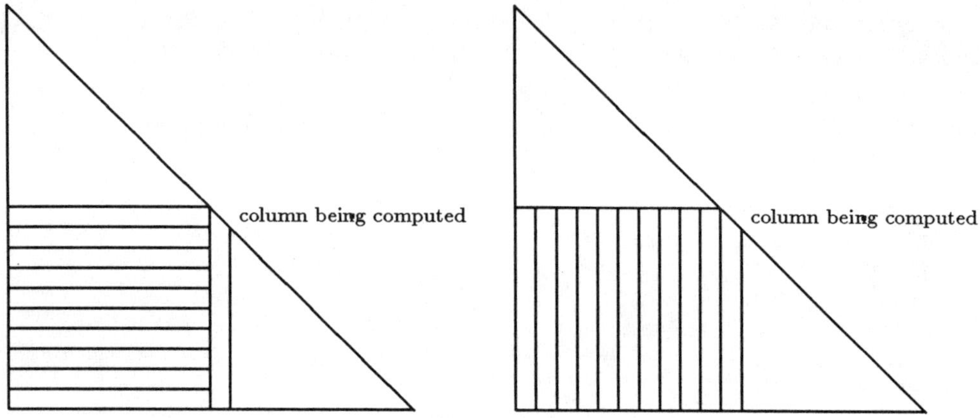

Figure 29: Two forms of column-Cholesky

The vector-sum form of column-Cholesky shares the same advantage as the scalar-product form in row-Cholesky. Indeed, the same Type 2 precedence relation exists among the n tasks

$$Tcol(1) \rightarrow Tcol(2) \rightarrow \cdots \rightarrow Tcol(n)$$

so that there is a high degree of potential concurrency among these tasks. It makes sense, therefore, to schedule these tasks for a number of parallel processors. The scalar-product form of the algorithm, however, suffers from the same difficulty as the vector-sum form of row-Cholesky, and imposes serial execution on the tasks. Thus, column-Cholesky will henceforth refer to the vector-sum version of the algorithm.

5.2.4 Submatrix-Cholesky

To apply the modification from column k of the Cholesky factor, we need to modify entries in the submatrix as given by the remaining $n - k$ columns of the matrix. Since the modification can be performed either by row or by column, we can again identify two different forms of submatrix Cholesky. Figure 30 illustrates these two forms.

The column-by-column form is more appropriate for self-scheduling loops, for the same reason we prefer the scalar-product row-Cholesky and vector-sum column-Cholesky forms. For our purpose, we shall use "submatrix-Cholesky" to refer to the column version, in which modification is applied column by column.

It is interesting to note that the n tasks $Tsub(*)$ are related in a rather different way. The task $Tsub(k)$ can start whenever the modifications to column k from the previous $k - 1$ tasks are done. It is therefore possible that the task $Tsub(k)$ can be

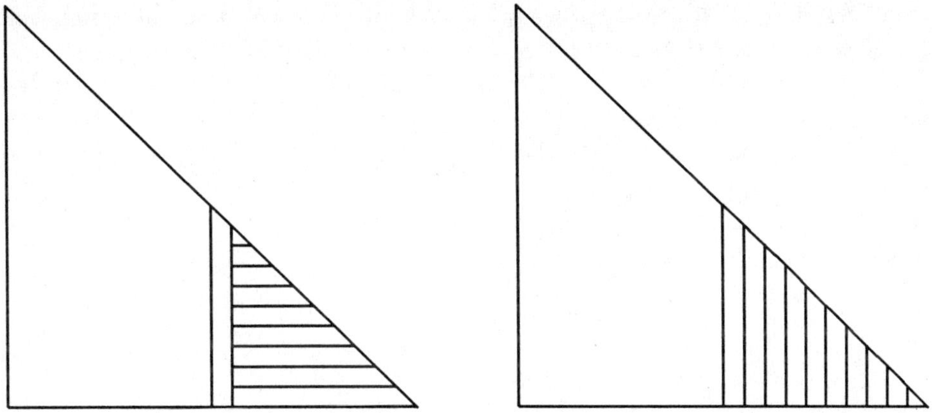

Figure 30: Two forms of submatrix-Cholesky

completed before a task $Tsub(c)$, where $c < k$. Thus the precedence relation

$$Tsub(1) \rightarrow Tsub(2) \rightarrow \cdots \rightarrow Tsub(n)$$

is of Type 3.

Another notable difference is the possibility of modifying the same column by different tasks at the same time. Some mechanism for mutual exclusion must therefore be incorporated to avoid simultaneous updates. This problem does not arise in row-Cholesky and column-Cholesky, since the modifications to a particular row or column are performed by only one task, so that the updates are done serially by one processor.

5.2.5 Work Profiles and Processor Utilization

In the previous section we have discussed three ways (row-Cholesky, column-Cholesky, and submatrix-Cholesky) of scheduling tasks in a parallel environment for the factorization of a symmetric positive definite matrix. For each of the three forms, we saw that a comparison of the precedence relations among the resulting tasks enabled us to identify the more inherently parallel of the two possible variations. In this section we turn to the problem of choosing from among the three basic forms the best for parallel execution. In order to do this we introduce the notion of *work profiles* of the different self-scheduling algorithms.

In general, consider the self-scheduling of the following loop:

for $t = 1, 2, \ldots, n$ **do**
  schedule $Task(t)$
end

where t is the task number or task id used for the purpose of task scheduling. In other words, if there are p processors available (assuming that $p < n$), the first p

tasks will be claimed by these processors. Whenever a processor becomes free, it will be responsible for the next task in the sequence, namely $Task(p+1)$.

For each task, we define $TaskWork(t)$ to be the amount of work required to complete it. The *work profile* is then the graph of $TaskWork(t)$ plotted against t. For simplicity, we assume that the amount of work for a task is the number of multiplicative operations required. Here, for uniformity, we shall regard a square root operation as another multiplicative operation. For the three basic forms of Cholesky it is easy to verify that:

$$TrowWork(i) = i(i+1)/2$$

$$TcolWork(j) = j(n-j+1)$$

$$TsubWork(k) = (n-k+1)(n-k+2)/2$$

With row-Cholesky, the relatively small tasks at the beginning should enable all processors to become fully utilized quickly, but saving the larger tasks to the end is likely to cause a significant number of processors to become idle while other processors finish with tasks involving the last few rows. Since these tasks require comparatively more time (proportional to $n^2/2$), the degradation in the overall efficiency of the scheme is non-trivial. Submatrix-Cholesky is rather the opposite: the relatively large tasks at the beginning may tend to inhibit full processor utilization early on, but its terminal behavior should be good because the tasks are getting smaller toward the end. Column-Cholesky has the best properties of both: task sizes taper up and then down in a smooth manner, leading to good processor utilization throughout the computation.

5.2.6 Implementation

We concluded in the previous section that the column-oriented Cholesky algorithm was the most desirable in terms of processor utilization. In this section we consider this algorithm in somewhat more detail, and discuss its implementation.

As before, let $Tcol(j)$ be the *task* that computes the j-th column of the Cholesky factor. Each such task consists of the following two types of subtasks:

1. $cmod(j,k)$: modification of column j by column k $(k < j)$;

2. $cdiv(j)$: division of column j by a scalar.

Thus, in terms of these sub-tasks, the basic algorithm can be expressed in the following condensed form.

```
for j = 1, 2, ..., n do
    for k = 1, 2, ..., j − 1 do
        cmod(j, k)
    cdiv(j)
```

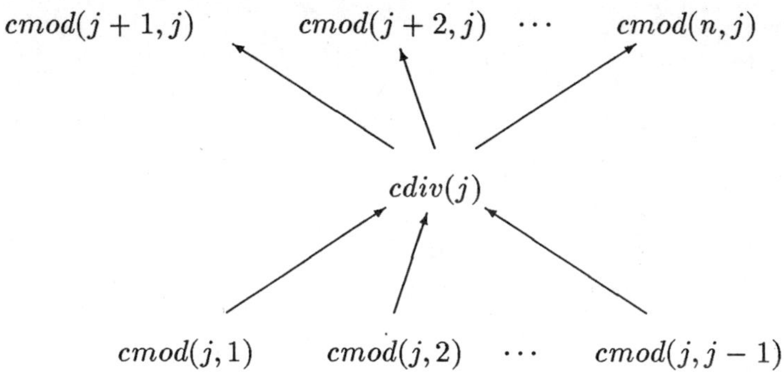

Figure 31: Subtask precedence graph for column-Cholesky.

We now consider the potential for parallelism in the above formulation of the algorithm. We implicitly assume throughout this paper that the *cmod* and *cdiv* operations are atomic in the sense that we do not attempt to exploit parallelism within them, although such exploitation is clearly possible.

Note first that $cdiv(j)$ cannot begin until $cmod(j,k)$ has been completed for all $k < j$, and column j can be used to modify subsequent columns only after $cdiv(j)$ has been completed. However, there is no restriction on the order in which the *cmod* operations are executed, and *cmod* operations for different columns can be performed concurrently. For example, after $cdiv(1)$ has completed, $cmod(2,1)$ and $cmod(3,1)$ could execute in parallel. These precedence relations are depicted in Figure 31.

The self-scheduling of the tasks $Tcol(j)$ can be implemented quite easily by maintaining a vector of flags "$ready[*]$", where $ready[j]$ indicates whether column j is ready to be used for modification of subsequent columns. Figure 32 gives an algorithmic description of the implementation.

We note that this implementation has the advantage of requiring no mutual exclusion or critical section other than that directly related to self-scheduling (i.e., processes picking up a unique task id).

5.3 Cholesky's Method - Sparse Case

5.3.1 A Modified Dense Shared-Memory Version

We now introduce a version of parallel dense factorization that differs somewhat from the one described in section 5.2.6. Its purpose is to motivate the parallel sparse column-Cholesky algorithm to be described in the section 5.3.2.

The new version, given in Figure 33, maintains a set of non-overlapping linked

```
Master Process
        for j = 1,2,...,n do
            ready[j] ← 0
        for j = 1,2,...,n do
            schedule Tcol(j)
Slave Process : Tcol(j)
        for k = 1,2,...,j − 1 do
            wait until ready[k] = 1
            cmod(j,k)
        cdiv(j)
        ready[j] ← 1
```

Figure 32: Cholesky Factorization: Shared-memory multiprocessor.

lists. Since they are non-overlapping, an n-vector $link[*]$ will be enough to implement it. The jth list is given by:

$$link[j], link[link[j]], \ldots, link^r[j],$$

where $link[link^r[j]] = 0$.

In this version, $nmod[j]$ is used to keep track of the number of column modifications that remain to be performed on column j; it is initialized to $j-1$. For column j, the linked list:

$$link[j], link[link[j]], \cdots$$

gives the columns that are currently ready to modify column j. After a column, say the kth one, is used to modify column j, it is removed from the linked list for j and passed on to column $j+1$. Moreover, after the $cdiv(j)$ operation, column j is now ready to modify subsequent columns, and it is placed in the linked list for column $j+1$.

Compared to the previous algorithm (using the *ready* vector), this version has several disadvantages. An extra n-vector is required, and more importantly, some form of critical section has to be set up during the update of the *link* vector. This is to avoid simultaneous update of the *link* vector by different *Tcol* tasks. In particular, since each update of the linked lists involves two different elements of the *link* vector, it is conceivable that these two elements may be modified by more than one processor simultaneously at any instance. To solve this problem, a *single* synchronization variable which is shared by *all* processors is used to protect the entire *link* vector from being modified by more than one processor at any given time. Such a mechanism is not necessary in the *ready*-version.

However, these changes to the basic scheme are required in order to take advantage of the parallelism derived from sparsity of L. Note that this new version of the factorization algorithm closely resembles the serial version discussed in section 2.2.5.

```
Master Process:
        for j = 1,2,...,n do
            link[j] ← 0
            nmod[j] ← j − 1
        for j = 1,2,...,n do
            schedule Tcol(j)
Slave Process: Tcol(j)
        while nmod[j] > 0 do
            wait until link[j] > 0
            k ← link[j]
            link[j] ← link[k]
            cmod(j,k)
            nmod[j] ← nmod[j] − 1
            if j < n then
                link[k] ← link[j + 1]
                link[j + 1] ← k
        cdiv(j)
        if j < n then
            link[j] ← link[j + 1]
            link[j + 1] ← j
```

Figure 33: A modified Cholesky algorithm for a shared-memory multiprocessor.

5.3.2 Parallel Sparse Column-Cholesky

The important difference between the sparse and dense versions of the algorithm is that column k or j should no longer be passed to the $(j+1)^{st}$ list; instead, they should be passed to the list as determined by the structure of the matrix.

The *link*-version for the dense factorization in the previous section is modified to factor sparse matrices by using the function $next(*,*)$. This is a function which denotes the row index of the nonzero immediately below the (j,k) position in L. Note that the subscript $next(j,k)$ depends on both the indices j and k. We adopt the convention that if $l_{j,k}$ is the last nonzero in column k, we put $next(j,k) = n+1$. The initialization remains the same except that $nmod[j]$ should be initialized to $|L_{j*}|-1$, the number of offdiagonal nonzeros in the jth row of L. The new algorithm is given in Figure 34.

```
Master Process:
          for j = 1,2,...,n do
              link[j] ← 0
              nmod[j] ← |L_j*| − 1
          for j = 1,2,...,n do
              schedule Tcol(j)
Slave Process: Tcol(j)
          while nmod[j] > 0 do
              wait until link[j] > 0
              k ← link[j]
              link[j] ← link[k]
              cmod(j,k)
              nmod[j] ← nmod[j] − 1
              nextnz ← next(j,k)
              if nextnz ≤ n then
                  link[k] ← link[nextnz]
                  link[nextnz] ← k
          cdiv(j)
          nextnz ← next(j,j)
          if nextnz ≤ n then
              link[j] ← link[nextnz]
              link[nextnz] ← j
```

Figure 34: Sparse Cholesky for a shared-memory multiprocessor.

Another implementational difference from the dense case is in the execution of $cmod(j,k)$. In the sparse case, the columns L_{*j} and L_{*k} are stored in a compact form. To perform $cmod(j,k)$, we need to unpack the compact form of the column L_{*j} so that modifications from other columns can be done efficiently. This implies that the processor executing the task $Tcol(j)$ requires a local working array of size n to facilitate the column update operations on column j. However, the data structure used for storing L in the parallel sparse Cholesky algorithm is the same as that in the sequential sparse Cholesky algorithm, since it is stored in global memory and is accessible by all processors.

The dense and the sparse cases differ in the order in which the *cdiv*'s are performed. In the dense case, the *cdiv*'s are performed in sequential order. However, in the sparse case, only a few *cmod*'s have to be applied to column j, and $cdiv(j)$ can be performed once all the necessary *cmod*'s have been applied. Thus not only can some of the *cmod*'s be carried out in parallel in the sparse case, some of the *cdiv*'s may also be completed simultaneously.

In the discussion above, we have not addressed the problem of assigning task id's. Our previous discussion on elimination trees and their relation to parallelism suggests that ordering the tasks in the task queue in a bottom-up manner on a balanced elimination tree ordering would be ideal.

5.3.3 Parallel Sparse Triangular Solutions

In this section we consider the parallel solutions of the triangular systems

$$Lv = u$$

and

$$L^T w = v,$$

where L and u are either given or computed elsewhere. It is assumed that the elements of L are stored column by column. We shall first describe a parallel algorithm for the backward solution and then present a parallel algorithm for the forward solve.

Let $Tw(j)$ denote the task that computes w_j. A parallel algorithm for computing w is given in Figure 35. It makes use of the vector *ready* described in the previous sections.

The functions $wmod(j,k)$ and $wdiv(j)$ are similar to *cmod* and *cdiv* in the Cholesky factorization, respectively. More precisely, $wmod(j,k)$ modifies v_j by w_k using L_{kj}, and $wdiv(j)$ divides v_j by L_{jj} to yield the result w_j. Note that once a component w_j is computed, it can be used to modify any $v_i, i < j$. The flag $ready[j]$ is used to signal when w_j is available. The parallel algorithm described in Figure 35 is efficient in terms of accessing the elements of L, since L is stored column by column.

When L is dense, the *wmod*'s can be performed in parallel, but the *wdiv*'s are completed sequentially. However, when L is sparse, $wmod(j,k)$ is performed only when L_{kj} is nonzero. As a result, not only can the *wmod*'s be carried out in parallel, some of the *wdiv*'s may also be completed simultaneously. Thus the algorithm described above exploits the parallelism inherent in the backward solution and that provided by the sparsity of L.

Now we consider the problem of solving the lower triangular system $Lv = u$ in parallel. We first recall that the elements of L are assumed to be stored column by

```
Master process:
        for j = 1,2,...,n do
            ready[j] ← 0
        for j = n,n-1,...,1 do
            schedule Tw(j)
Slave Process: Tw(j)
        for k ∈ {Ω(L_{*j}) - {j}} do
            wait until ready[k] ≠ 0
            wmod(j,k)
        wdiv(j)
        ready[j] ← 1
```

Figure 35: Sparse backward solution for a shared-memory multiprocessor.

column. Let $Tv(j)$ denote the task that computes v_j. In the following discussion, *vmod*(j,k) modifies u_j by v_k using L_{jk}, and *vdiv*(j) divides u_j by L_{jj} to yield the result v_j.

The algorithm is shown in Figure 36. Note that it makes use of a vector *nmod* and is column-oriented.

```
Master process:
        for j = 1,2,...,n do
            nmod[j] ← |L_{j*}| - 1
        for j = n,n-1,...,1 do
            schedule Tv(j)
Slave Process: Tv(j)
        wait until nmod[j] = 0
        for k ∈ {Ω(L_{*j}) - {j}} do
            vmod(k,j)
            nmod[k] ← nmod[k] - 1
```

Figure 36: Sparse forward solution for a shared-memory multiprocessor.

In this version, once v_j is computed, it can be used, together with column j of L, to modify the right hand side vector. Thus accessing the elements of L is very efficient.

Obviously, some form of critical section must be set up when performing the *vmod*'s and decrementing *nmod*. This is particularly important in the sparse case since the sparsity of L may cause a component of u to be modified simultaneously by several components of v. Thus, in general, there must be some form of synchronization lock for each component of u. This will be a difficulty if the multiprocessor system provides only a small number of such locks.

6 Algorithms for Distributed-Memory Machines

6.1 Introduction

As we noted earlier, on shared-memory machines, global communication is a relatively minor concern, since all processors have access to a common memory. Results computed by one processor can be thought of as being immediately available to all other processors. However, on distributed-memory muliprocessors, this is no longer true. Communication over the interconnection network tends to cause delays in which some processors must wait for necessary prior results to be communicated by other processors. Since communication speeds, at least currently, are relatively slow compared to the computational speed of the processors, communication becomes a critical factor affecting computational efficiency. In a distributed-memory environment, we must usually be content with a static load balance that is determined in advance of the computation, in contrast to a dynamic load balance that is easily implemented in shared memory by means of a pool of tasks.

In our discussion of algorithms for distributed-memory computers, we implicitly assume the presence of a host that can communicate with all of the node processors. The role of the host is to distribute the rows or columns of the matrix to the node processors in advance of the computation, along with any other information needed by them, and to receive the results of the computations from the node processors.

We make frequent use of an n-vector called *map*, whose elements specify in which of the p node processors the columns or rows of the matrix reside. Thus, $map[i] = j$ would mean that row i is contained in processor j. Each processor has a copy of *map*. A prototypical host program is shown in Figure 37.

```
Compute map
for s = 1,2,...p do
    send map to processor s
for j = 1,2,...,n do
    send A_{*j} to processor map[j]
repeat n times
    receive a column of L
```

Figure 37: Sample host program.

Similarly, in a typical node program, there will be an initialization part involving receipt of *map* and its set of rows or columns. In order to avoid redundancy, we will not explicitly include this initialization in our pseudo-code descriptions of algorithms. We will often make use of the sets *mycols* or *myrows*, which are the indices of the columns or rows assigned to a node processor. These sets are assumed to have a natural order associated with them, and when we indicate a loop over the elements of a set, the elements are to be taken in that natural order.

A typical node program is shown in Figure 38, where for completeness, we have included the receipt of the map and the data, and sending the results to the host. However, in general we will omit these items in our descriptions.

```
receive map
for j ∈ mycols do
    receive A_{*j}
for i = 1,2,...,n do
    for j ∈ mycols do
        x_i ← x_i + A_{ij}
    send x_i to host
```

Figure 38: Sample procedure for a node processor.

6.2 Cholesky's Method - Dense Case

In this section we study the implementation of the column-oriented Cholesky algorithm that we considered in detail in an earlier section for shared-memory architectures. We will adopt the same level of task granularity. That is, the *cmod*(j,k) and *cdiv*(j) operations will be the basic subtasks to be scheduled on the processors. In the shared-memory environment, when a processor is assigned the task of computing a given column of L, it first performs all of the necessary modifications by previous columns, then performs the column division. All of the required previous columns are directly accessible to each processor because they are stored in the common global memory. Moreover, by sharing the pool of tasks among all of the processors and assigning columns to processors dynamically, the computational load tends to be automatically balanced among the processors. In a message-passing, distributed-memory system, the pool of tasks cannot easily be shared, so the load balance is entirely dependent on the static assignment of columns to processors. Since all memory is private, each column computed by a given processor must be explicitly sent to other processors that need it for updating the columns they have been assigned. Thus, in general, global communication is required, and this may or may not be well supported by a particular interconnection network.

Let us assume that each processor has been assigned a subset of the columns of L to compute and that the corresponding columns of the original matrix A reside in the local memory of the processor. Later we will discuss ways to make this assignment effectively. For now, we let the indices of the columns assigned to a processor be contained in the set *mycols*. Then the program that runs on each processor in the message-passing version of the column-Cholesky algorithm might be as shown in Figure 39.

```
for j = 1, 2, ..., n do
    if j ∈ mycols then
        cdiv(j)
        send L_{*j} to processors {map[s] | j < s ≤ n}
    else
        receive L_{*j}
    for k ∈ mycols, k > j do
        cmod(k, j)
```

Figure 39: Cholesky factorization for a local-memory multiprocessor

Note that the *cdiv* operation on a given column will be done only after all of its necessary modifications have taken place. Thus, the proper synchronization of the algorithm is implicit in the flow of completed columns through the network.

It is possible to overlap the computation and communication to a greater extent than shown here, in order to mask some of the communication cost with computation. The philosophy is to send out results at the earliest possible time in the hope that this will minimize any subsequent waiting for them. In particular, as soon as any column has had all of its modifications completed, the *cdiv* operation should be carried out immediately so that the broadcast of the resulting column of L can be started.

The overlapping of communication with computation enables the modified algorithm to maintain a higher level of processor utilization than the "synchronous" algorithm. Of course the amount of improvement will depend on the relative speeds of communication and computation, so the difference in performance on a given machine may or may not be significant. This strategy is usually referred to as "pipelining"[35]. A processor, rather than continuing its current computation, sends already computed values to other processors. A more detailed version of the algorithm, exploiting this notion, is shown in Figure 40. Note that it uses an array *nmod* to keep track of how many times a column must be modified.

At this level of specification, the algorithm is independent of the specific type of message-passing multiprocessor architecture. Only the details of the send and receive procedures depend on the communication supported by a particular interconnection network.

In addition to the communication technique, the other main feature affecting performance of the algorithm is the mapping of the matrix onto the processors. We have already considered this problem to some extent for sparse matrices in our discussion of elimination trees. In the next section we consider the problem for dense matrices.

for $j = 1, 2, \ldots, n$ **do**
  $nmod[j] \leftarrow j - 1$
if $1 \in mycols$ **then**
    $cdiv(1)$
    send L_{*1} to processors $\{map[s] \mid 1 < s \leq n\}$
    $mycols \leftarrow mycols - \{1\}$
while $|mycols| > 0$ **do**
  receive a column, say column L_{*k}
  for $j \in mycols$ **do**
    $cmod(j, k)$
    $nmod[j] \leftarrow nmod[j] - 1$
    if $nmod[j] = 0$ **then**
      $cdiv(j)$
      send L_{*j} to processors $\{map[s] \mid j < s \leq n\}$
      $mycols \leftarrow mycols - \{j\}$

Figure 40: Improved Cholesky factorization algorithm.

6.2.1 Mapping Columns to Processors

The manner in which the matrix is mapped onto the processors will affect the communication requirements, the degree of concurrency, and the load balance among the processors. We would like to minimize communication, maximize concurrency, and have a uniform work load across the processors. These tend to be conflicting desiderata, however, so acceptable compromises must be sought.

One possibility is to map a contiguous block of n/p columns onto each processor; this is referred to as *block* mapping. Another possibility, already mentioned in a previous section, is to assign the columns to the processors in the same manner one would deal cards, assigning one column to each processor and then wrapping back to the beginning with further columns; this is usually called *wrap* mapping.

Which of these mappings should give the best performance? Regardless of the mapping used, after completing the last of its assigned columns, each processor becomes idle for the remainder of the factorization. The block mapping therefore causes the processors containing the earlier blocks to be idle much of the time, whereas the wrap mapping tends to keep all processors busy as long as possible. Thus, we would expect the wrap mapping to yield much higher concurrency and processor utilization than the block mapping. On the other hand, the block mapping has potentially smaller communication requirements, since each completed column needs to be sent only to higher numbered processors, rather than to all processors as in the wrap mapping.

Thus, it would appear that which mapping is best will depend on the relative speeds of communication and computation. However, for machines that are currently available, even though they have rather slow communication, the poor concurrency of block mapping leads to performance that is uniformly inferior to that of the wrap mapping.

6.3 Cholesky's Method - Sparse Case

6.3.1 Numerical Factorization

As before, we let $Tcol(j)$ be the task of computing the jth column of the sparse Cholesky factor L, where this task consists of the two types of subtasks $cmod(j,k)$ and $cdiv(j)$.

Recall that the task $Tcol(j)$ can be expressed in the following algorithmic form, where the number of *cmod* operations required in the task $Tcol(j)$ is given by the number of off-diagonal nonzeros in the jth row of L.

for $k \in \{\Omega(L_{j*}) - \{j\}\}$ **do**
    $cmod(j,k)$

$cdiv(j)$

As we did for the shared-memory version of the algorithm, we introduce the vector *nmod*, where the value $nmod[j]$ is the number of column modifications *cmod* required in the execution of $Tcol(j)$. This vector can be obtained by simply counting the number of off-diagonal nonzeros in each row of L.

We assume that an assignment of the column tasks $Tcol(*)$ to the computational nodes of the multiprocessor has been given. (We implicitly assume here and with other column-oriented algorithms that the matrix has been ordered to reduce fill, and reordered so that its elimination tree is balanced, as discussed in sections 4.2 and 4.3. Additionally, we assume that the columns have been allocated to the processors by subtrees, as discussed in section 4.3.) As usual, let *map* be the mapping of these n tasks into the p processors. The work to be performed by the node processors is described in Figure 41.

As usual, we implicitly assume that the host processor is responsible for the initiation of the tasks by sending the relevant information to each node processor, and then for the collection of the computed columns of L. Note that the computed columns are sent only to those processors that actually need them during the factorization. Moreover, only *one* copy of the column L_{*j} should be sent to a processor even though that processor may use the column to modify more than one column in the processor.

Note that as soon as a column L_{*j} is completely formed, it is immediately sent to the other processors that need this column. This will tend to reduce wait time on node processors. In order to make the presentation simple, our pseudo-code may also send column j to itself, even though it already has it. Whether this would be done in the actual implementation would probably involve a tradeoff between efficiency and program simplicity.

for $j \in mycols$ **do**
  if $nmod[j] = 0$ **then**
    $cdiv(j)$
    send L_{*j} to processors $\{map[r] \mid r \in \{\Omega(L_{*j}) - \{j\}\}\}$
    $mycols \leftarrow mycols - \{j\}$
while $|mycols| > 0$ **do**
  receive a column of L, say L_{*k}
  for $j \in \{mycols \cap \{\Omega(L_{*k}) - \{k\}\}\}$ **do**
    $cmod(j, k)$
    $nmod[j] \leftarrow nmod[j] - 1$
    if $nmod[j] = 0$ **then**
      $cdiv(j)$
      send L_{*j} to processors $\{map[r] \mid r \in \{\Omega(L_{*j}) - \{j\}\}\}$
      $mycols \leftarrow mycols - \{j\}$

Figure 41: Procedure for the node processor.

Note also that the main loop is driven by the incoming columns of L. This implies that the parallel algorithm is working at the granularity level of the subtasks $cmod(j, k)$ and $cdiv(j)$, rather than at the level of the tasks $Tcol(j)$.

Another important characteristic of this formulation is that it is independent of the interconnection network topology. For different processor interconnections, it may be desirable to choose a different task-to-processor mapping function $map[*]$ or a different message routing strategy. But the basic algorithm remains unchanged.

6.3.2 Symbolic Factorization

In this section we consider a parallel algorithm for symbolic Cholesky factorization of sparse symmetric matrices. It makes heavy use of the notions already introduced in the previous section, so our discussion will be brief. The reader might find it useful to review section 2.2.4 and particularly the algorithm displayed in Figure 7. We will adopt the usual notation from that section as well.

To begin the discussion, let us consider the example in Figure 6 again. Note that the structures of the first 3 columns of L are independent of each other in the sense that the structure of each of them depends only on the corresponding columns of A. Thus, if 3 processors are available, the structures of the first three columns of L can be computed simultaneously. Using the results in section 2.2.4, the structures of columns 1 and 2 of L will be used to modify the structures of columns 5 and 4 of A, respectively. Clearly these can also be performed simultaneously. (Note that column 4 of A will also be modified by the structure of column 3 of L, but this has to be performed either after or before column 4 of A has been modified by column 2 of L.) This small example illustrates the fact that there is a certain degree of parallelism associated with the sequential symbolic factorization algorithm which we can exploit. In terms of the elimination tree, columns at the same level in the elimination tree can be computed independently. Our goal in this section is to present an algorithm that exploits such parallelism.

The algorithm is presented in Figure 42. We assume that the number of structure modifications required to compute column k of L, which we denote by *smod*$[k]$, is known. This information allows us to determine when the structure of a column of L has been fully computed. Note that *smod*$[k]$ is simply the number of columns of L whose first off-diagonal nonzeros are in row k. The computation of *smod* will be described after we have presented the parallel symbolic factorization algorithm. The

for $k \in mycols$ **do**
    $\mathcal{S}_k \leftarrow \phi$
    $\mathcal{L}_k \leftarrow \Omega(A^{\triangleright}_{*k})$
    if $smod[k] = 0$ **then**
        $mycols \leftarrow mycols - \{k\}$
        if $|\mathcal{L}_k| > 1$ **then**
            determine m_k
            send $\mathcal{L}_k$ to processor $map[m_k]$
while $|mycols| > 0$ **do**
    receive a column, say $\mathcal{L}_i$
    determine m_i
    $\mathcal{L}_{m_i} \leftarrow \mathcal{L}_{m_i} \cup \mathcal{L}_i - \{i\}$
    $smod[m_i] \leftarrow smod[m_i] - 1$
    if $smod[m_i] = 0$ **then**
        $mycols \leftarrow mycols - \{m_i\}$
        if $|\mathcal{L}_{m_i}| > 1$ **then**
            determine m_{m_i}
            send $\mathcal{L}_{m_i}$ to processor $map[m_{m_i}]$

Figure 42: Sparse Symbolic Factorization Algorithm.

parallel symbolic factorization algorithm is basically the same as the sequential algorithm presented in section 2.2.4, except that the computation has been reorganized. The structures of the columns of the Cholesky factor may now not be computed in the natural order. Structure modifications to a particular column arrive asynchronously, and they are applied to the target column as they are received. As a result, the data structure for implementing the parallel algorithm has to be flexible. This differs from the sequential algorithm, in which the structure modifications to a particular column are applied in the natural order, and the structures of the columns of L are computed sequentially.

In the description of the parallel symbolic factorization algorithm, we have assumed that *smod* is given. On the surface, this does not make much sense since the definition of *smod* suggests that the structure of L, which we are trying to compute, is already known. However, as noted earlier in section 2.2.6, an efficient sequential algorithm is available to compute the elimination tree of L from the structure of A. Moreover, a parallel algorithm for computing the elimination tree on a distributed-memory multiprocessor has been developed by Gilbert and Hafsteinsson [33]. Given the existence of these algorithms, it is reasonable to assume that *smod* can be made available to the symbolic factorization algorithm. Note that $smod[j]$ is the number of immediate descendents (child vertices) of vertex j in the elimination tree.

References

[1] *HEP Fortran 77 User's Guide.* Denelcor, Inc., Aurora, CO, February 1982.

[2] R.P. Brent and F.T. Luk. *A systolic architecture for the singular value decomposition.* Technical Report TR-82-522, Dept. of Comput. Sci., Cornell University, Ithaca, 1982.

[3] T.F. Coleman, A. Edenbrandt, and J.R. Gilbert. Predicting fill for sparse orthogonal factorization. *J. Assoc. Comput. Mach.*, 33:517–532, 1986.

[4] J.J. Dongarra, F.G. Gustavson, and A. Karp. Implementing linear algebra algorithms for dense matrices on a vector pipeline machine. *SIAM Review*, 26:91–112, 1984.

[5] I.S. Duff. Full matrix techniques in sparse Gaussian elimination. In G. A. Watson, editor, *Lecture Notes in Mathematics (912)*, Springer-Verlag, 1982.

[6] I.S. Duff. *MA28 - A set of FORTRAN subroutines for sparse unsymmetric linear equations.* Technical Report AERE R-8730, Harwell, 1977.

[7] I.S. Duff. Parallel implementation of multifrontal schemes. *Parallel Computing*, 3:193–204, 1986.

[8] I.S. Duff and L. Johnsson. The effect of orderings on the parallelization of sparse code. Technical Memorandum. Mathematical and Computer Science Division, Argonne National Laborary, 1986.

[9] I.S. Duff and J.K. Reid. The multifrontal solution of indefinite sparse symmetric linear equations. *ACM Trans. on Math. Software*, 9:302–325, 1983.

[10] I.S. Duff and J.K. Reid. The multifrontal solution of unsymmetric sets of linear equations. *SIAM J. Sci. Stat. Comput.*, 5:633–641, 1984.

[11] S.C. Eisenstat, M.H. Schultz, and A.H. Sherman. Applications of an element model for Gaussian elimination. In J. R. Bunch and D. J. Rose, editors, *Sparse Matrix Computations*, pages 85–96, Academic Press, 1976.

[12] S.C. Eisenstat, M.H. Schultz, and A.H. Sherman. Minimal storage band elimination. In A.H. Sameh and D. Kuck, editors, *Proceedings of the Symposium on Sparse Matrix Computations*, pages 273–286, 1977.

[13] S.C. Eisenstat, M.H. Schultz, and A.H. Sherman. Software for sparse Gaussian elimination with limited core storage. In I.S. Duff and G.W. Stewart, editors, *Sparse Matrix Proceedings*, pages 135–153, SIAM Press, 1978.

[14] G.E. Forsythe and C.B. Moler. *Computer solution of linear algebraic systems.* Prentice-Hall Inc., Englewood Cliffs, N.J., 1967.

[15] G.A. Geist and M.T. Heath. *Parallel Cholesky factorization on a hypercube multiprocessor.* Technical Report ORNL-6211, Oak Ridge National Laboratory, Oak Ridge, Tennessee, 1985.

[16] J.A. George and M.T. Heath. Solution of sparse linear least squares problems using Givens rotations. *Linear Algebra and its Appl.*, 34:69–83, 1980.

[17] J.A. George, M.T. Heath, and J.W-H. Liu. Parallel Cholesky factorization on a shared-memory multiprocessor. *Linear Algebra and its Appl.*, 77:165–187, 1986.

[18] J.A. George, M.T. Heath, J.W-H. Liu, and E.G-Y. Ng. *Solution of sparse positive definite systems on a shared memory multiprocessor.* Technical Report CS-86-10, Dept. of Computer Science, York University, 1986. (to appear in Internat. J. Parallel Programming).

[19] J.A. George, M.T. Heath, J.W-H. Liu, and E.G-Y. Ng. *Sparse Cholesky factorization on a local-memory multiprocessor.* Technical Report CS-86-02, Department of Computer Science, University of Waterloo, 1986. (to appear in Siam J. Sci. Stat. Comput.).

[20] J.A. George, M.T. Heath, J.W-H. Liu, and E.G-Y. Ng. Symbolic Cholesky factorization on a local-memory multiprocessor. *Parallel Computing*, 1987. (special issue, to appear).

[21] J.A. George and J.W-H. Liu. Compact structural representation of sparse Cholesky, QR and LU factors. In R. Glowinski and J.-L. Lions, editors, *Computing Methods in Applied Sciences and Engineering, VII*, Elsevier Publishers B.V. (North-Holland), 1985.

[22] J.A. George and J.W-H. Liu. *Computer Solution of Large Sparse Positive Definite Systems.* Prentice-Hall Inc., Englewood Cliffs, New Jersey, 1981.

[23] J.A. George and J.W-H. Liu. Householder reflections versus Givens rotations in sparse orthogonal decomposition. *Linear Algebra and its Appl.*, 1986. (To appear.).

[24] J.A. George and J.W-H. Liu. On the evolution of the minimum degree algorithm. 1987. in preparation.

[25] J.A. George and J.W-H. Liu. An optimal algorithm for symbolic factorization of symmetric matrices. *SIAM J. Comput.*, 9:583–593, 1980.

[26] J.A. George, J.W-H. Liu, and E.G-Y. Ng. Communication reduction in parallel sparse Cholesky factorization on a hypercube. In M.T. Heath, editor, *Hypercube Multiprocessors*, SIAM Publications, Philadephia, PA., 1987.

[27] J.A. George, J.W-H. Liu, and E.G-Y. Ng. Communication results for parallel sparse Cholesky factorization on a hypercube. 1987. (submitted to Parallel Computing).

[28] J.A. George, J.W-H. Liu, and E.G-Y. Ng. *A data structure for sparse QR and LU factors.* Technical Report CS 85-16, Dept of Computer Science, University of Waterloo, 1985. (to appear in SIAM J. Sci. Stat. Comput.).

[29] J.A. George and E.G-Y. Ng. An implementation of Gaussian elimination with partial pivoting for sparse systems. *SIAM J. Sci. Stat. Comput.*, 6:390–409, 1985.

[30] J.A. George and E.G-Y. Ng. On the complexity of sparse QR and LU factorization of finite element matrices. 1987. (submitted to SIAM J. Sci. Stat. Comput.).

[31] J.A. George and E.G-Y. Ng. Orthogonal reduction of sparse matrices to upper triangular form using Householder transformations. *SIAM J. Sci. Stat. Comput.*, 7:460–472, 1986.

[32] J.A. George and E.G-Y. Ng. *Symbolic factorization for sparse Gaussian elimination with partial pivoting.* Technical Report CS-84-43, Department of Computer Science, University of Waterloo, 1984. (to appear in SIAM J. Sci. Stat. Comput.).

[33] J.R. Gilbert and H. Hafsteinsson. *A parallel algorithm for finding fill in a sparse symmetric matrix.* Technical Report TR 86-789, Dept. of Computer Science, Cornell University, Ithaca, New York, 1986.

[34] J.R. Gilbert and T. Peierls. *Sparse partial pivoting in time proportional to arithmetic operations.* Technical Report TR 86-783, Dept. of Computer Science, Cornell University, Ithaca, New York, 1986.

[35] M.T. Heath and C.H. Romine. *Parallel solution of triangular systems on distributed memory-multiprocessors.* Technical Report ORNL/TM-10384, Mathematical Sciences Section, Oak Ridge National Laboratory, Oak Ridge, Tennessee 37831, 1987.

[36] J.A.G. Jess and H.G.M. Kees. A data structure for parallel L/U decomposition. *IEEE Trans. Comput.*, C-31:231–239, 1982.

[37] H.F. Jordan. Experience with pipelined multiple instruction streams. *Proc. IEEE*, 72:113–123, 1984.

[38] J.W-H. Liu. A compact row storage scheme for Cholesky factors using elimination trees. *ACM Trans. on Math. Software*, 12:127–148, 1986.

[39] J.W-H. Liu. Computational models and task scheduling for parallel sparse Cholesky factorization. *Parallel Computing*, 3:327–342, 1986.

[40] J.W-H. Liu. *Equivalent sparse matrix reordering by elimination tree rotations.* Technical Report CS-86-12, Dept. of Computer Science, York University, November 1986.

[41] J.W-H. Liu. Modification of the minimum degree algorithm by multiple elimination. *ACM Trans. on Math. Software*, 11:141–153, 1985.

[42] J.W-H. Liu. On general row merging schemes for sparse Givens transformations. *SIAM J. Sci. Stat. Comput.*, 7:1190–1211, 1986.

[43] J.W-H. Liu. *Reordering sparse matrices for parallel elimination.* Technical Report CS-87-01, Dept. of Computer Science, York University, 1987.

[44] D.P. O'Leary and G.W. Stewart. Data-flow algorithms for parallel matrix computations. *Comm. Assoc. Comput. Mach.*, 28:840–853, 1985.

[45] S.V. Parter. The use of linear graphs in Gaussian elimination. *SIAM Review*, 3:364–369, 1961.

[46] F.J. Peters. *Sparse Matrices and Substructures.* Mathematisch Centrum, Amsterdam, The Netherlands, 1980. Mathematical Centre Tracts 119.

[47] R. Schreiber. A new implementation of sparse Gaussian elimination. *ACM Trans. on Math Software*, 8:256–276, 1982.

[48] A.H. Sherman. Algorithm 533. NSPIV, a FORTRAN subroutine for sparse Gaussian elimination with partial pivoting. *ACM Trans. on Math. Software*, 4:391–398, 1978.

[49] A.H. Sherman. *On the efficient solution of sparse systems of linear and nonlinear equations.* Technical Report 46, Dept. of Computer Science, Yale University, 1975.

[50] G.W. Stewart. The economical storage of plane rotations. *Numer. Math.*, 25:137–138, 1976.

[51] G.W. Stewart. *Introduction to Matrix Computations.* Academic Press, New York, 1973.

[52] R.E. Tarjan. Efficiency of a good but not linear set union algorithm. *J. ACM*, 22:215–225, 1975.

[53] Z. Zlatev, J. Wasniewski, and K. Schaumburg. *Y12M: Solution of large and sparse systems of linear algebraic equations.* Lecture Notes in Computer Science, Vol. 121, Springer-Verlag, 1981.

LEVEL-INDEX ARITHMETIC : AN INTRODUCTORY SURVEY

C.W.Clenshaw
Mathematics Department, University of Lancaster
Lancaster, LA1 4YL, UK

F.W.J.Olver
Institute for Physical Science and Technology
University of Maryland, MD 20742, USA

P.R.Turner
Mathematics Department, University of Lancaster
Lancaster, LA1 4YL, UK
and
Mathematics Department, US Naval Academy
Annapolis, MD 21402, USA

Foreword

During the past few years, two teams have been pursuing research into the development and application of level-index arithmetic, one in Maryland, USA, and one in Lancaster, UK. The long-term members of the Maryland team are F.W.J. Olver and D.W. Lozier, their counterparts at Lancaster being C.W. Clenshaw and P.R. Turner: several others have made contributions, including S. Cramer, F. Golam-Hossen, R.E. Kaylor, I. Reid and C. Sims.

Each of the following seven lectures was prepared and presented by its named author, but it has been our intention that this set of seven should form a unified whole which will provide an introduction to the subject; we have collaborated in an endeavour to produce a compatibility of approach and to secure a uniformity of notation. The first two lectures prepare the way for their successors by demonstrating the motivation for seeking arithmetic systems other than floating point, and by describing some of the early responses to the need. The next four lectures describe the level-index system and its most important properties, show how it can be implemented and how it performs on some numerical examples. Finally, Lecture 7 points to the future by showing ways in which implementation may be made more efficient.

We should like to acknowledge the value of many helpful discussions, and particularly those with D.W.Lozier (National Bureau of Standards, Gaithersburg, Maryland), A. Feldstein (Arizona State University) and D. Parkinson (QMC, London).

We also acknowledge with gratitude the support we have received, through grants and contracts over several years, from the US Army Research Office, the National Science Foundation, the Science and Engineering Research Council and the Lancaster University Research Fund.

1: ALTERNATIVES TO FLOATING POINT - THE NEED

P.R. Turner

1.1 Introduction

In the first three lectures of this course, we shall be introducing various alternatives to the floating-point system of arithmetic and number representation within a computer. From Lecture 3 onwards the focus will be firmly on just one of these, the level-index system.

Before dealing with any particular proposed scheme, we should ask why we need to seek alternatives to the floating-point system. The primary purpose of this first lecture is to establish this need by considering the fascinating topic of the distribution of the leading significant digits of numbers and the implications of this for floating-point arithmetic.

After observing the well-used pages at the beginning of books of log. tables and then conducting an analysis of 20,229 numbers gathered from a varied host of sources, Benford [3] first suggested in 1937 that numbers are logarithmically distributed. Specifically, he conjectured that the frequency of leading significant digit (lsd), n, in radix r is given by

$$F^{(r)}(n) = \log_r (n+1) - \log_r (n).$$

This implies, for example, that 30% of decimally represented numbers have lsd 1 and that the frequency of lsd n decreases steadily to about 4.6% for lsd 9.

Since Benford's original work many authors have attempted to establish his conjecture. It is widely accepted as being fact by statisticians who allow for this distribution in the design of random number generators. Knuth [40] dedicates some nineteen pages to the phenomenon. Other justifications have been based on assumptions such as invariance to scaling or on the expected frequencies of lsd for the integers assuming some cut-off point. (See [20], [28], [53] and [54]). Hamming in [28] established that, once obtained, the logarithmic distribution is not disturbed by other arithmetic operations. He also showed that any reasonable distribution is moved closer, in a relative L_1 metric, to the logarithmic one by multiplication. The relevance of the logarithmic distribution to the study of errors in floating-point arithmetics has been studied, amongst others, by Barlow and Bareiss [2] and Goodman, Feldstein and Bustoz [22,23] and are discussed in Feldstein's lectures in this volume.

Section 1.2 will be concerned with establishing the validity of the logarithmic distribution in a way which seems more natural than the earlier approaches. Also in that section we that the logarithmic distribution implies an immediate preference as to the choice of radix for floating-point arithmetic.

In Section 1.3 we go on to consider the consequences of the logarithmic distribution for the potential frequency of overflow and underflow in floating-point arithmetic. We see that in the absence of scaling of, or structure in, the problem being solved this risk can be alarmingly high. However any scaling or structure - which is typically inherent in physical problems - only replaces the overflow/underflow difficulty with a comparably worrying frequency of catastrophic cancellation causing serious loss of precision.

In Section 1.4 some conclusions are drawn and suggestions are made as to how a computer arithmetic should be designed to overcome these, apparently dual, hazards of the floating-point system.

1.2 The logarithmic distribution

We begin this section with a justification of the logarithmic distribution based on the simple observation that the leading significant digit of a number is always potentially affected by multiplication or division whereas the lsd of a sum of two numbers is typically just that of the larger of the summands. It is therefore reasonable to consider the distribution of lsd resulting from repeated multiplication. We shall assume that the factors are taken from a uniform distribution. A full treatment of this can be found in [59].

For simplicity of description and notation we shall use the decimal base throughout and denote the frequencies of lsd n resulting from i multiplications by $F_i(n)$ for $n=1,2,...,9$ and $i=0,1,2,...$. The corresponding continuous density functions will be denoted by f_i so that

$$F_i(n) = \int_n^{n+1} f_i(t)\, dt. \tag{1.2.1}$$

It is not hard to obtain the following recurrence relation.

$$f_{i+1}(t) = \frac{1}{9}\int_1^{10} \frac{1}{x} f_i(x)dx + \int_t^{10} \frac{1}{x} f_i(x)dx$$

$$= \frac{10}{9}\int_1^{10} \frac{1}{x} f_i(x)dx - \int_1^{t} \frac{1}{x} f_i(x)\, dx \quad . \tag{1.2.2}$$

THEOREM 1.2.1

If $f_i \to f_\infty$ uniformly on [1,10] as $i\to\infty$ then $f_\infty(t) = 1/(t \ln 10)$. (1.2.3)

Proof

If (f_i) converges uniformly then the limit function satisfies the limit version of (1.2.2). It follows that f_∞ is differentiable and satisfies the separable differential equation

$$f_\infty{}'(t) = - f_\infty(t)/t. \tag{1.2.4}$$

Hence $f_\infty(t) = A/t$. Also

$$\int_1^{10} f_\infty(t)\,dt = 1$$

from which it follows that $A = 1/\ln 10$.

From (1.2.2) we may deduce, by induction and integration by parts, that f_n is a polynomial of degree n in $(-\ln t)$. We can expand this polynomial using truncated exponential series for $1/t = \exp(-\ln t)$. Thus we write

$$f_n(t) = \sum_{i=0}^{n} c_i^{(n)} E_i(-\ln t) \tag{1.2.5}$$

where

$$E_i(x) = 1 + x + \ldots + x^i/i! \tag{1.2.6}$$

Using (1.2.2) we find, for $0 \leqslant i \leqslant n$, that

$$c_{i+1}^{(n+1)} = c_i^{(n)} = c_{n-i}\text{, say} \tag{1.2.7}$$

and

$$c_{n+1} = \Sigma\, c_{n-i}\,[1 - 10E_{i+1}(-\ln t)]/9\,. \tag{1.2.8}$$

Rearranging the series for f_n in increasing powers of $(-\ln t)$ we deduce that (f_n) is uniformly convergent provided that $\Sigma\, c_i$ is bounded and this in turn follows from the boundedness of the sequence $\{f_n(1)\}$. Full details of this argument can be found in [59]. Combining this result with Theorem 1.2.1 establishes

THEOREM 1.2.2

The sequence (f_n) converges to the limit f_∞ and

$$F_\infty(n) = \log_{10}(n+1) - \log_{10}(n). \tag{1.2.9}$$

In a subsequent paper [60] it was shown that none of the other arithmetic operations disturb this distribution. Thus, for example, multiplication of all numbers by a constant leaves the logarithmic distribution unchanged. One consequence is that the logarithmic distribution implies invariance to scaling, and so the earlier work establishing the logarithmic distribution under that assumption was simply half of a circular argument. It is also established in [60] that forming powers, taking roots or reciprocals (and, therefore, division) all leave the logarithmic distribution unchanged *as do addition and subtraction.* The observation that the arguments used to establish convergence above hold for any other (continuous) initial distribution implies that any number of other arithmetic operations could be included in the sequence of multiplications without affecting that convergence.

The combination of all these remarks amounts to a very strong justification of Benford's original conjecture that numbers are indeed logarithmically distributed. The frequencies of lsd n for the decimal system are given in the following table.

Table 1.2.1 Frequencies (%) of lsd for the decimal system.

n	1	2	3	4	5	6	7	8	9
$F_\infty(n)$	30.1	17.6	12.5	9.7	7.9	6.7	5.8	5.1	4.6

Of course it follows from all the arguments presented above or elsewhere - and others! - that 100% of binary represented numbers have leading significant digit 1.

We conclude this section with a brief look at one immediate consequence of this distribution with respect to the choice of radix for floating-point representation. A more detailed description of these considerations can be found in [57]; we present here a simple comparison between the hexadecimal and binary representations.

A number X is represented in the (normalised) floating-point, base β, system by

$$X = \pm f * \beta^{\pm E} \tag{1.2.10}$$

where $\beta^{-1} \leqslant f < 1$ and E is an integer.

We shall compare the hexadecimal and binary systems using a hypothetical 32-bit word with a positive representable range of $2^{-256} \leqslant X < 2^{256}$. At this stage we are not concerned with detailed electronic implementation of this representation and so may assume that the exponent range is balanced. (See also Feldstein's lectures in this volume).

Now $2^{256} = 16^{64}$ and $64 = 2^6$ so that the hexadecimal word must allocate 6 bits to the exponent E (and one for its sign). Allowing for the sign of the number X this leaves 24 bits for the *mantissa* or *fraction* f. Similarly, since $256 = 2^8$, the binary word must have 8 bits (and a sign) allocated to the exponent. This leaves just 22 for the binary mantissa.

However since the leading significant digit of a floating-point binary number is 1 this need not be stored explicitly. Thus the binary format has an effective 23 bits for the mantissa.

Consider the representation errors associated with the storage of X in these two forms. This can be gauged by the relative mesh sizes of the two systems, that is the gap between successive representations. Suppose the hexadecimal exponent of X is n, so that

$$16^{n-1} \leqslant X < 16^{n} \text{ or } 2^{4n-4} \leqslant X < 2^{4n}. \qquad (1.2.11)$$

The difference between successive hexadecimal fractions is 16^{-6} (since the 24 mantissa bits represent 6 hexadecimal digits) and so the hexadecimal gap between X and its successor is $16^{n-6} = 2^{4n-24}$.

From (1.2.11), it follows that the binary exponent of X is 4n-3, 4n-2, 4n-1 or 4n and the difference between successive binary fractions is 2^{-23}. Now the binary exponent is 4n-3 if and only if the leading significant hexadecimal digit, $lsd_{16}(X)$, is 1 which occurs according to the logarithmic distribution with probability 1/4. Similarly the binary exponent is 4n-2 if and only if $lsd_{16}(X) = 2$ or 3 and so on for the other cases. The following table summarizes the various possibilities. We see immediately that the probability that the binary representation having greater error than the hexadecimal one is only 1/4 or, conversely the binary representation is more accurate than hexadecimal 50%, and at least as good 75%, of the time.

Table 1.2.2 Comparison of hexadecimal and binary representation errors.
n = hexademical exponent of X.

$lsd_{16}(X)$	binary exponent	binary gap	hexadecimal gap	probability
1	4n-3	2^{4n-26}	2^{4n-24}	1/4
2,3	4n-2	2^{4n-25}	2^{4n-24}	1/4
4,5,6,7	4n-1	2^{4n-24}	2^{4n-24}	1/4
8,. . . , 15	4n	2^{4n-23}	2^{4n-24}	1/4

There are several arguments based on probabilistic analyses of floating-point arithmetic which demonstrate advantages of the binary system relative to other bases. Some of these are presented at this meeting by Feldstein while others can be found in, for example, [21] and [23].

1.3 Overflow, underflow and cancellation

In this section we consider two fundamental difficulties which can arise in floating-point computation and demonstrate that both can occur with disquieting frequency. We begin by considering the probabilities of overflow or underflow resulting from the add-magnitude and subtract-magnitude operations in a floating-point arithmetic system using the representation (1.2.10), that is

$$X = \pm f * \beta^{\pm E}.$$

We denote by X^+ and X_- the greatest and least representable positive numbers in the system. In a balanced system of floating-point arithmetic, we would have

$$X^+ * X_- \doteqdot 1. \tag{1.3.1}$$

Now the assumption of the logarithmic distribution implies that the fractions of floating-point numbers are logarithmically distributed and, assuming continuity of the distribution, that their exponents are uniformly distributed. Under this assumption it can be shown that the probability of overflow from add-magnitude is given by

$$P^+ = \pi^2/6(\ln(X^+/X_-))^2 \tag{1.3.2}$$

and the probability of underflow from subtract-magnitude is

$$P_- \doteqdot P^+.$$

(Detailed derivations can be found in [19]).

Define $E^+ = \log_2(X^+)$ and $E_- = \log_2(X_-)$. Then E^+ and E_- are approximately the maximum and minimum *binary* exponents of representable numbers. For a balanced system $E_- \doteqdot -E^+$ and so it follows that

$$P^+ \doteqdot K / (E^+)^2 \tag{1.3.3}$$

where

$$K = (\pi / \ln 2)^2 / 24 \doteqdot 0.856.$$

Typically E^+ is of the order of 2^7 or 2^8. Table 1.3.1, which is a shorter version of Table 2.4 of [19], demonstrates clearly the risk of overflow under this assumption of the uniform distribution of exponents for machines with various speeds and binary exponent ranges.

In practice, it would normally be the case for scientific computation that the problem is scaled by the choice of units. The effect is that the assumption of uniform distribution of exponents may well be invalid in such situations. Even in this case however the choice of units is made so that the input and final results are reasonably scaled rather than the intermediate results which could well have exponents near the extremes of the range. Sweeney's paper [58] presented a detailed statistical survey of major scientific computing programs. The assumption in Table 1.3.1 that just 5% of run-time is spent on add-magnitude operations is based on Sweeney's data.

Table 1.3.1. Probability of overflow resulting from add-magnitude operations.

(Addition time = T; 1μs = 10^{-6} and 1 ns = 10^{-9} seconds; assume 5% of run-time is used on add-magnitude).

One - minute run

T \ E^+	2^8	2^{12}	2^{16}	2^{20}	2^{26}
10 μs	0.98	0.02	0.00	0.00	0.00
100 ns	1.00	0.78	0.01	0.00	0.00
1 ns	1.00	1.00	0.45	0.00	0.00

One - hour run

T \ E^+	2^8	2^{12}	2^{16}	2^{20}	2^{26}
10 μs	1.00	0.60	0.00	0.00	0.00
100 ns	1.00	1.00	0.30	0.00	0.00
1 ns	1.00	1.00	1.00	0.13	0.00

One of the subjects Sweeney addressed was the distribution of exponent differences in add- and subtract-magnitude operations. This topic was considered further by Feldstein and Goodman [17] who again concluded that the (hidden bit) binary system was the optimal floating-point arithmetic from the point of view of cancellation. We now address the problem of the frequency of *catastrophic cancellation*, that is where at least half of the significant digits are lost through cancellation in subtract-magnitude operations. The conclusions of [19] are summarised in Table 1.3.2 in which we consider the likelihood of losing 13 bits from a 27-bit mantissa. (This is half the precision since the 27-bit mantissa is equivalent to 26 bits together with the implicit or hidden leading bit).

Table 1.3.2 Probability P of catastrophic cancellation occuring at least once in N subtract-magnitude operations.

N	5	15	50	100
P	0.21	0.51	0.91	0.99

We see that the frequency of such cancellation happening with fractions of length only 27 bits is very high. However the frequency of the loss of 13 bits would be unchanged for a much longer mantissa of say, 100 bits.

1.4 Conclusions and suggestions

In the last section we saw that, under the assumption of the logarithmic distribution of lsd, the risk of overflow can be alarmingly high - especially for the unwary programmer who performs no precautionary prescaling of the problem. This danger can be alleviated greatly by the use of a much greater exponent range using 26 bits - or 27 to account for the exponent sign or bias.

In the common situation where the problem is scaled either by the programmer or by the choice of units, the danger of catastrophic cancellation and a resulting loss of precision in the final output becomes a real worry. This can be countered by the use of a much greater mantissa length of perhaps 100 bits.

The combined effect of these possible safeguards would be a proposal for the use of a standard floating-point word for scientific computing of 128 bits consisting of the sign, the 100-bit fraction and the biassed 27-bit exponent. This proposal has substantial associated costs - both computational and financial. The storage requirement would be double to quadruple those for most current scientific computation. Arithmetic operations would inevitably be significantly slower than for conventional single or double length floating-point arithmetic and yet the benefits would probably be relevant only to large scale number-crunching.

The essential message is that in the floating-point environment improved machine speeds should be accompanied by the simultaneous increase of the exponent range and of the available relative precision. The alternative is to use some other arithmetic system in which the representable range is increased so as to reduce or eliminate the risk of overflow and underflow and in which the precision used is appropriate to the magnitude of the number.

The next lecture deals with several of the arithmetic systems which have been proposed with these ends in mind. One possibility is to retain the standard 32 or 64 bit word but with overflow/underflow from a reduced floating-point range into the level-index arithmetic first proposed by Clenshaw and Olver [8]. This possibility is discussed in [62]. The level-index system is the subject developed in the later lectures of this course.

It is worth commenting here that Benford in his original paper on the logarithmic distibution [3] suggested that nature does not count 0,1,2,3, . . . but instead uses $0, 1, e, e^e, \ldots$. These numbers, which we could term "Benford integers", are just the break points between different levels of the level-index system. In his 1985 address to the British Mathematical Colloquium [27], Hammersley argued that the Arabic number system itself was perhaps reaching its limits of usefulness in areas of physical and computation science and that some more compact notation was necessary for the representation of very large numbers. His most promising suggestion was for something similar to a base 2 level-index system.

From all the foregoing it is plain that alternatives to floating-point arithmetic are needed at least within special purpose and experimental scientific computing environments. The rest of the course is devoted to considering some of the possibilities, and level-index arithmetic in particular.

2: ALTERNATIVES TO FLOATING POINT - SOME CANDIDATES

P.R. Turner

2.1 Introduction

In the previous lecture we discussed some of the reasons for seeking alternatives to floating-point arithmetic for scientific computing. In particular, we highlighted the problems of overflow, underflow and cancellation. These considerations suggest the desirability of systems with a greater range of representable numbers and an appropriate precision for numbers of all magnitudes.

One major benefit of any new arithmetic which meets these requirements will be the saving in program development time and even perhaps run-time. These savings result from the removal of the need for programmers to scale problems so as to avoid difficulties such as overflow. Similarly, scientific software designers would no longer need to incorporate special code to handle exceptional situations which might result in overflow at an intermediate stage. Even if the arithmetic itself were to run more slowly, it may well be the case that these simpler programs would accomplish their tasks more quickly.

There are of course other reasons why alternative arithmetic systems may be wanted; these include special precision requirements in some particular range or some special purpose arithmetic for handling special data types. In this very brief survey of some of the recently suggested candidates, we shall concentrate largely on the number representations (and their ranges and precisions) of systems which are designed to overcome the problems of overflow and underflow. Time and space permit only an outline description of these systems and there are notable omissions. For instance there is no section on interval arithmetic; the review article [43] provides a full description and bibliography of this system. A good starting point for the interested reader on many of the systems would be the proceedings of the various IEEE Symposia on Computer Arithmetic, for example [38].

The next sections of this lecture contain brief descriptions of a number of systems which are typical of the differing approaches to computer arithmetic. In section 2.2, we consider the CADAC system due to Hull and his coworkers ([32], [33] and [34]) which uses a decimal floating-point representation with dynamically variable range and precision and built-in exception-handling.

Section 2.3 is devoted to modified binary floating-point representations due to Matsui and Iri [45] and Hamada [25] both of which are designed to reduce the risks of overflow and underflow while retaining some of the familiarity of floating-point binary arithmetic. We also consider a proposal of Kahan to utilise wrap-around characteristic with a counter to extend the range of conventional floating-point arithmetic.

In Section 2.4, we consider one of the rational arithmetics, namely the lexicographic continued fraction system of Kornerup and Matula [41] and [42]. One of the attractions of this system is the possibility of very rapid evaluation of complicated rational expressions in two variables.

Section 2.5 concludes this lecture with a brief comparison of the ranges and local precisions of some of these systems and the level-index and symmetric level-index systems [8], [9] and [10].

2.2 The CADAC system

The CADAC (or Clean Arithmetic with Decimal base And Controlled precision) system has been developed over the past several years by Hull and his coworkers. It is much more than just an alternative computer arithmetic system since it has its own programming language which includes special facilities for exception handling and even user-defined exceptions for particular special situations. This last feature allows the possibility of automatic switching among different algorithms for the solution of a problem. For full details of the system see [32], [33] and [34].

CADAC uses decimal floating-point arithmetic with dynamically variable precision and range. The precision and range are related so that as the range is increased so is the number of significant digits - and vice versa. Specifically, a number is represented at precision p as a normalised decimal floating-point number with p digits in the mantissa, and exponent in the range [-10p, 10p]. The value of p can be changed at any point in the program.

The programming language, *Numerical Turing,* is PASCAL-like and the precision of a variable is determined by the variable declaration. For example, the statement

```
var a,b : real (10)
```

declares the variables a and b to be 10-digit normalised decimals with exponents in the interval [−100, 100] while

```
var a,b : real
```

sets the precision of a and b to the current working precision; this is stored as the variable "currentprecision" which may be accessed and used within subsequent precision declarations.

The variability of the precision allows different parts of a routine to use different precisions so as to control the precision of the final result.

For example, the following routine has the effect of producing the inner product of two n-dimensional vectors with only one rounding error. This is effected by performing all the multiplications and accumulating the partial sums at the appropriate higher precision as explained below. This example is taken from [33].

Example 2.2.1 Computation of dot product in CADAC.

```
var dotproduct : real
begin
     precision:=  currentprecision + getexp(n) +1
     var temp : real := 0.0
     for i := 1...n
               temp := temp + a(i)*b(i)
     endfor
     dotproduct := temp
end
```

Note that the variable dotproduct is given the current working precision (of the calling program) on entry to the routine and that this precision is immediately increased by getexp(n)+1 for the internal working of the procedure. The function getexp obtains the (normalised decimal) floating-point exponent of its argument - which in this case is first converted to type real. The increased precision is exactly what is required to allow for all possible overflows, underflows or shifts within the accumulation of n products of real numbers. The single rounding, at the precision level of a,b and dotproduct, occurs in the abbreviation statement

```
dotproduct := temp.
```

We see that in some senses CADAC is not so much a new system of computer arithmetic as a familiar one being used together with dynamically controlled precision so as to reduce the risks of overflow, underflow and cancellation.

The other main feature of CADAC is its exception-handling. There are just three built-in exceptions - *overflow, underflow* and *domainerror*. The last of these is used for situations such as the square-root of a negative argument. However, the user has the freedom to define other exceptions and handlers to create, for example, automatic switches between different algorithms. Similarly, handlers can be attached to any routine so that appropriate action can be taken in the event of any of the built-in exceptions arising.

For example, it may be desirable in some applications simply to increase the precision - and therefore the range - in the event of overflow or underflow. There is not space to include details of these aspects here and the interested reader is referred to [32].

2.3 Overflow, underflow-free binary floating-point systems

In this section we consider some of the suggestions that have been made for eliminating, or at least reducing, the incidence of overflow and underflow from binary floating-point arithmetic. Specifically we shall consider the representations of Matsui and Iri [45] and Hamada [25]. Before describing these in detail, however, we give brief consideration to a suggestion of Kahan made in a panel discussion at the IEEE Symposium on Computer Arithmetic in Como, May 1987.

The idea is simply to have a one-byte (8 bit) counter which represents as a binary integer the number of times a quantity has "wrapped around". When a result exceeds the floating-point range this counter is activated and, essentially, just registers the carries from the most significant end of the floating-point representation which itself wraps round so that the next representable number beyond X^+ would be represented by 00000001 in the counter and a string consisting entirely of zeros in the original part of the word (allowing for the hidden bit). There seem to be several drawbacks to this. In terms of the storage requirement, it is necessary to know *in advance* which quantities are likely to overflow the standard range or to allocate this extra byte to every real variable. The latter course is equivalent to increasing the wordlength of the floating-point representation by eight bits all of which are allocated to the exponent. Such an increase in wordlength can, of course, be incorporated into any computer arithmetic system *but it does not address the underlying problem.*

If an extra eight bits were to be incorporated into, say, IEEE standard single-length floating-point format, it is also questionable as to whether they should all be allocated to increasing the exponent range. The use of 16 out of 40 bits for the sign and the biassed exponent seems excessive - especially in the light of the cancellation problem discussed in the previous lecture.

Even if it is known in advance which variables are likely to cause overflow or underflow, there are remaining difficulties: all internal routines must be able to support both data types, and the accumulators and temporary storage must all be designed for the 40-bit word. For example, any one of the products or partial sums computed in the evaluation of a scalar product may overflow or underflow even though the final result is moderate in size. Effectively therefore all internal calculation would be in terms of the extended format as would all library function routines.

A suggestion that extra bits be incorporated into the representation of real numbers to increase range, precision or both can equally well be made for any system; it is not particularly relevant to the present discussion.

Matsui and Iri [45] advanced an extended floating-point format in which a variable portion of the computer word is allocated to storage of the exponent. Figure 2.3.1 shows the structure of the Matsui and Iri, "MI", 64-bit word.

Figure 2.3.1 Structure of a 64-bit MI word.

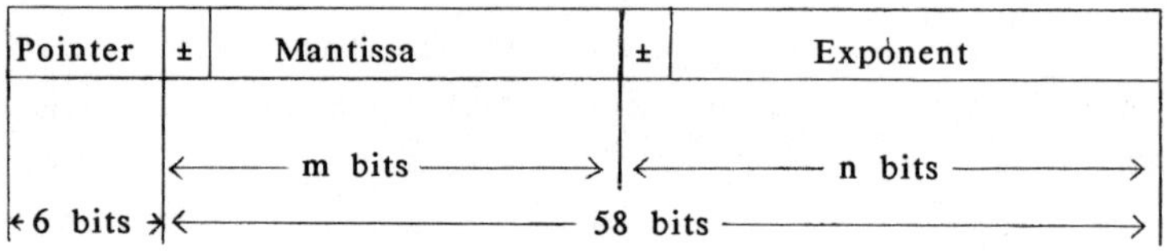

The pointer registers the number of bits used for the signed exponent of a normalised binary floating-point representation.

For |X| close to unity, the exponent is small and is represented exactly with a small value of n. Consequently most of the 58 bits are available for the mantissa. As |X| increases more bits are required for the (still exact) exponent and fewer are available for the mantissa. Thus the relative precision decreases as the magnitude of the number increases. Eventually the exponent occupies the full 58 bits and all relative precision is lost. At this point, Matsui and Iri's "level 1" is used and the binary exponent of X is represented as a normalised floating-point number. That is, X is represented as $\pm 2^{f * 2^E}$. Unless extra bits are allocated to the pointer to incorporate a "level indicator" very little of level 1 is available in the word described in Figure 2.3.1 which would only enable the pointer values 59, 60, 61, 62 and 63 to be used to determine how many bits are allocated to this exponent E and the sign of f.

One question which arises immediately out of Figure 2.3.1 concerns the structure of the corresponding single-length MI format. Because the pointer would still require 5 bits, the single-length version would have only 27 for the two signs, the mantissa and the exponent. This renders the single-length format of MI insufficiently accurate for many purposes.

Although Matsui and Iri described MI arithmetic as overflow/underflow-free, this is not strictly true since the set of representable numbers is not closed under the basic arithmetic operations. For example, the product of two level-0 numbers could easily exceed the available range of the representation; however, with even a one-bit level indicator the range is, of course, increased greatly. Continuing this process leads effectively to a version of base 2 level-index arithmetic in which the index is represented in floating-point form.

The computational experience reported in [45] does not include any results which use representations beyond level 0, and the arithmetic processes for handling such numbers are not described.

The final extended floating-point format to be considered here is Hamada's "Universal Representation of Real Numbers", or URR, [25,26]. This representation uses the same principle as MI arithmetic in that the mantissa and exponent are of variable length; but in URR the pointer is used as part of the exponent information. Specifically, we express a representable number X as

$$X = \pm f * 2^{\pm(2^m + n)} \tag{2.3.1}$$

where, after an initial sort,

m is stored as a *unary* integer,

n is a *binary* integer, m bits long and

f $\in$ [1,2) is the (hidden bit) binary mantissa.

The following example of the URR representation should help to clarify this.

Example 2.3.1 The URR representation of 57 is obtained as follows.

Step 1

$X > 0$ => leading bit **0**

$|X| \geqslant 1$ => second bit opposite to first => **01**

$X \geqslant 2$ => third bit 1 => **011**

Step 2

For each i such that $X \geqslant 2^{2^i}$ (i=1,2, . . .) add another 1 and finally a 0 to mark the termination of this step. Denote the number of 1's by m. In this case

$$2^{2^2} = 16 \leqslant X < 256 = 2^{2^3}$$

so m = 2 and representation is now **011 110**.

Step 3

Find the m-bit integer n such that

$$2^{2^m+n} \leqslant X < 2^{2^m+n+1}$$

(Note that $n < 2^m$.)

For our example,

$$2^{2^2+1} = 32 \leqslant X < 64 = 2^{2^2+2}$$

so n = 1 and so we now have **011 110 01**.

Step 4 Find the mantissa in [1,2) which is then represented using hidden bit binary.
For X= 57, we get the mantissa 1.1100100...0 and so the final representation is given by
011 110 01 1100100. . . 0
where we see that the 1 before the binary point of the mantissa is not stored explicitly.

The first three bits are the result of the initial sort, the next group (of three in this case) are the unary representation of m (= 2) followed by the 0 terminator. The third group is the binary integer representation of n (this is m bits long). The final group represents the mantissa in hidden bit binary form. (The spaces in the representation are here included only for clarity of demonstration).

Step 1 is an initial sort in which the sign and exponent sign of the representation are obtained. The number of consecutive 1's after the leading three bits indicates the length of the exponent as well as forming part of it. For X ϵ (0,1), m is represented as a string of 0's followed by a 1 to mark its termination, while negative numbers are stored as the two's complement of their magnitude.

The above example and remarks also show how to interpret URR data which can then be processed by extensions of standard floating-point arithmetic - but with added complications, as follows.

Firstly, the exponent is not in simple binary form : this complicates the computation of shifts. Also, since mantissas are not necessarily of the same length their alignment and manipulation is even less straightforward. There is, as yet, no direct algorithm for URR arithmetic without separating data into two full words to represent the exponent and mantissa of the standard floating-point form.

Perhaps most importantly, URR does not eliminate overflow and underflow. However, the overflow level in Hamada's original description is precisely at the point at which all relative precision in the representation is lost. This is seen by some advocates of floating-point arithmetic as being a significant advantage. This "advantage" only pertains because Hamada's choice was to cut the representation off at the point where the exact exponent could no longer be represented in the computer word. Thus if 2n+3 is greater than the number of bits available, then there is insufficient wordlength to represent n exactly and the quantity overflows the URR system. Because of this, URR has many redundant representations which Hamada suggests should be used for NAN's (Not-A-Numbers) which, as with IEEE standard floating-point [37], can behave like infinity symbols or denote indeterminates. On the principle that some information is better than none, it seems to be to be more desirable to continue the URR representation by simply carrying as much of the exponent as can be accommodated. The fact that such quantities would have no relative precision does not in itself render them worthless. (See Lectures 3 and 7, for example.) If necessary, the final few possible representations could still be reserved for the small number of NAN's needed by any system.

2.4 LCF arithmetic

The LCF or lexicographic continued fraction system of computer arithmetic was developed by Kornerup and Matula [41]. This rational arithmetic grew out of their earlier work on fixed- and floating- slash arithmetic. Of course the fact that LCF is a rational arithmetic does not set it apart from the floating-point forms we have discussed hitherto; what is different is that it is overtly rational.

The representation is based on the fact that a nonnegative rational number p/q can be uniquely expressed in the form

$$p/q = a_0 + \cfrac{1}{a_1 + \cfrac{1}{a_2 + \cfrac{1}{\ddots \; a_{m-1} + \cfrac{1}{a_m}}}} \tag{2.4.1}$$

where the integers $\{a_i\}$ satisfy $a_0 \geqslant a_i \geqslant 1$ for $i \geqslant 1$ and $a_m \geqslant 2$ if $m \geqslant 1$. Denote this expression by $[a_0,a_1,a_2,...,a_m]$.

To enable compact representation of such a rational expression the *lexicographic* encoding of the integers is used. That is, the leading bit of the binary representation is replaced by the appropriate number of 1's followed by a 0; so that if the leading bit represents 2^3, say, it is replaced by 1110. The terminal 0 here acts as a separation between this leading "bit" and the rest of the binary representation whose length is determined by the number of 1's. For example, 27 = 16 + 8 + 2 + 1 and so has the binary representation 11011. The lexicographic code replaces the leading 1 (which represents 2^4) by 11110 and so 27 is stored as 111101011. The four leading 1's indicate that the remainder of the number occupies four more bits after the first 0 and so no further separator or indicator is needed to show that a_i terminates here and that a_{i+1} starts with the next bit.

Thus the representation (2.4.1) can be achieved by simply concatenating the lexicographic encoding of the integers a_0, a_1, a_2, . . . , a_m. It is also the case that the natural ordering of the integers is preserved by this encoding and so discriminations between LCF numbers are straightforward. The representation is readily extended to "signed" LCF to incorporate negative rationals, too. The LCF representation has no redundancy since there is a one-one correspondence between the irreducible fractions and the (theoretically infinite) set of LCF representations.

The arithmetic operations of the LCF system have been thoroughly researched by Kornerup and Matula who now have detailed hardware designs for LCF arithmetic chips. The operation of reciprocation is particularly simple since it only entails forming the 2's complement of the original quantity. The other arithmetic operations can be performed in a bit-serial manner so that, after a small start-up time, the result is produced bit-by-bit as the data is read bit-by-bit. Thus arithmetic can be performed in a time which is a linear function of the string length.

Recently in [42], Kornerup and Matula have extended this bit-serial algorithm so that rational expresssions of the form

$$f(x,y) = \frac{axy + bx + cy + d}{exy + fx + gy + h}$$

where x and y are in LCF form, can be evaluated in the same sort of way - again in linear time relative to the string lengths.

The LCF system can be used to provide a variable-precision, variable-range "real" computer arithmetic by allowing the string lengths to grow and using the ideas of "best rational approximation" to real numbers with the sequence of *convergents* $[a_0]$, $[a_0,a_1]$, $[a_0,a_1,a_2]$, ... Such a sequence has quadratic oscillating convergence.

There is insufficient space to give more than the briefest of summaries of some of the important facts and results that Kornerup and Matula have obtained for this system - see [41], [42] for further details.

2.5 Summary and comparisons

In the previous sections we have introduced various approaches to some of the problems of computer arithmetic which have included complete computing environments such as CADAC, modifications of the familiar floating-point systems and a completely different approach in the rational LCF arithmetic of Kornerup and Matula. In the first lecture we also mentioned the level-index system. The principal reason advanced for our study is the desire to overcome overflow and underflow while preserving a satisfactory measure of precision. In order to understand how well the various systems achieve these goals, we present in Table 2.5.1, a comparison of their overflow and underflow levels together with their mesh sizes at

1 and (approximately) 2^{255}. These mesh sizes μ and M are defined as the gaps between 1 and the next representable number, and between 2^{255} and its successor. LCF arithmetic is omitted from this table since its primary motivation is different from those of the other systems.

Table 2.5.1

	32-bit word				64-bit word			
	X_-	X^+	μ	M	X_-	X^+	μ	M
CADAC	2^{-200}	2^{203}	2^{-20}	over-flow	2^{-431}	2^{434}	2^{-43}	2^{207}
MI	$2^{-2^{25}-1}$	$2^{2^{25}-1}$	2^{-25}	2^{238}	$2^{-2^{56}-1}$	$2^{2^{56}-1}$	2^{-56}	2^{207}
URR	$2^{-2^{15}}$	$2^{2^{15}+1}$	2^{-29}	2^{240}	$2^{-2^{31}}$	$2^{2^{31}+1}$	2^{-61}	2^{208}
ℓi	2^{-28}	$>2^{2^{65536}}$	2^{-28}	2^{237}	2^{-60}	$>2^{2^{65536}}$	2^{-60}	2^{206}
sℓi	$<2^{-2^{2^{65536}}}$	$>2^{2^{2^{65536}}}$	2^{-27}	2^{238}	$<2^{-2^{2^{65536}}}$	$>2^{2^{2^{65536}}}$	2^{-59}	2^{207}
IEEE	2^{-129}	2^{127}	2^{-23}	over-flow	2^{-1025}	2^{1023}	2^{-52}	2^{207}

As in Lecture 1 the symbols X_- and X^+ are used to represent the smallest and largest representable positive numbers. The values given here for X^+ for the IEEE floating-point system should actually all be multiplied by $(1-2^{-k})$ where 2^{-k} is the precision of the mantissa. The values for the CADAC system are all approximate binary equivalents of the correct decimal quantities; the approximation is made for ease of comparison. (It is also the case that the word lengths chosen are not necessarily the most suitable for the decimal-based CADAC system).

Again, the ℓi and sℓi figures are approximate values, but they demonstrate that although their mesh sizes do grow rapidly the relative precision at 2^{255} remains comparable with the systems of floating-point type. The Hamada (URR) and Matsui and Iri (MI) schemes also lose relative precision steadily as the magnitude increases but they, too, are comparable at this level. For the MI scheme the figures quoted for X_- and X^+ are for level 0 of their system for the reasons given in Section 2.3. For the URR system, likewise, the limits given are based on Hamada's suggestion that the exponent must be represented exactly.

3: LEVEL-INDEX ARITHMETIC

C.W. Clenshaw

3.1 Large numbers

The aim of this lecture is to give a general and elementary introduction to level-index arithmetic. It is likely that questions will be raised more often than answered, but many of the answers should be forthcoming as the succeeding lectures unfold.

At the heart of this work, just as in some of the other work on new arithmetics which Peter Turner has outlined, is a concern with the two basic properties of the real numbers which we seek to represent and manipulate. These properties are magnitude and precision which, for computer arithmetic, have a crucial interdependence which has often been ignored in the past. In particular, floating-point arithmetic (flp), for all its attractive simplicity, has serious deficiencies in extreme (though not necessarily rare) situations, as has been shown here by both Alan Feldstein and Peter Turner.

On the one hand, any practical implementation of flp has a cut-off point. There is a largest representable number X^+ which can be easily exceeded by carrying out ordinary, simple arithmetic operations on numbers less than X^+. On the other hand, the precision of flp numbers is always associated with relative error, even when this measure is utterly inappropriate. So ingrained has this measure become, that even experienced numerical analysts are inclined to assert that if a number has no correct significant figures then it is meaningless. It seems unfortunate that the expression "most significant figures" has been appropriated to describe the leading figures in the mantissa of a flp number, when it is clear that the information carried in the exponent is more important.

We shall return to this topic shortly; for the present we note the example given in [8]. Dirac estimated the ratio M, say, of the mass of the universe to the mass of a single proton to be about 10^{78}, though the last digit of the decimal exponent is very uncertain. Clearly there are no correct significant figures in this estimate of M, and it would be meaningless to speak of its relative error. Nevertheless there is, equally clearly, important information in the estimate; we simply need a new error measure which will deal with very large numbers as readily as with those near unity. (We might indeed by tempted to refer to the relative error of the exponent in this example. In doing so, we should be setting off on the road to level-index arithmetic).

These, then, are the problems we address - a satisfactory treatment of magnitude and precision. It is our contention that level-index arithmetic (ℓi) provides just the framework we seek. First we define the ℓi system.

Given any positive number X, we may represent it within the computer by its ℓi-image x, which is obtained from X by taking natural logarithms as many times as necessary (ℓ times, say) to bring the result (f, say) into the interval [0,1). For example, if X=1234567 then lnX=14.02623, ln(lnX)=2.640929 and ln(ln(lnX))=0.9711308. The image x is then given by ℓ+f; in this example it is 3.9711308. (Here $dX/dx=4.6\times10^7$, so if X is correct to the nearest integer then it is appropriate to record x to seven decimal places). The numbers ℓ and f are the *level* and the *index* respectively, of the number X.

More concisely, we define the mapping function to be the *generalized logarithm* (gl) ψ, given by

$$x = \psi(X) = 1+\psi(\ln X),$$

with x=X for $0 \leqslant x < 1$.

The inverse function (whose existence is obvious) is the *generalized exponential* function (ge) ϕ, given by

$$X = \phi(x) = e^{\phi(x-1)}$$

with X=x for $0 \leqslant X < 1$.

It may be noted that we could define other ge and gl functions (which would be equally suitable for ℓi purposes) by changing the nature of ψ and ϕ within [0,1]. In certain contexts this may be advantageous, as Lecture 7 will indicate; for the present, however, we use the functions defined above, by virtue of their formal simplicity.

It is easy to see that ψ and ϕ are continuous, as are their first derivatives, but that there are discontinuities in their second derivatives at integer values of x.

The ability of ℓi arithmetic to deal with very large numbers is apparent from the above example; we may also note that the constant of Dirac to which we referred might be written as M $\doteqdot$ $\phi(4.50)$, and we shall return to this example shortly when considering precision.

This ability cannot be achieved without paying some price, however. After all, in a fixed word-length of, say 32 bits, there are just 2^{32} different configurations of 0's and 1's. If some are to be used for representing very large numbers, then there will be fewer available for more moderate numbers. However, we argue, and we shall bring evidence to support our argument, that the redistribution of magnitude and precision brought about by ℓi (as compared with flp) is quite natural and wholly desirable.

3.2 Precision

We now turn to the question of precision, and in this discussion we shall use the definitions given by Olver [48]. In his notation we have

$$\bar{x} \simeq x \ ; \ ap(\alpha) \quad \Leftrightarrow \quad \bar{x} = x+u \ ; \ |u| \leqslant \alpha \ ,$$

and

$$\bar{x} \simeq x \ ; \ rp(\alpha) \quad \Leftrightarrow \quad \bar{x} = xe^{v} \ ; \ |v| \leqslant \alpha \ .$$

We say that $\bar{x}$ approximates x with a precision (absolute in the first case, relative in the second) of α. The former is conventional, while the latter differs from the conventional $\bar{x} = x(1 + v)$ in second-order terms only. Olver's definition has the great advantage of symmetry ($\bar{x} \simeq x \iff x \simeq \bar{x}$) and, because it is a metric, it is often more convenient in analysis.

When we are dealing with numbers of moderate magnitude (say unity or less) then the recording of numbers to fixed absolute precision is appropriate and convenient. (Consider, for example, familiar tables of elementary functions likes sines and cosines). With somewhat larger numbers we tend to use flp; that is to say we record to fixed relative precision. But when our numbers become very large then, as we have seen, this measure is inappropriate. However, with ℓi arithmetic we have a ready-made alternative.

Let $X = \phi(x)$, where, for all nonnegative X, x is recorded to a fixed absolute precision α, say. Then, if $x < 1$ (that is, X is a level-0 number), X=x and is therefore itself given to absolute precision α . If $1 \leqslant x < e$ then X is a level-1 number (in fact, $X = e^{x-1}$) and is given to relative precision α . The generalisation beyond x=e is obvious. We say that X is given to *generalized precision* α when x is given to absolute precision α. This uniform measure embraces the familiar measures for small and moderate numbers, and provides a logical extension for large numbers.

Returning briefly to Dirac's constant, we might say that $M = \phi(4.50)$ to that number of figures. Conventionally, this implies that M is bounded below by $\phi(4.495) \doteqdot 2.93 \times 10^{75}$, and above by $\phi(4.505) \doteqdot 1.68 \times 10^{82}$. We do not know whether these bounds are in fact valid for this example: the point is that we are able to express any number, however large, in a concise and standard form that carries an indication of precision as well as of magnitude. Generally we shall write

$$X \simeq \phi(x) \ ; \ gp(\alpha) \quad \iff \quad X = \phi(\bar{x}) \ ; \ |\bar{x}-x| \leqslant \alpha.$$

3.3 Arithmetic operations

The substantial theoretical advantages offered by ℓi arithmetic lead us to consider the possibility of its implementation: that is, we examine methods for performing arithmetic operations on numbers represented by their ℓi images. To perform addition, for example, we shall need to be able to compute z, given x and y, where

$$\phi(z) = \phi(x) + \phi(y).$$

This is a nontrivial task which we shall explore, but first we note that if we can devise acceptable methods for both addition and subtraction, then our task is complete. Other basic operations are essentially equivalent. For example, if

$$\phi(z) = (\phi(x))^{\phi(y)}$$

then, taking logarithms, we have

$$\phi(z-1) = \phi(y)\ \phi(x-1)$$

and again

$$\phi(z-2) = \phi(y-1) + \phi(x-2).$$

Thus powering, multiplication and addition are equivalent operations at different levels.

It is also relevant to note that the set of numbers for which we must construct addition and subtraction algorithms is limited by the fact that a very large ℓi number is unchanged by the addition to it (or subtraction from it) of a smaller number, to any finite working precision. For example, to the implied precision,

$$\phi(4.45678) + \phi(3.87654) = \phi(4.45678).$$

(Indeed, this result still holds with a gp of 10^{-9}). We say that in this case, addition is trivial. This phenomenon is of fundamental importance, in theory and in practice, and it is treated in depth in Lecture 4. For now we quote just two pertinent results:

$$\phi(5) + \phi(x) \approx \phi(5)\ ;\ gp(2^{-27}) \text{ when } x \leqslant 5$$
$$\phi(6) + \phi(x) \approx \phi(6)\ ;\ gp(2^{-5,500,000}) \text{ when } x \leqslant 6.$$

We may readily check that even to the moderate precision of 28 bp in the ℓi image, some addition is nontrivial at level 5, whereas even to 5½ million bp, all addition is trivial at level 6.

It is easy to conclude that, in implementing ℓi arithmetic, it is both necessary and sufficient to allocate just 3 bits to the level.

3.4 Comparison of ℓi with flp

An early point to be made for the flp system is that it represents small numbers as readily, and as accurately, as large numbers. We see that ℓi may be given a similar property by allocating a single special bit to denote reciprocation; we express any number less than unity in the form $X=(\phi(x))^{-1}$. This bit plays a role similar to that of the sign of the exponent in flp. Where it is desirable to distinguish this modified ℓi system from the original scheme, we shall call it the *symmetric level-index* (sℓi) system.

An alternative notation for sℓi introduces a new mapping function Ψ defined by

$$\Psi(X) = \begin{cases} \psi(\ln X) & \text{for } X \geqslant 1 \\ -\psi(-\ln X) & \text{for } X < 1 \; . \end{cases}$$

The corresponding inverse function is Φ, where

$$\Phi(x) = \begin{cases} e^{\phi(x)} & \text{for } x \geqslant 0 \\ e^{-\phi(-x)} & \text{for } x < 0 \; . \end{cases}$$

To make a specific comparison with flp, we now suppose that we have a computer word of 32 bits, and we weigh orthodox flp against sℓi. The flp system will need two of its bits for signs, and will typically allocate 7 to the exponent, leaving 23 for the mantissa - which effectively becomes 24 when we take the "hidden" bit into account. The largest representable number is $2^{127}(1-2^{-24})$.

The sℓi system will also need the two sign bits (one of these being the reciprocation bit mentioned above) and, as we have seen, 3 are needed for the level. Thus 27 are available for the index. The largest representable number is $\phi(8-2^{-27})$, which is much too large to be written in flp form. (If high precision is not needed in our very small numbers, then we can use ℓi, as opposed to sℓi, and 28 bits are available for the index).

It is not difficult to show that the precision offered by the two systems is similar for numbers around 1000; sℓi is rather better for smaller numbers, flp for larger numbers until, of course, they become as large as 2^{127}. At that point flp fails abruptly, while sℓi continues to operate, and to provide the uniform precision $gp(2^{-27})$.

An argument might be raised to the effect that the abolition of overflow is not to be desired: certainly some writers of scientific programs tend to regard the appearance of overflow in flp computations as a useful indicator of error, either in the coding or in the underlying numerical analysis. Such diagnostic considerations have an undeniable validity, but it should be noted that ℓi arithmetic permits their more logical resolution. Whenever any computed quantity exceeds some bound B, say, then this fact may be brought to the user's attention. This number B could reasonably be given the default value $\phi(6)$ perhaps, and of course the user would have the option of changing it. He would be able to abort the calculation when B was exceeded if he so wished; otherwise the program would continue. The user could also, in any case, be given an indication of the largest number (in modulus) generated by his program. This situation seems preferable to the familiar one in which a bound is set arbitrarily by the system for all programs, most of which will fail completely when it is exceeded. (Also see [16]).

A deeper comparison will be described in Lecture 4 and further details may be found in [44].

3.5 Addition and subtraction

Later lectures will treat details of implementation, and possibilities for the future. We now deal with the fundamental algorithms for addition and subtraction in broad outline, first in ordinary ℓi (not $s\ell i$) arithmetic.
We take the basic problem to be the computation of z from x and y, where

$$\phi(z) = \phi(x) \pm \phi(y) \ ; \ x \geqslant y > 0. \tag{3.5.1}$$

We shall use ℓ, m, and n to denote the levels of $X = \phi(x)$, $Y = \phi(y)$, and $Z = \phi(z)$ respectively, while f, g, and h denote the corresponding indices. Then we define sequences

$$a_j = 1/\phi(x-j), \qquad j = \ell-1, \ell-2, \ldots, 0, \tag{3.5.2}$$

$$b_j = \phi(y-j) \ / \ \phi(x-j), \qquad j = m-1, m-2, \ldots, 0, \tag{3.5.3}$$

$$c_j = \phi(z-j) \ / \ \phi(x-j), \qquad j = 0, 1, \ldots . \tag{3.5.4}$$

The sequences $\{a_j\}$ and $\{b_j\}$ are computed to a fixed absolute precision. Efficient routines are needed for this purpose; they may be built upon the relations

$$a_{j-1} = e^{-1/a_j} \quad , \qquad a_{\ell-1} = e^{-f} \quad , \tag{3.5.5}$$

$$b_{j-1} = e^{-(1-b_j)/a_j} \quad , \qquad b_{m-1} = a_{m-1}\, e^{g} \quad , \tag{3.5.6}$$

Then we may compute the value of c_o from the equation

$$c_o = 1 \pm b_o \tag{3.5.7}$$

which is an immediate consequence of (3.5.1) on division by $\phi(x)$. The remainder of the sequence $\{c_j\}$ may then be calculated, from

$$c_j = 1 + a_j \ln c_{j-1}. \tag{3.5.8}$$

(We note that the "c-relation" (3.5.8) is the same as the "b-relation" (3.5.6) except that it proceeds in the opposite direction). The c-relation is continued until there is a value such that $c_j < a_j$. Then

$$z = n + h, \qquad \text{where } n=j,\ h=c_j/a_j.$$

Several special cases need to be treated, and several important points need further investigation; not least the question of the precision to be carried in the various quantities. Full details can, however, be found in [9]. This reference includes a detailed error analysis of the algorithm, and shows that by carrying a specific number of extra bits in the sequences we can ensure that the accumulation of round-off error is small compared with the unavoidable errors caused by roundings in the x and y. (Indeed, we can make the error arbitrarily small.)

The implementation of sℓi requires a few simple variations; we find that in detail some of these are simplifications rather than complications.
We must consider operations like

$$\phi(z) = \phi(x) + \phi(y)$$
$$\phi(z) = \phi(x) + 1/\phi(y)$$
$$1/\phi(z) = 1/\phi(x) + 1/\phi(y)$$

In each case we suppose that the first term on the right is the larger. For simplicity, we now refer to these three cases as "large", "mixed" and "small" arithmetic respectively. The approach is as before, but the sequences generated vary. The large case is the one already treated by ℓi, of course, and we need not refer to it again.

In the mixed case we use the same $\{a_j\}$ and $\{c_j\}$ sequences as before, but instead of the b-sequence we compute

$$\alpha_j = 1/\phi(y-j),$$

and in this case the linking equation is

$$c_0 = 1 + a_0\alpha_0$$

For small arithmetic we again compute the $\{a_j\}$ and $\{c_j\}$ as before, but now the $\{b_j\}$ are replaced by

$$\beta_j = \phi(x-j)/\phi(y-j),$$

and the linking equation by

$$1/c_0 = 1 + \beta_0.$$

Reference [10] contains a detailed analysis of the fundamental sℓi algorithm. The purpose of the present brief account is to indicate that it is possible to carry out ℓi and sℓi arithmetic by using algorithms which generate numbers of moderate magnitude in fixed precision. Further research will be aimed at producing faster algorithms.

4: CLOSURE AND PRECISION

F.W.J. Olver

4.1 Closure

In preceding lectures we have heard how the ℓi and $s\ell i$ systems were introduced to combat the problems of overflow and underflow, and seen how arithmetic operations can be carried out in these systems. We have also seen that with quite modest values of the ℓi image x it is possible to represent numbers $X = \phi(x)$ that lie far beyond the ranges of representable numbers of all other existing and proposed systems, including those of Matsui and Iri [45] and Hamada [25]. Nevertheless, it might be argued that even with the ℓi systems overflow is still possible. This is indeed so. However, overflow is impossible within the framework of arithmetic operations (other than division by zero) in finite-precision arithmetic. We begin with an explanation of this statement.

To fix ideas, let us suppose that the internal arithmetic base of the computer is r, and the fractional parts of the ℓi images are stored to d r-nary places. Then if we add any positive representable number $\phi(x)$ to itself the stored ℓi image of the sum will not exceed x, provided that

$$\phi(x+cr^{-d}) > 2\phi(x). \tag{4.1.1}$$

Here c is a positive constant: if the ℓi arithmetic processor computed to infinite precision, then we would have $c = 1$ for chopping, and $c = \frac{1}{2}$ for rounding. By using a sufficient number of guard digits in the processor we can approach these values arbitrarily closely, but the precise value of c is unimportant. On taking logarithms the inequality (4.1.1) becomes

$$\phi(x-1+cr^{-d}) > \phi(x-1) + \ell n\ 2.$$

If $x > 2$, then $\phi'(x-1)$ is increasing. Hence by application of the mean-value theorem we see that the last inequality is satisfied when

$$cr^{-d}\phi'(x-1) \geqslant \ell n\ 2.$$

This is equivalent to

$$x \geqslant \chi(c^{-1}r^{d}\ell n\ 2) + 1,$$

where $x = \chi(t)$ is the inverse function to $t = \phi'(x)$.

Sample values of $\chi(t)$, computed with the aid of Newton's rule, are as follows:

$$\chi(2^{32}) = 4.06..., \qquad \chi(2^{64}) = 4.26..., \qquad \chi(2^{5,502,869}) = 5.00... \ .$$

Consider now the set A of all numbers generated by the addition or subtraction of any two numbers, beginning with any pair of numbers whose ℓi images are numerically less than $\chi(c^{-1}r^{d}\ell n\ 2) + 1$. Assume that the ℓi images of the members of A are generated by arithmetic operations in the ℓi system and stored to d r-nary places. In consequence of the result just established, the absolute values of the members of A are bounded by $\phi\{\chi(c^{-1}r^{d}\ell n 2) + 1\}$. In a sense $\phi\{\chi(c^{-1}r^{d}\ell n\ 2) + 1\}$ represents "infinity", or rather, an "upper bound for infinity", for the operations of addition and subtraction. If we extend the arithmetic processes to include multiplication and division, other than division by zero, then the set of numbers so obtained is bounded in absolute value by $\phi\{\chi(c^{-1}r^{d}\ell n\ 2) + 2\}$. This is a consequence of the equivalence of multiplication and division to addition and subtraction at one level below; for example,

$$\phi(z) = \phi(x)\phi(y) \qquad \Leftrightarrow \qquad \phi(z-1) = \phi(x-1) + \phi(y-1),$$

when $x,y \geqslant 1$.

We have therefore, shown that when the ℓi system is used with any finite-precision arithmetic, there is a subset of its set of representable numbers that is closed under the operations of addition, subtraction, multiplication and division, excluding division by zero. Obviously, the same conclusion also applies to the $s\ell i$ system.

From the numerical values of $\chi(t)$ quoted above, it follows that with $r = 2$ and $d = 32$, that is, with 32 bits assigned to the storage of the fractional part of the ℓi image, an upper bound for the absolute values of numbers generated by addition and subtraction is $\phi(5.07)$, whether the abbreviation mode be chopping or rounding. For multiplication and division the corresponding bound is $\phi(6.07)$. To raise the overall upper bound from $\phi(6.07)$ to $\phi(7)$, or higher, we would need a wordlength in excess of 5,500,000 bits. Thus, in practice, levels beyond 6 will not be entered; in consequence *it will always suffice to allocate 3 bits to the storage of the integer part of the ℓi image.*

Once again, it could be argued that the closure property is not peculiar to the ℓi and the $s\ell i$ systems. When the floating-point system is augmented by the additon of NAN's ("Not-A-Numbers") [37] it can be regarded as being closed. However, NAN's are artificial and their occurrence signals a complete loss in precision. In contrast, the approach to infinity in the ℓi and the $s\ell i$ systems is completely natural, and the appropriate measure of precision (gp) is maintained.

4.2 Precision

For the fixed-point system the appropriate error measure is absolute error. For the floating-point system the appropriate error measure is relative error, or more satisfactorily, relative precision. For the ℓi and the $s\ell i$ systems the appropriate error measure is generalized precision, that is, the absolute error in the ℓi or $s\ell i$ image. (See Lecture 3.)

How can we compare, say, relative precision and generalized precision? Suppose that $X \geqslant 1$ and the ℓi image x of X is correct to d places in base r. (By "correct" we mean correctly chopped or correctly rounded.) The maximum relative error in X is $\{\phi'(x)/\phi(x)\}r^{-d}$ Therefore a measure m, say, of the number of correct digits in the mantissa of the corresponding floating-point form of X is given by

$$m = d - \log_r\{\phi'(x)/\phi(x)\}.$$

Since

$$\phi'(x)/\phi(x) = \phi'(x-1) = \phi(x-1)\phi(x-2). . . \phi(x-\ell+1),$$

where $\ell(\geqslant 1)$ is the integer part of x, we have

$$m \doteqdot d - \log_r e.\phi(x-2). \tag{4.2.1}$$

Now let λ and μ denote the exponent and mantissa, respectively, of the normalized floating-point form of X; thus

$$\phi(x) = r^\lambda\mu \text{ , with } 1/r \leqslant \mu < 1.$$

Then the number of digits, n say, in λ is represented by $\log_r\lambda$. On taking logarithms we have

$$\phi(x-1) = \lambda \ln r + \ln \mu,$$

and, as long as $x \geqslant 2$,

$$\phi(x-2) = \ln \lambda + \ln\ln r + \ln\left(1 + \frac{\ln \mu}{\lambda \ln r}\right) \doteqdot \ln \lambda.$$

Hence

$$\log_r e.\phi(x-2) \doteqdot \log_r\lambda = n. \tag{4.2.2}$$

Combination of (4.2.1) and (4.2.2) yields

n + m ≑ d.

In words, *in the corresponding floating-point form of* X, *the number of digits in the exponent plus the number of correct digits in the mantissa approximately equals the number of correct places in the* ℓi *image of* X.

The result just derived provides a rough, but useful, practical guide. However, for very large values of X it breaks down. More accurate information can be arrived at as follows.

Suppose, for example, that r, the internal arithmetic base, is 2 and the computer word has 32 bits. According to the IEEE standard [37] one of these bits is allocated to the sign of the represented number X, one is allocated to the sign (or bias) of the exponent, and 7 are allocated to the exponent itself. This leaves 24 for the mantissa (after allowing for the hidden bit). This arrangement is indicated schematically by the trapezium OABC in Figure 4.2.1. In this diagram, the vertical scale is $\xi \equiv \log_2\log_2 X$. Since $\xi \geqslant 0$ this means we are considering values of X such that $X \geqslant 2$. The triangular part to the left of the ξ-axis represents the digits actually used for this exponent, whereas the rectangular part to the right of the ξ-axis represents the digits assigned to the mantissa. Overflow takes place at the horizontal boundary BC.

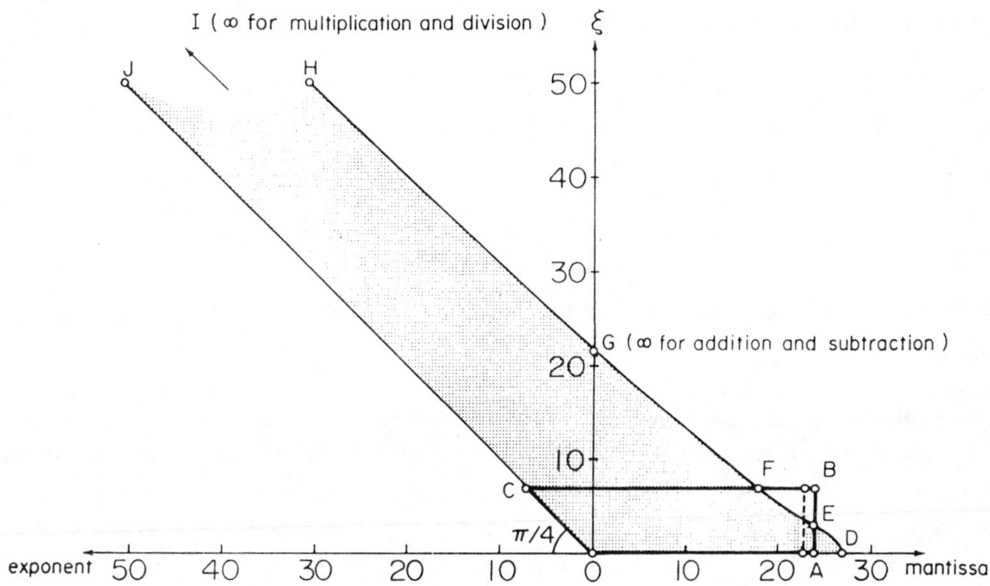

Figure 4.2.1. Accurate bits, in floating-point form, of the IEEE standard word and the corresponding sℓi word. (Approximately to scale).

With a 32-bit word in the sℓi system we allocate one bit to the sign of X and 4 bits to the integer part and sign (or bias) of the sℓi image x. This leaves 27 bits for the fractional part of x. The absolute change δX, say, in X resulting from an increase of 1 unit in the last place of x is therefore given by

$$\delta X = \Phi(x+2^{-27}) - \Phi(x) = \phi(x+1+2^{-27}) - \phi(x+1).$$

Provided that $\delta X \leqslant X$, we may take

$$m \equiv \log_2(\delta X/X)$$

as our measure of the number of correct digits in the mantissa of X. In Figure 4.2.1, m is the horizontal distance of the curve DEFG from the ξ-axis. At the intersection G of this curve with the ξ-axis we have $\delta X = X$ and $\xi = 21.70$, approximately.

At G no digits remain in the mantissa of the equivalent floating-point form; in other words, all of the information afforded by the sℓi representation has passed into the exponent. Not surprisingly, G is indistinguishable from the point on the ξ-axis that represents the upper bound for the closed set of numbers generated by the operations of addition and subtraction, given by $X = \phi\{\chi(2^{27}\ell n\ 2) + 1\}$; compare Section 4.1 above. We also observe that by the time G has been reached both the Matsui-Iri [45] and Hamada systems [25] have overflowed.

When $\xi > 21.70$ only part of the exponent is meaningful. In fact, the number of correct digits in the exponent is measured by

$$n \equiv -\log_2\left\{\frac{\delta(\log_2 X)}{\log_2 X}\right\} = -\log_2\left\{\frac{\phi(x+2^{-27})}{\phi(x)} - 1\right\} .$$

In Figure 4.2.1, n is the horizontal distance from the $3\pi/4$ ray OCJ to the extrapolated curve GH.

One further point, I, associated with Figure 4.2.1 deserves attention. This is the point off the diagram at which the extrapolations of the ray OCJ and the curve DEFGH intersect. At I, $\xi = 3{,}405{,}000$ (approximately), and no correct digits remain in the exponent. I is also on the same horizontal level as the point of the ξ-axis that corresponds to $X = \phi\{\chi(2^{27}\ell n\ 2) + 2\}$, and therefore represents "infinity" for the operations of multiplication and division.

We can proceed to still higher values of ξ in the sℓi system, but the corresponding floating-point form becomes meaningless.

Part of the enormous increase in representation range afforded by the sℓi system over the floating-point system can be seen by comparing the shaded area of Figure 4.2.1, and its extension to the point I, with the trapezium OABC. The only part of the trapezium that is not contained in the shaded region is the curvilinear triangle EBF, corresponding to $2.85 \leq \xi \leq 7$, approximately. In this region the floating-point system gains and the maximum gain amounts to 5 or 6 bits in the mantissa. Elsewhere the sℓi system gains: for $0 \leq \xi \leq 2.85$ there are up to 3 or 4 extra bits in the mantissa, and for $\xi > 7$ there is no overflow failure. On the scale employed in the diagram the point I is about 6 kilometers away from O.

It should also be noted that the region where the floating-point system achieves its greatest gain in local precision abuts the region in which it fails completely owing to overflow. In other words, gains in accuracy by the floating-point system over the ℓi and sℓi systems are likely to be realizable only at the price of living dangerously.

4.3 Example

The discussion of precision and range just concluded is interesting from the mathematical standpoint but does it have practical value? In particular, can approximations that are completely devoid of relative precision be useful in computational algorithms?

We first comment that the erosion of relative precision (and *a fortiori* absolute precision) that accompanies an increase in the magnitudes of stored numbers in the ℓi and sℓi systems is analogous to the erosion of absolute precision that accompanies an increase in the magnitudes of stored approximations in the floating-point system. Just as there was a shift in emphasis from absolute error to relative error on the introduction of floating-point hardware, so there may well be a corresponding shift towards the use of generalized precision as error measure if the use of ℓi and sℓi systems becomes widespread.

Secondly, we provide an example to show how relative precision can disappear and subsequently be recovered in the course of a computational algorithm.

Consider the iterative method

$$x_n = Ax_{n-1}, \qquad n = 1,2, \ldots$$

for computing the spectral radius λ, say, of a matrix A. For any vector norm and almost any choice of the initial vector x_0 we have [18]

$$\|x_n\|^{1/n} \to \lambda, \qquad n \to \infty .$$

Suppose, for example, that $n = 8192$ and

$$\|x_{8192}\| = \phi(4.87654) \quad ,$$

where the ℓi image is correct to ±0.00001, say. Since

$$\{\phi'(4.87654)/\phi(4.87654)\} \times 10^{-5} = \phi'(3.87654) \times 10^{-5} = 16.73.\ .\ .\ .\ ,$$

there are no correct significant digits in the corresponding floating-point mantissa, and only four of the five figures in the decimal exponent (27380) are reliable. However,

$$\ell n^{(2)} \{\|x_{8192}\|^{1/8192}\} = \phi(2.87654) - \ln 8192,$$

and on evaluating the right-hand side and exponentiating twice, we find that

$$\|x_{8192}\|^{1/8192} = \phi(3.71327).$$

Almost four decimals in the ℓi image here are accurate, and the corresponding floating-point form of $\|x_{8192}\|^{1/8192}$ has three correct digits in its mantissa.

What happened in this example is that the taking of a high-order root restored the relative precision. Further examples of this kind will appear in Lecture 6.

5: IMPLEMENTATION SCHEMES FOR LI AND SLI ARITHMETIC

P.R.Turner

5.1 Introduction

In this lecture we turn our attention to the vital question of how to implement ℓi or $s\ell i$ arithmetic in an efficient manner so that the loss of speed is reduced as far as possible. In the final lecture of this series, Olver will introduce an alternative scheme to those we discuss here. The aim of the present study is to find fast direct implementations of the algorithm described by Clenshaw in Lecture 3.

Recall that the basic arithmetic problem is that of finding z such that

$$\phi(z) = \phi(x) \pm \phi(y) \qquad (x \geq y) \tag{5.1.1}$$

and the computation of z is based on the construction of the sequences

$$a_j = 1/\phi(x-j) \qquad b_j = \phi(y-j)/\phi(x-j) \qquad c_j = \phi(z-j)/\phi(x-j) \tag{5.1.2}$$

according to the recurrence relations

$$a_{j-1} = \exp(-1/a_j) \tag{5.1.3}$$

$$b_{j-1} = \exp((b_j-1)/a_j) \tag{5.1.4}$$

$$c_{j+1} = 1 + a_{j+1}\ln c_j \tag{5.1.5}$$

with the appropriate initial values $a_{\ell-1}$, b_{m-1} and c_0.

There is a clear need therefore for efficient (direct) algorithms for the evaluation of functions such as $\exp(-1/a)$ and $1 + a\ln c$. In order to achieve this it will be necessary to make maximal use of parallelism.

We shall consider two different approaches to this problem, each of which yields a potential implementation for which arithmetic speeds should not be many times slower than for floating-point long multiplication. The first of these is based on using, with a high degree of parallelism, the CORDIC algorithms for evaluation of elementary functions. This parallelism will include all the forms which have been discussed elsewhere in this meeting. Task-, vector- and bit- parallelism all play important roles. The second approach will again make use of parallelism wherever possible within a basic framework of table look-up implementation. In both of these approaches, we make full use of Carry Save Adders which have the advantage that numbers can be added in a bit-parallel manner without regard to carries. The effect of this is that we store internal quantities, whenever possible, as *double numbers*; that is as the sum of two numbers. We shall describe the basic idea of the use of these CSA's in the next section; for a complete description see, for example [35], and [65].

The CORDIC algorithm was originally developed by Volder [63] for rapid solution of trigonometric problems for early in-flight navigation computers. The basic algorithm has since been extended to all the elementary functions and is now widely used within electronic hand-held calculators; see [56] and [64] for example. A brief account of the CORDIC algorithms is given in Section 5.3. Their modification and use in the implementation of ℓi airthmetic is described in Section 5.4.

In Section 5.5 we turn to the table-look-up approach in which we combine small tables with parallelism and the use of the CSA's to arrive at an implementation which, like that of Section 5.4, yields potential computing times not much slower than conventional floating-point long multiplication. We conclude this lecture with a comparison of these timings in Section 5.6.

5.2 Carry Save Adders

In this section we give a little background to the use of the Carry Save Adder, or CSA, and the philosophy of representing internal quantities as the sum of two numbers; that is as *double numbers.* The basic principle of the CSA is that the fundamental operation of addition is that *three* numbers are added to produce the sum in the form of a double number rather than the conventional operation of adding just two numbers to produce a single number output. This conventional addition is usually performed by either a Carry Propagate Adder, CPA, or a Carry Look-Ahead Adder.

The big advantage of the CSA lies in the fact that the addition can be performed in a bit-parallel manner with no regard to, or delay from, carries. Thus three binary numbers x, y and z are added to produce the answer s+c where c contains the carry bits. The following simple example demonstrates both the idea and the availability of bit-parallel addition; that is that the ith bits x_i, y_i, z_i of x, y, z can be added simultaneously for each i.

Example 5.2.1 Let

x = 0	1	0	0	0	1	1	1	=	71	
y = 0	0	1	0	1	0	1	1		+43	
z = 0	0	0	1	1	1	0	1		+29	= 143
s = 0	1	1	1	0	0	0	1	=	113	
c = 0	0	0	1	1	1	1			+30	= 143

Here the carry bits have been shifted one place to the left for simplicity but the principle is plain; each column can be added simultaneously.

In what sense does this yield a computational advantage?

To answer this question we consider the multiplication of two 32-bit numbers. The first step is to obtain (simultaneously) 32 numbers which are each either shifted copies of the first factor or zero depending on whether the corresponding bit of the multiplier is 1 or 0. These terms must be added to produce the product. We can use a tree of CSA's as in Figure 5.2.1 below. The symbol ⊕ is used to signify the CSA addition of the three terms to its left to produce the two to the right; → indicates that the corresponding term is just moved to the next stage of the addition.

Figure 5.2.1 CSA tree for multiplication

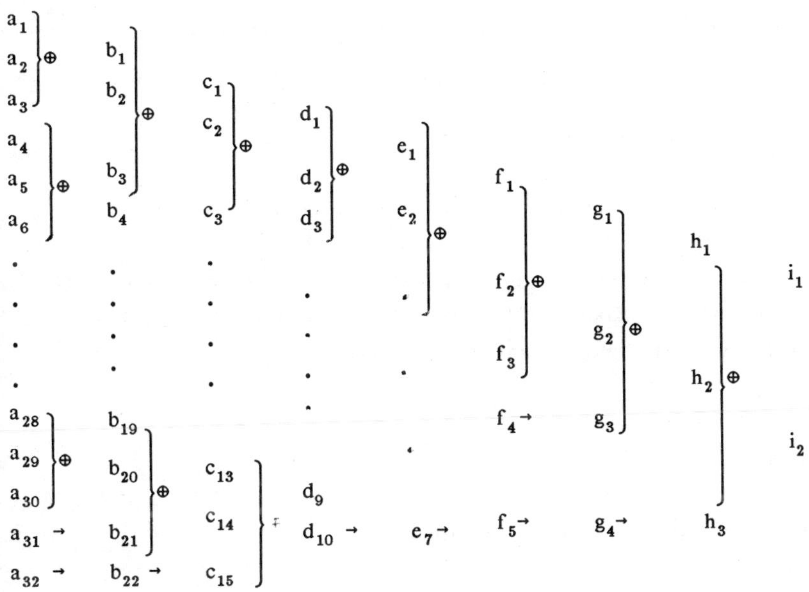

For the purpose of computing timings for the various arithmetic operations, we make the following assumptions:

One CSA operation takes	a time units (t.u.)
One shift operation takes	b t.u.
One CPA operation takes	c t.u.

Furthermore we shall suppose that

$a \doteqdot b$ and $c \doteqdot 32a$.

These are reasonable assumptions since the shift can also be executed in a bit-parallel manner while the CPA must of course operate bit-serially and we are considering 32-bit words.

For the above multiplication, the terms $a_1, \ldots, a_{32}$ can be obtained simultaneously in b t.u. Next there are eight stages of simultaneous CSA operations and finally the quantities i_1 and i_2 must be added by the CPA to produce the final product. The total time for this operation is therefore (8a + b) + c t.u. which is less than 1.5c t.u. The corresponding time for the ordinary "long multiplication" algorithm is 31c t.u. The potential saving is apparent.

For many of our purposes, the final CPA may be eliminated and the double number result used for the next stage of the algorithm. The multiplication of two such double numbers requires the addition of 128 terms. This requires 11 stages of simultaneous CSA operations to reduce the number of terms remaining from

128 to 86 to 58 to 39 to 26 to 18 to 12 to 8 to 6 to 4 to 3 to 2

and so the total time to produce the product as a double number is 11a + b t.u.

The potential usefulness of these ideas within arithmetic hardware is obvious. This is explored in great detail in, for example, [35]. In the next sections we see how these advantages can be exploited in the implementation of ℓi arithmetic.

5.3 CORDIC algorithms

The COordinate Rotation DIgital Computer algorithms developed by Volder [63] for the evaluation of the elementary functions are based on the idea of computing f(x) by decomposing the argument x in the form

$$x = \sum_{k=1}^{\infty} \delta_k \epsilon_k \tag{5.3.1}$$

where $\delta_k = \pm 1$ and $\Sigma\, \epsilon_k$ is a convergent series of positive terms chosen so as to combine with addition rules for $f(a \pm b)$ to yield a convenient decomposition of $f(x)$. The CORDIC algorithms use the following decomposition theorem which in this form is due to Walther [64] and Schelin [56].

Theorem 5.3.1 Suppose $\epsilon_0 \geq \epsilon_1 \geq \ldots \geq \epsilon_n > 0$ satisfy

$$\epsilon_k \leq \sum_{j=k+1}^{n} \epsilon_j + \epsilon_n \qquad (k=0,1,\ldots,n) \tag{5.3.2}$$

and let

$$|r| \leq \sum_{j=0}^{n} \epsilon_j.$$

Define $s_0 = 0$ and $s_{k+1} = s_k + \delta_k \epsilon_k$ $(k=0,1,\ldots,n)$ where $\delta_k = \operatorname{sgn}(r - s_k)$. Then

$$|r - s_k| \leq \sum_{j=k}^{n} \epsilon_j + \epsilon_n \qquad (k=0,1,\ldots,n+1) \tag{5.3.3}$$

In particular, we deduce that

$$|r - s_{n+1}| \leq \epsilon_n.$$

Note that the choice $\epsilon_k = 2^{-k}$ satisfies the conditions of the theorem and so any quantity $r \in [-2,2]$ can be decomposed in the form

$$r = \pm 1 \pm 1/2 \pm 1/4 \pm \ldots \pm 2^{-n} + E$$

where $|E| \leq 2^{-n}$. Thus we see that a prescribed absolute precision is obtainable with a known finite number of steps in this decomposition. Convenient addition rules exist for all the trigonometric and hyperbolic functions; these may be combined with the above theorem to yield algorithms for their evaluation. One of the major advantages of these procedures is that the computation then involves only shift and addition operations. This lack of any multiplicative operations is what gives them their speed by comparison with series or rational function approaches. Decimal versions of these algorithms are commonly employed in hand calculators for the evaluation of the elementary functions and for multiplication and division. The remarkable aspect of the CORDIC algorithm is that essentially the same algorithm with different choices of the parameters, signs and starting values achieves all of these. A general description of this algorithm follows.

CORDIC Algorithm (Binary version)

Compute three sequences according to the recurrence relations

$$\begin{aligned} x_{k+1} &= x_k - m\delta_k y_k 2^{-k} \\ y_{k+1} &= y_k + \delta_k x_k 2^{-k} \\ z_{k+1} &= z_k - \delta_k \epsilon_k \end{aligned} \tag{5.3.4}$$

We use:

$m = 0$	with $\epsilon_k = 2^{-k}$	for multiplication and division,
$m = 1$	with $\epsilon_k = \tan^{-1}2^{-k}$	for the trigonometric functions,
$m = -1$	with $\epsilon_k = \tanh^{-1}2^{-k}$ $(k\geq 1)$	for the hyperbolic functions.

(Note that the equations (5.3.4) use no operations other than addition or subtraction of known constants or shifted copies of current values).

The choice of the signs δ_k and of the starting values depends on the individual function to be evaluated. There are two "modes" of operation - the *rotation* mode in which, typically, y_0 is set to zero and δ_k is chosen to drive z_k toward zero by setting $\delta_k = \mathrm{sgn}(z_k)$, and the *vectoring* mode in which $z_0 = 0$ and $\delta_k = -\mathrm{sgn}(y_k)$. We shall consider just one example. (A more complete treatment can be found in [56] and [64].)

The third case, with $m = -1$, which is used for the hyperbolic, exponential and logarithmic functions requires a little more explanation before we proceed to the example. The choice $\epsilon_k = \tanh^{-1} 2^{-k}$ does *not* satisfy the condition (5.3.2) of the theorem. However if the steps for $k = 4, 13, 40, \ldots, j, 3j+1, \ldots$ are repeated the resulting sequence $\{\epsilon_k\}$ is suitable.

Standard addition formulae for the hyperbolic functions yield

$$\begin{aligned} \cosh(s_k+\delta_k\epsilon_k) &= \cosh s_k \cosh \epsilon_k + \delta_k \sinh s_k \sinh \epsilon_k \\ &= \cosh \epsilon_k \{\cosh s_k + \delta_k \tanh \epsilon_k \sinh s_k\} \\ &= \cosh \epsilon_k \{\cosh s_k + \delta_k 2^{-k} \sinh s_k\} \end{aligned} \tag{5.3.5}$$

and similarly,

$$\sinh(s_k+\delta_k\epsilon_k) = \cosh \epsilon_k \{\sinh s_k + \delta_k 2^{-k} \cosh s_k\}. \tag{5.3.6}$$

As we pointed out earlier, the number of steps of the CORDIC algorithm required to obtain a specified absolute precision is predeterminable. It follows that the factors $\cosh \epsilon_k$ of (5.3.5) and (5.3.6) are known *in advance*, as is the number of them, N say. All that is necessary therefore is to store one additional constant

$$K = \prod_k \cosh \epsilon_k$$

where the product includes the repetitions mentioned above. The initial values x_1 and y_1 can be premultiplied by this constant so that the above relations then resemble the first two of (5.3.4) with $m = -1$. This multiplication by K does not contravene the desire to use only shifts and additions since the quantities to be premultiplied are either 0 or 1.

With the initial values $x_1 = K$, $y_1 = 0$ and $|z_1| \leqslant \Sigma\, \epsilon_k$ and the signs δ_k given by $\mathrm{sgn}(z_k)$ the values z_k approach zero and

$$|z_1 - \sum_{k=1}^{n} \delta_k \epsilon_k \,| \leqslant \epsilon_n \qquad (n=1,2,...,N) \tag{5.3.7}$$

while the recurrence relations (5.3.4) yield

$$|x_{N+1} - \cosh z_1| \leqslant 2^{-N} \quad \text{and} \quad |y_{N+1} - \sinh z_1| \leqslant 2^{-N}. \tag{5.3.8}$$

5.4 Implementation of ℓi arithmetic using CORDIC algorithms

Our primary aim here is to develop versions of the CORDIC scheme for the particular functions needed for the implementation of level-index arithmetic. The relation (5.1.3) for generating the sequence $\{a_j\}$ uses the function given by $\exp(-1/x)$ with the argument x lying in (0,1]. A similar function is needed to generate the sequence $\{b_j\}$.

Firstly, consider the exponential function itself. From equations (5.3.8) we see that

$$x_{N+1} \pm y_{N+1} \doteqdot \exp(\pm z_1) \tag{5.4.1}$$

with an error not greater than 2^{-N+1}. The recurrence relations (5.3.4) for this case are

$$\begin{aligned} x_{k+1} &= x_k + \delta_k y_k 2^{-k} \\ y_{k+1} &= y_k + \delta_k x_k 2^{-k} \\ z_{k+1} &= z_k - \delta_k \epsilon_k \end{aligned}$$

Adding the first two of these and denoting $x_k + y_k$ by u_k we obtain

$$u_{k+1} = u_k + \delta_k u_k 2^{-k}, \tag{5.4.2}$$

which still only uses shifts and additions or subtractions to give

$$|u_{N+1} - \exp(z_1)| \leqslant 2^{-N+1}$$

for any $|z_1| \leqslant \Sigma\ \epsilon_k \doteqdot 1.13$.

But, of course, we require evaluation of quantities $\exp(-1/x)$ and must therefore adapt this procedure to this special situation. Instead of decomposing the argument z_1 in the form $\Sigma\delta_k\epsilon_k$ as in the above algorithm, we obtain the decomposition

$$-1 \doteqdot \Sigma\ \delta_k\epsilon_k x = \Sigma\ \delta_k\xi_k \tag{5.4.3}$$

so that

$$-1/x \doteqdot \Sigma\ \delta_k\epsilon_k.$$

Here the new constants ξ_k. defined by $\xi_k = x\epsilon_k$, can all be computed simultaneously in an arithmetic unit with sufficient parallelism. The CORDIC-like algorithm for the function $\exp(-1/x)$ thus becomes:

$$\begin{aligned} u_1 &= K, \qquad\qquad v_1 = 1 \\ \delta_k &= -\mathrm{sgn}(v_k) \\ u_{k+1} &= u_k + \delta_k u_k 2^{-k} \\ v_{k+1} &= v_k + \delta_k\xi_k \end{aligned} \tag{5.4.4}$$

For the sequence $\{b_j\}$ the recurrence relation relies on evaluations of the function $\exp((y-1)/x)$. This is achieved by simply setting

$$v_1 = 1 - y$$

in the algorithm defined by equations (5.4.4).

There is a serious and obvious drawback to this algorithm as we have it thus far. The restriction of the range to $|z_1| \leqslant 1.13$ implies that the above algorithm is only useful for $|x| \geqslant 1/1.13$ and since we have $0 < x \leqslant 1$ we clearly must extend this range. This extension of the range is a common problem for elementary function routines. In this case the inclusion of two additional steps using $\epsilon_1 = \tanh^{-1} 1/2 \doteqdot 0.549$ extends the range of the original algorithm to $|z_1| \leqslant 2.228$ and therefore the range for x above to $x \geqslant 0.45$.

Now the analyses of the ℓi arithmetic algorithms [9,10] show that (for single precision working) the values a_j are required to an absolute precision of 2^{-37}. For any $x < 2^{-5}$ we have

$$\exp(-1/x) < e^{-32} < 2^{-46} \simeq 0 \; ; \; ap(2^{-37}) \qquad (5.4.5)$$

so that if $a_j < 1/32$ we simply return the value $a_{j-1} = 0$. If $x \geqslant 1/32$ then doubling x at most four times results in an argument greater than 1/2 and so within the range of our algorithm. This is compensated by squaring the result the same number of times since

$$\exp(-1/x) = (\exp(-1/2x))^2.$$

We turn now to considerations of the operation time for computation of the sequence $\{a_j\}$ using this approach in combination with the "double-number" philosophy discussed in Section 5.2. First, we must compute $a_{\ell-1} = e^{-f}$ *as a double number* with the required accuracy 2^{-37}. This needs 39 steps of the CORDIC algorithm allowing for the necessary repetitions. (Note that there is no range extension needed for this part of the calculation). Thus we set

$$u_1 = K, \qquad v_1 = f, \qquad \delta_k = \mathrm{sgn}(v_k)$$

and, *in parallel with each other,*

$$u_{k+1} = u_k - \delta_k u_k 2^{-k} \quad \text{and} \quad v_{k+1} = v_k - \delta_k \xi_k.$$

For $k > 1$ the choice of δ_k involves finding the sign of a double number. We assume this operation takes d t.u. Each of the parallel steps takes 2a + b t.u. since the u_k are double numbers. Hence the total time for this stage is 39(2a + b + d) t.u.

For the computation of a_{j-1} from a_j, we have a maximum of four "doublings" of a_j to obtain the argument $x > 0.45$. This takes a maximum of 4b t.u. since multiplication by 2 is simply a shift of one place. Next the quantities $\{\xi_k\}$ are obtained as double numbers by parallel multiplication of the stored single-number constants $\{\epsilon_k\}$ by x. This step takes (10a + b) t.u. The necessary 41 steps of CORDIC algorithm are accomplished in 41(2a + b + d) t.u. and must be followed by up to four "squarings" of the result. These "double × double" multiplications take 12a + b t.u. each. The maximum time for computing a_{j-1} from a_j is therefore

$$140a + 50b + 41d \text{ t.u.}$$

With a simple modification of the way in which the sequence $\{b_j\}$ is started this can be computed in parallel with the $\{a_j\}$ in the same time.

The recurrence relation for the sequence $\{c_j\}$ can be implemented with the standard CORDIC algorithm for the logarithm function modified to use the constants

$$\alpha_k = a\epsilon_k$$

in order to evaluate the function a × ln c. Again some initial scaling is required but this is readily accomplished. (A complete treatment of this and the timings involved can be found in [61]).

The overall times for this operation are critically dependent on the time, d, needed for obtaining the sign of a double number. In Section 5.6 we discuss and compare these times with those of the table look-up approach of the next section. For a "typical" case where the levels of both x and z are 3, we find that this operation time ranges from about twice to six or seven times that of a floating-point long-multiplication as d varies from about 2a to about 16a.

5.5 The table look-up approach

In this section we turn to a different approach to the implementation of level-index arithmetic in which the double number philosophy is combined with a partial table look-up scheme. The idea is to use several small tables for the appropriate exponential functions. The argument is divided up into subwords corresponding to the various tables which can then be consulted simultaneously.

Consider first the computation of a_{j-1} to the required working precision of 2^{-37}. For this approach this calculation has two stages the first of which is the efficient reciprocation of a_j. We require

$$t \approx 1/a_j \;;\quad ap(2^{-37}). \tag{5.5.1}$$

By (5.4.5) we may again deduce that

$$a_j \leqslant 1/32 \;\Rightarrow\; a_{j-1} \approx 0;\ ap(2^{-37})$$

and hence $t < 32$. Storage of t thus requires 42 bits including five before the binary point. To facilitate the reciprocation, we first shift a_j to the interval [1/2,1) and subtract the result from unity so that we seek $1/(1-\delta)$ with $\delta \in (0,1/2]$. (Of course, in forming t we must shift the result the appropriate number of places to compensate for this initial shift. This has been accounted for in the timings below).

The necessary accuracy in $1/(1-\delta)$ can be obtained from the geometric series using the terms

$$1 + \delta + \delta^2 + \ldots + \delta^{41}.$$

(Alternatively, we may use the approximation

$$1/(1-\delta) \doteqdot (1+\delta)(1+\delta^2)(1+\delta^4)\ldots(1+\delta^{32})$$

which may achieve a further marginal speed-up.) Now, obtaining δ consists of a shift and the forming of the two's complement. We assume the latter takes the same time b t.u. so that finding δ takes 2b t.u. Next we form δ^2 as a double number; this takes a further 8a + b t.u.

For the remaining powers of δ, parallel computation yields significant gains. For k = 1,2,3,4,5, we can form $\delta^{2^k+1}, \ldots, \delta^{2^{k+1}}$ as double numbers *in parallel.* Each such stage takes 12a + b t.u.

There are now 40 double and two single numbers to be summed and, finally, shifted to form t as a single number. This part of the operation takes 10a + b + c t.u. which yields a total time for the reciprocation of

$$78a + 10b + c \text{ t.u.}$$

Note that further slight gains can be achieved by forming some of the partial sums simultaneously with the computation of the next group of terms in the expansion.

To complete the calculation of a_{j-1} we must evaluate e^{-t}. Write

$$t = t_1 + t_2 + \ldots + t_7 \tag{5.5.2}$$

where each t_i is the ith 6-bit subword of the 42-bit word t. This can be achieved by simultaneous shift operations and takes b t.u. We can now use parallel table look-up for the values $\exp(-t_i)$, and we shall assume that this look-up operation takes e t.u.

There are now seven factors to be multiplied. This can be accomplished using a tree of parallel multiplications each of which makes maximal use of the CSA and double numbers.
The time for this part of the calculation is

$$32a + 3b + c \text{ t.u.}$$

and so the total time for the computation of a_{j-1} from a_j is

$$110a + 14b + 2c + e \text{ t.u.}$$

In parallel with this calculation we can compute b_{j-1} using a_j and b_j with just a small time-lag caused by the need to compute

$$t' = (1 - b_j)t.$$

This multiplication requires $10a + b$ t.u. and so *assuming that* $e > 10a + b$, we can compute a_{j-1} and b_{j-1} in

$$110a + 14b + 2c + 2e \text{ t.u.}$$

The calculation of $a_{\ell-1}$ (and, perhaps, $b_{\ell-1}$) is a much quicker process for the table look-up approach since no reciprocation step is needed. This can be achieved in

$$32a + 4b + c + 2e + \max(c,e) \quad \text{t.u.}$$

Full details can be found in [52].

For the computation of the sequence $\{c_j\}$ using the recurrence relation

$$c_{j+1} = 1 + a_{j+1}\ln c_j$$

we begin by forming $c_0 = 1 \pm b_0$ which entails either the insertion of a 1 before the binary point or the taking of the two's complement of b_0. This operation takes b t.u. Write

$$c_j = s(1 + \sigma) \tag{5.5.3}$$

where s comprises the first six significant bits of c_j. It follows that

$$\sigma < 2^{-5}.$$

Dividing c_j by s using a similar approach to the reciprocation above, we can obtain σ to the required absolute precision of 2^{-32} in a total of $57a + b$ t.u.

Since $\sigma < 2^{-5}$, the necessary accuracy for $\ln(1+\sigma)$ can be obtained from the approximation

$$\sigma - \sigma^2/2 + \sigma^3/3 - \sigma^4/4 + \sigma^5/5.$$

The terms of this expansion can now be computed in three stages as follows: first σ^2, $\sigma/3$ and $\sigma/5$ are computed as double numbers; from these $\sigma^3/3$ and σ^4 and then, finally, $\sigma^5/5$, $\sigma^2/2$ and $\sigma^4/4$ can be obtained. Each of these steps takes 12a + b t.u.

This leaves nine terms to be added together with ln s. This last term is obtained by table look-up taking account of the initial shift required to obtain the leading six *significant* bits. The total time to produce ln c_j is therefore 98a + 11b t.u. and hence we may deduce that the overall time required for computing c_{j+1} is

109a + 12b + c t.u.

In this timing we have assumed that the table look-up time for the six bit table, e t.u., does not exceed 100a ÷ 3c t.u. and therefore that ln s can be obtained simultaneously with the rest of the calculation described above.

In the next section we compare the timings for the implementation of the ℓi arithmetic algorithms adopting this approach with those for the CORDIC techniques. The assumptions made for the table look-up time bracket this between (approximately) c/3 and 3c and, for this range of values of e, we find that "typical" arithmetic times are of the order of twice those of floating-point long multiplications.

In this case it is also necessary to address the question of the storage requirement of the tables suggested. For the sequence $\{a_j\}$ and $\{b_j\}$ we need seven tables, each with $2^6 = 64$ entries, while for $\{c_j\}$ only one such table is necessary. In total therefore, some 512 entries each of substantially fewer than 60 bits would be needed. A generous upper bound for the storage requirement is thus around 30,000 bits which would occupy only a small corner of a level-index arithmetic chip. The feasibility of on-chip storage of the tables renders the table look-up times cited above attainable.

5.6 Overall arithmetic times and conclusions

The steps needed to complete the ℓi arithmetic algorithms can be accomplished in essentially the same ways as have been described in the preceding sections.

For subtraction when $n < \ell$, that is when the level of the result z is less than that of x, a single division is required. For the case $n = \ell$ and for addition the quantity f + ln $c_{\ell-1}$ must be computed. The addition algorithm may require yet one further logarithm. Whether the CORDIC or the table look-up approach is to be adopted these final stages can be readily accomplished with minor variations - often simplifications - of the procedures already described.

The "worst case" for which level-index arithmetic ever need be performed is with $\ell = m = n = 5$ since all arithmetic beyond that level is trivial in the sense discussed elsewhere in this course. The overall time for the CORDIC implementation for this case is

$$C_1 = 1168a + 467b + c + 409d \text{ t.u.}$$

while for the table look-up method it is

$$L_1 = 1006a + 120b + 14c + 9e + \max(c,e) \text{ t.u.}$$

The major attraction of the CORDIC scheme is the retention of the double numbers throughout the calculation to the extent that only one CPA operation is used. The associated cost is plain - the coefficient of d is large. The dependence of the table look-up method on e is less critical. Under the assumptions made earlier that

$$a \doteqdot b \doteqdot c/32,$$

these times are given by

$$C_1 \doteqdot 52c + 409d \qquad \text{and} \qquad L_1 \doteqdot 49c + 9e + \max(c,e).$$

For a more realistic assessment of the potential of these implementations which is not unfairly favourable to the ℓi system consider a "typical" operation with $\ell = n = 3$. The corresponding times are then

$$C_2 = 722a + 293b + c + 259d \doteqdot 33c + 259d \text{ t.u.}$$

and

$$L_2 = 568a + 68b + 8c + 6e + \max(c,e) \doteqdot 28c + 6e + \max(c,e) \text{ t.u.}$$

Single precision floating-point long multiplication would be expected to take of the order of 25 c t.u. and it is this figure which is used in the following table to estimate level-index times in terms of floating-point multiplications, fpm.

TABLE 5.5.1

CORDIC			Table look-up		
d	Typical time	Worst case time	e	Typical time	Worst case time
2a or c/16	49c or 2fpm	77c or 3fpm	11a or c/3	34c or 1.5fpm	53c or 2fpm
4a or c/8	65c or 2.5fpm	103c or 4fpm	32a or c	37c or 1.5fpm	59c or 2.5fpm
8a or c/4	98c or 4fpm	154c or 6fpm	96a or 3c	47c or 2fpm	79c or 3fpm

It would be unwise at this stage to draw firm conclusions as to what will be the best way in which to implement ℓi arithmetic. What can be safely concluded from the study of these approaches is that it will be possible to implement it in ways which will not be much slower than floating-point computation. The final implementation may well use combinations of all the ideas considered here and others too. The surface-fitting approach to be considered in the final lecture of this series is also highly promising.

It is relevant to point out here that the evidence of Sweeney [58] is that only a very small proportion of computer run-time is spent on floating-point operations even within large scientific computing programs. Thus the effect of slowing down the arithmetic operations themselves is much diluted in its effect on run-times. In combination with the fact that a robust arithmetic system will allow much simpler programs to be used, the overall effect on program development and run times is likely to be beneficial.

6: APPLICATIONS

C.W.Clenshaw

6.1 Introduction

In this lecture we examine the results obtained from the application of sℓi arithmetic to two particular problems in numerical analysis. Programs were prepared for the VAX at Lancaster and for the UNIVAC at the NBS, based on software implementations of the algorithms already described. These implementations use flp arithmetic internally, and they also use standard machine routines for the exponential and logarithmic functions. Thus they are slow, and they are not amenable to the kind of error analyses which we explored in [9] and [10]. Nevertheless, they are genuine sℓi programs, in the sense that their results will be similar to those produced by a fully polished sℓi implementation which has similar limitations of precision. Since our first aim is to confirm the practical usefulness of sℓi, we now apply these programs to (i) the calculation of the binomial probability distribution; (ii) the solution of real polynomial equations by Graeffe's method of root squaring.

6.2 The binomial probability distribution

The function which we wish to calculate is defined by

$$I(n,r,p) \;=\; \sum_{s=0}^{r} \binom{n}{s} p^s q^{n-s}$$

where $q = 1-p$, $0 \leqslant p \leqslant 1$, $0 \leqslant r \leqslant n$.

The problems associated with magnitude here are well known, and widely discussed in the literature. If calculation is undertaken directly in flp, the numerator and denominator of $\binom{n}{s}$ overflow, and p^s and q^{n-s} underflow, for quite moderate values of the parameters, even though the resulting I always lies between 0 and 1.

It is possible to overcome the difficulties by careful analysis and scaling, but the resulting algorithm is inevitably somewhat complicated. Sterbenz [57] devotes several pages to this problem. In contrast, we can implement almost any algorithm, however simple, in sℓi. For example, given p, n, and r, we might first evaluate

$$q^n = q \times q \times q \times q. \;.\;.\quad .$$

Then for each s from 1 to n form

$$\begin{aligned} n!/(n-s)! &= \{n!/(n-s+1)!\} \times (n-s+1) \\ s! &= (s-1)! \times s \\ p^{s} &= p^{s-1} \times p \\ q^{n-s} &= q^{n-(s-1)}/q \end{aligned}$$

following which each term of the sum can be calculated directly, and I found by summation.

In the case p = 0.1, n = 2000, r = 200 (which is used by Sterbenz) we find, for the last term in the sum, the values

$$\begin{aligned} n \times (n-1) \times \ldots \times (n-r+1) &= \phi\,(4.68842\ 667) \\ r! &= \phi(4.64769\ 156) \\ p^{r} &= 1/\phi(4.59530\ 170) \\ q^{n-r} &= 1/\phi(4.50519\ 488) \end{aligned}$$

Then

$$\binom{n}{r} p^{r} q^{n-r} = 1/\phi(3.22893\ 471)$$

and we find, on adding this to the other similarly-calculated terms,

$$I = 1/\phi(1.65619\ 393) = 0.51882\ 226 \quad .$$

There is some loss of accuracy here through cancellation (the correct 8D result is 0.51882 040) but we clearly have a useful result, even in this rather extreme case. (We may note, for example, that $\phi(4.6) \doteqdot 6.3 \times 10^{210}$ while $\phi(4.7) \doteqdot 3 \times 10^{778}$).

This loss is interesting: it tends to provoke a quick reaction from flp defenders who will say that this is evidence that sℓi is subject to loss of precision in circumstances where flp cannot lose precision. We have performed multiplications and divisions, followed by addition of positive quantities. There has been no subtraction, so there can be no cancellation.

This plausible-sounding argument is based upon the assumption that the proper measure of precision is *relative* precision. As we have already seen, this is not the case for large numbers. When we divide two such numbers there is cancellation in the exponent, and information is lost. If the above example could be carried out by the same method in flp (which is not possible for any existing orthodox system) then it would suffer similar cancellation.

Not least of the advantages of ℓi and sℓi is that they make manifest some properties of ordinary arithmetic operations that are concealed by flp; in this case, the essential similarity of the nature of subtraction and division from the standpoint of precision.

6.3 Root squaring

Graeffe's method of root squaring for the solution of polynomial equations presents greater difficulties because the very large (or very small) numbers which occur naturally in its execution are much more difficult to remove by scaling. Nevertheless, the theoretical attractions of the method are such that in pre-computer days, when the result of each arithmetic operation could be observed and acted upon, root squaring featured prominently in the literature. Up until the 1960's many books on numerical analysis included an account of root squaring (among them Hildebrand [31] and Ralston [55]) and a comprehensive description of its implementation on desk machines was given by Olver [47]. In 1960 Bareiss [1] showed how a computer might be programmed to perform root squaring, and more recently, a discussion of this method appeared in Henrici [30]. Generally speaking, however, numerical analysts have sought methods which are more readily programmable. The attractive simplicity of the root-squaring method tends to be lost in the devices needed to avoid catastrophic overflow and underflow in a flp system.

Thus this method is an obvious test vehicle for any new system of arithmetic which claims to free computations from these phenomena. We now give a very brief outline of the method. Suppose that the polynomial whose zeros are required is given by

$$p(z) = K(z-\alpha_1)(z-\alpha_2)(z-\alpha_3)\ .\ .\ .\ (z-\alpha_n). \tag{6.3.1}$$

Then

$$(-1)^n\, p(z)p(-z) = K(z^2-\alpha_1^{\,2})(z^2-\alpha_2^{\,2})\ .\ .\ .\ (z^2-\alpha_n^{\,2}),$$

so the multiplication of $p(z)$ by $p(-z)$ gives a new polynomial whose zeros are the squares of those of the original. This process may be repeated as often as we wish.
We write

$$f_0(z) = p(z)$$
$$f_{r+1}(z^2) = (-1)^n f_r(z)\ f_r(-z), \qquad r = 0,1,2,... \tag{6.3.2}$$

with

$$f_r(z) = \sum_{j=o}^{n} (-1)^{n-j}\, a_j^{(r)} z^j. \qquad (6.3.3)$$

Substitution of (6.3.3) in (6.3.2) gives

$$a_j^{(r+1)} = (a_j^{(r)})^2 + 2 \sum_{k=1}^{\min(j,n-j)} (-1)^k a_{j-k}^{(r)}\, a_{j+k}^{(r)} \qquad (6.3.4)$$

The mechanical process of root squaring is therefore very simple. What it achieves is, of course, the increased separation of real, distinct zeros (we shall see what happens to other types of zero later) to the point at which they can be deduced very easily from the coefficients.

For simplicity, let us examine a cubic.

Let α_1, α_2 and α_3 be the real zeros of

$$f_0(z) = p(z) = a_3^{(0)} z^3 - a_2^{(0)} z^2 + a_1^{(0)} z - a_0^{(0)},$$

and let us, for the sake of brevity, denote by t_1, t_2 and t_3 the zeros of

$$f_r(z) = a_3^{(r)} z^3 - a_2^{(r)} z^2 + a_1^{(r)} z - a_0^{(r)} .$$

(Thus we have $t_j = \alpha_j^{2^r}$).

The ratios of successive coefficients yield

$$\frac{a_2^{(r)}}{a_3^{(r)}} = t_1 + t_2 + t_3 ,$$

$$\frac{a_1^{(r)}}{a_2^{(r)}} = \frac{t_1 t_2 + t_1 t_3 + t_2 t_3}{t_1 + t_2 + t_3}$$

$$\frac{a_0^{(r)}}{a_1^{(r)}} = \frac{t_1 t_2 t_3}{t_1 t_2 + t_1 t_3 + t_2 t_3}$$

If $|t_1| >> |t_2| >> |t_3|$, these are approximately t_1, t_2 and t_3 respectively. If r is large enough for the separation of zeros to be acceptable, then we can then recover the wanted α_j by taking the appropriate root of t_j (except that the sign will be undetermined as yet, since this will have been lost in the first squaring).

So in all cases we define

$$\alpha_j^{(r)} = \left| a_{n-j}^{(r)} \;/\; a_{n-j+1}^{(r)} \right|^{2^{-r}} .$$

If $|\alpha_1| > |\alpha_2| > |\alpha_3| \ldots$ then $\alpha_j^{(r)} \to |\alpha_j|$ as $r \to \infty$.

As a simple example, consider the cubic $z^3 - 6z^2 + 11z - 6$. In this case we find (to 5D)

$$\begin{aligned}
\alpha_1^{(0)} &= 6 & & \\
\alpha_1^{(1)} &= 14^{2^{-1}} &&= 3.74166\\
\alpha_1^{(2)} &= 98^{2^{-2}} &&= 3.14635\\
\alpha_1^{(3)} &= 6818^{2^{-3}} &&= 3.01444\\
\\
\alpha_2^{(0)} &= 11/6 &&= 1.83333\\
\alpha_2^{(1)} &= (49/14)^{2^{-1}} &&= 1.87083\\
\alpha_2^{(2)} &= (1393/98)^{2^{-2}} &&= 1.94170\\
\alpha_2^{(3)} &= (1686433/6818)^{2^{-3}} &&= 1.99143\\
\\
\alpha_3^{(0)} &= 6/11 &&= 0.54545\\
\alpha_3^{(1)} &= (36/49)^{2^{-1}} &&= 0.85714\\
\alpha_3^{(2)} &= (1296/1393)^{2^{-2}} &&= 0.98212\\
\alpha_3^{(3)} &= (1679616/1686433)^{2^{-3}} &&= 0.99949
\end{aligned}$$

Further similar cycles of the squaring process will demonstrate rapid convergence of the three sequences to the limits 3, 2 and 1 respectively.

Several questions are raised at once: for example

(i) How do we determine the sign of each real zero?

(ii) What happens when not all zeros are well separated? (In particular, how are complex zeros and repeated zeros determined?)

(iii) How do we deal with the large numbers, and the consequent large errors, that appear after only a few cycles?

The first two questions are readily answered. When a real zero has been isolated, we simply calculate $p(\pm\alpha_k)$; the magnitude of the two residuals will indicate which is the correct sign.

Now suppose that two dominant zeros of the cubic share the same modulus $|\alpha|$, and let us write the zeros of $f_r(z)$ as

$$t_1 = pe^{i\theta},\ t_2 = pe^{-i\theta},\ t_3\ .$$

In this case we have

$$\alpha_1^{(r)} = |2p\cos\theta + t_3|^{2^{-r}} \rightarrow |2p\cos\theta|^{2^{-r}} \text{ if } \cos\theta \neq 0,$$

$$\alpha_2^{(r)} = \left|\frac{p^2 + 2pt_3\cos\theta}{2p\cos\theta + t_3}\right|^{2^{-r}} \rightarrow \left|\frac{p}{2\cos\theta}\right|^{2^{-r}}$$

$$\alpha_3^{(r)} = \left|\frac{p^2 t_3}{p^2+2pt_3\cos\theta}\right|^{2^{-r}} \rightarrow t_3^{2^{-r}} \rightarrow |\alpha_3|\ .$$

We see that the distinct root is still exposed as before, but that the pair has become relatively complicated. However, if we define

$$\beta_j^{(r)} = \left(\alpha_j^{(r)}\ \alpha_{j-1}^{(r)}\right)^{½}$$

we see that here $\beta_2^{(r)} \rightarrow p^{2^{-r}} \rightarrow |\alpha|$, so the modulus of the pair is exposed by the β-sequence just as is that of the real root by the α-sequence.

In solving an arbitary polynomial equation with real coefficients, we must expect real roots and complex conjugate pairs; higher multiplicities are much less common. We therefore envisage a standard output from a root-squaring program which will give values of $\alpha_j^{(r)}$ and $\beta_j^{(r)}$, together with the residuals $p(\pm\alpha_j^{(r)})$ and $p(\pm\beta_j^{(r)})$. These should serve to identify isolated real zeros, and pairs of zeros sharing the same modulus. (A pair may be complex conjugates, real and equal, or real and equal-opposite).

6.4 Precision

It is easy to appreciate that for even quite moderate values of r, the coefficients in $f_r(z)$ will be subject to large relative errors. (In our example above, the coefficents were exact for r = 0,1,2,3, but even here their sheer magnitude will force rounding, and therefore rounding errors, in succeeding cycles). These errors grow so rapidly that there may soon be no correct significant figures in the $a_j^{(r)}$. However, this is not in itself a matter for concern; the information that is apparently lost in the process of root squaring is patently recovered in the calculation of the α_j. This phenomenon is described more fully in [11]: see also [44], and Olver's Lecture 4.

It is important to notice that we have not yet invoked sℓi arithmetic. We have merely observed that relative error is an inappropriate measure for calculations of this nature, where results of moderate magnitude are extracted from intermediate calculations where numbers may be very large (or small). We now note that the use of sℓi arithmetic, and the naturally associated notion of generalized precision, enables this problem to be treated without difficulty.

6.5 The application of sℓi to root squaring

In [11] we give an account of the results obtained when a simple implementation of sℓi is applied directly to four separate polynomials of widely differing nature. Here we merely sketch the salient points, and describe some results from just two of those problems.

We have used sℓi (rather than ℓi) in order that we need not address the problem of scaling. It is true that, whether we use ℓi or sℓi, the simple multiplication of the coefficients of a polynomial by a constant will affect the conditioning of the problem of root-finding, and there is some interest in determining the optimal scaling procedure of this kind. However, our present purpose is to investigate the efficacy of level-index arithmetic in dealing with magnitude and precision in extreme circumstances; we therefore proceed with an algorithm which is immediately applicable to any polynomial with real coefficients, even though end-figure errors may be larger than necessary.

The calculation of the phase of complex zeros also raises questions which, for similar reasons, we shall not pursue here. Our aim will be to calculate the modulus of each zero, and we shall incidentally determine (in most cases) whether it is real or complex. In the case of real zeros, we shall determine their signs also.

The routine used is a straightforward implementation in sℓi of the algorithm described in Section 6.3. (This implementation uses double-precision floating-point arithmetic internally, and yields a gp of about 10^{-13} where there is no cancellation.) The output after each iterative cycle consists of the four vectors mentioned above: it contains the information that will in general yield on inspection the required distinct real zeros, and the moduli of complex zeros (as well as the equal/opposite real pairs). Higher multiplicities may not be revealed at once. However, their occurrence should be rare, and in any case it is not difficult to extract them once their presence is suspected. (See [11] for further details, including a more detailed presentation of numerical results.)

The first example which we present here is a simple one used by Matsui and Iri [45]. It is a polynomial of degree four, given by

$$p(x) = x^4 - a_3x^3 + a_2x^2 - a_1x + a_0$$

where

a_3 = 10.43857 59302 0614 $\quad$ a_1 = 69.60408 57054 5396

a_2 = 40.58740 56758 7410 $\quad$ a_0 = 44.36715 61490 6059 .

This is a fairly ill conditioned problem: to 14D the smallest zero is 2, the largest is 3, and the other two are the rounded values of e and √(7.4), namely

2.71828 18284 5904 and 2.72029 41017 5647

respectively.

It transpires that after 15 cycles the α–values in the first output vector have settled to within a unit or two in the 14th decimal place. We give below the approximations to the second largest zero for iterations 14 through 17.

Iteration	Approximation
14	2.72029 41026 3166
15	2.72029 41017 3066
16	2.72029 41017 3064
17	2.72029 41017 3065

The approximation is remarkably stable - after 50 cycles we obtain the value 2.72029 41017 3055 - but it is clear that stability is no guarantee of accuracy. In this case the ill-conditioning overlaying the limited precision of the arithmetic is sufficient to cause the loss of the last four figures given. (The other three zeros are similarly indicated, but with rather greater accuracy: the second zero is the worst conditioned).

We may note that in this example, which is by no means an extreme one, our computation has dealt with numbers that are very large by conventional standards. After r cycles the coefficient $a_3^{(r)}$ exceeds 3^{2^r}, since all the zeros are positive. With n=15 this is about 2×10^{15634}. (The value with n=50, where we examined meaningful results, exceeds $10^{10^{14}}$).

Our second example, also treated at greater length in [11], is a polynomial of degree 14 with zeros of varying character. It is given by

k	$\alpha_k^{(10)}$	$p(\pm\alpha_k^{(10)})$		$\beta_k^{(10)}$	$p(\pm\beta_k^{(10)})$	
1	2.31237647787152	[4.01576326389222]	1			
		[3.90307296757064]	-1			
2	1.95977873402839	[3.91434499764033]	1	2.12878985491760	[3.97360414323500]	1
		[3.927902060761220]	1		[-3.96974834023348]	1
3	2.00016078777021	[3.93137606887548]	1	1.97986680777004	[3.92311703703905]	1
		[-3.67512474462557]	1		[3.90713538488153]	1
4	2.04088266744541	[3.94641513170820]	1	2.02041913569046	[3.93909242828714]	1
		[-3.93444724555156]	1		[-3.91089253762885]	1
5	1.66779521821806	[2.33112195692192]	-1	1.84493207292560	[3.84646463238930]	1
		[3.95029146536699]	1		[3.9558596726904]	1
6	1.66553887877075	[2.34854857504526]	-1	1.66666666666437	[4.06921586867569]	-1
		[3.95004512185302]	1		[3.95016880628101]	1
7	1.41517116999438	[2.66976438637781]	1	1.53525978379593	[3.62559718897801]	1
		[3.69800654487441]	1		[3.92587594535703]	1
8	1.41325660274334	[-2.66799164372710]	1	1.41421356237542	[4.05673599717207]	-1
		[-3.69772407136798]	1		[3.92261804103378]	-1
9	1.20126536676056	[-4.08922902562647]	-1	1.30295671885956	[-3.59769879290383]	1
		[-3.92213072487545]	1		[-3.90946181799193]	1
10	1.00000000000000	[3.70533444232518]	1	1.09602252110099	[3.66550681884250]	1
		[-3.92358170908464]	1		[-3.92504514072812]	1
11	1.00000000000000	[3.70533444232518]	1	1.00000000000000	[3.70533444232518]	1
		[-3.92358170908464]	1		[-3.92358170908464]	1
12	0.74999999999934	[-4.06003154461810]	-1	0.86602540378406	[3.70084734296961]	1
		[-3.91023431656315]	1		[-3.91769717929753]	1
13	0.70758558490781	[-3.66204731725957]	1	0.72848417188048	[-3.60370966135972]	1
		[-3.90711796420531]	1		[-3.90867534223799]	1
14	0.70662830145921	[-3.66384329068752]	1	0.70710678118685	[-3.66295147349345]	1
		[-3.90704566387311]	1		[-3.90708181205614]	1

Table 6.5.1: Output from the polynomial of degree 14. [±x] ±1 denotes $(\pm\phi(x))^{\pm 1}$.

$$p(x) = a_{14}x^{14} + a_{13}x^{13} + a_{12}x^{12} \dots + a_1 x + a_o$$

where

a_{14} =	+684	a_9 =	−34563	a_4 =	−45782
a_{13} =	−1278	a_8 =	+66896	a_3 =	+103410
a_{12} =	−3570	a_7 =	−797	a_2 =	−65400
a_{11} =	+13655	a_6 =	−34629	a_1 =	+58000
a_{10} =	−9495	a_5 =	−15151	a_0 =	−3000

In Table 6.5.1 we give the complete output array for iteration 10. Again, the last few figures of the α_k and β_k are subject to error. (Errors in the function values p may be much larger, but this is of little consequence.) In order to deduce the required zeros we should compare this array with the corresponding array at iteration 11, but we do not reproduce this here. In any case, Table 6.5.1 itself indicates real zeros at $-\alpha_1$, α_9 and α_{12}; their small residuals are conspicuous. When we compare the two arrays, we shall see that there is a pair of real zeros with modulus close to 5/3; $\beta_6^{(10)}$ is close to $\beta_6^{(11)}$, even though $\alpha_5^{(10)}$ and $\alpha_6^{(10)}$ are far from $\alpha_5^{(11)}$ and $\alpha_6^{(11)}$, and the fact that this pair is real is betrayed by the small residual $p(\beta_6^{(10)})$. Next we observe a pair with modulus 1.41421 . . . ; again this is indicated by the closeness of $\beta_8^{(10)}$ and $\beta_8^{(11)}$, but this time the residuals $p(\pm\beta_8^{(10)})$ show that the two are equal and opposite. Similar agreement between the iterations shows another pair at β_{11}, but the absence of a small residual suggests that this is a complex pair, and another complex pair is indicated by β_{14}.

This simple investigation leaves the 2nd, 3rd and 4th zeros undetermined. If we perform several more cycles, we shall find that there are three zeros with modulus 2, with at least one real zero at −2. A brief discussion of how such an example might be unravelled may be found in [11]; we pursue it no further here.

6.6 Conclusion

We have presented some of the results obtained from the application of an sℓi implementation to standard problems in numerical analysis which tend to generate large numbers. These results demonstrate that the sℓi system does indeed free the user from concern about overflow and underflow; they also confirm that the associated measure of generalized precision is both useful and convenient.

Our examples suffice to make a case for the implementation of sℓi arithmetic in those extreme circumstances in which it may be difficult to scale a problem so as to avoid very large and very small numbers. The case of a more general use of sℓi must depend to some extent upon future success in achieving fast implementations.

7: GENERALIZED EXPONENTIALS AND LOGARITHMS; SURFACE FITTING; CONCLUSIONS

F.W.J. Olver

7.1 Generalized exponentials and logarithms

Let us return to the definitions of the generalized exponential and logarithmic functions. For nonnegative arguments we have

$$\phi(x) = x, \quad 0 \leqslant x < 1 \quad ; \quad \phi(x) = e^{\phi(x-1)}, \quad x \geqslant 1 \quad ;$$
$$\psi(x) = x, \quad 0 \leqslant x < 1 \; ; \quad \psi(x) = 1 + \psi(\ell n\ x), \quad x \geqslant 1.$$

As we saw in Lecture 3, both $\phi(x)$ and $\psi(x)$ are C^1 functions when $x \in [0,\infty)$. An immediate benefit of this smoothness is that the ℓi and sℓi systems are free from the annoying phenomenon of wobbling precision associated with the floating-point system [12], p.7 and [49], Section 5. However, for reasons given in the next section it is likely that we shall be needing even smoother mapping functions of the same type. To construct these, we adhere to the recurrence relations

$$\phi(x) = e^{\phi(x-1)}, \quad \psi(x) = 1 + \psi(\ell n\ x), \quad x \geqslant 1,$$

and the initial conditions $\phi(0) = \psi(0) = 0$, but we modify the definitions when $x \in (0,1)$. We must maintain the inverse relationship of ϕ and ψ, of course, which means that each function has to be increasing on (0,1).

Can we set for example,

$$\phi(x) = a_1x + a_2x^2 + \dots + a_{N+1}x^{N+1}, \quad 0 \leqslant x \leqslant 1,$$

and arrive at a generalized exponential function that is in $C^N[0,\infty)$? Different forms of $\phi(x)$ apply on opposite sides of $x = 1$, namely

$$a_1x + a_2x^2 + \dots + a_{N+1}x^{N+1} \text{ and } \exp\{a_1(x-1) + a_2(x-1)^2 + \dots + a_{N+1}(x-1)^{N+1}\}.$$

We have only to choose the coefficients a_n in such a way that ϕ and its first N derivatives are continuous at $x = 1$. In the case $N = 2$ we find that

$$\phi(x) = ax - a^2x^2/4 + a^2x^3/6, \qquad 0 \leqslant x \leqslant 1,$$

with $a = 6 - 2\surd 6 = 1.10\ \dots$. Also, since $\phi'(x) > 0$ on [0,1] the corresponding $\psi(x)$

is definable as the inverse function.

In principle, the same method can be used to construct generalized exponentials and logarithms that are in $C^3[0,\infty)$, $C^4[0,\infty)$, . . . , but the equations to be solved for the coefficients a_n are nonlinear and increasingly complicated. Nevertheless, the process suggests that generalized exponentials and logarithms of arbitrarily high smoothness might exist. Indeed, we are led to wonder whether there are generalized exponentials and logarithms that are in $C^\infty[0,\infty)$. Somewhat remarkably this problem had already been resolved by the German analyst H. Kneser in a paper published in 1950 [39]. This was discovered by Clenshaw in the course of a chance conversation with P.L.Walker whilst Walker was a member of the Mathematics Department at Lancaster University.

Kneser actually proved a stronger result. He showed that there exists a generalized logarithm that is *analytic* on a complex domain that includes the interval $[0,\infty)$. This was achieved by application of a fixed-point theorem due to G.Koenigs. The final result is purely existential: no formula or algorithm is supplied for constructing the analytic generalized logarithm Kneser also indicated that the solution to his problem is not unique, and this is easily verified directly. Given one analytic generalized logarithm $\psi(x)$, another is furnished by

$$\psi_1(x) = \psi(x) + kF(\psi(x))$$

where F is any periodic function with period unity, and k is a constant which secures the increasing character of ψ_1. (A simple example is $\psi_1 = \psi + k\sin 2\pi\psi$ with $-1<2\pi k<1$).

Given the existence of an infinite set of generalized logarithms, is there a distinguished member that possesses an important property? If so, how can we compute this function? We do not know the answers to these questions. Fortunately, from our standpoint it will suffice to be able to compute one of the analytic generalized logarithms, and of course its inverse, but even this more limited requirement is not easy to fulfil.

Walker considered the problem of computing analytic generalized logarithms in 1982, before he was aware of possible applications in computer arithmetic. If the Maclaurin expansion of $\psi(x)$ is denoted by

$$\psi(x) = c_1x + c_2x^2 + c_3x^3 + \ldots, \tag{7.1.1}$$

then formally

$$\psi(e^x) = \sum_{n=1}^{\infty} c_n e^{nx} = \sum_{n=1}^{\infty} \left[c_n \sum_{m=0}^{\infty} \frac{(nx)^m}{m!} \right].$$

Substituting these series expansions in the fundamental recurrence relation

$$\psi(e^x) = \psi(x) + 1, \tag{7.1.2}$$

and equating coefficients, we immediately arrive at an infinite set of linear algebraic equations for the coefficients c_j, given by

$$c_1 + c_2 + c_3 + \ldots = 1,$$

$$\frac{c_1}{1!} + \frac{2c_2}{1!} + \frac{3c_3}{1!} + \ldots = c_1$$

$$\frac{1^2 c_1}{2!} + \frac{2^2 c_2}{2!} + \frac{3^2 c_3}{2!} + \ldots = c_2,$$

$$\cdots\cdots\cdots\cdots$$

Walker approached this problem numerically by setting $c_n = 0$ for $n \geqslant N + 1$, for various values of N. Numerical values of the nonzero coefficients for the case $N = 40$ are supplied in Table 7.1.1. Subsequently Walker computed $\psi(x)$ and $\psi(e^x)$ by summing the expansion (7.1.1), and then compared the two sides of equation (7.1.2). Extracts from these results are given in Table 7.1.2, again for the case $N = 40$. It is evident that satisfactory agreement was obtained.

If the numerical values of the coefficients calculated by Walker's process were to settle down as N increases, then there would be a strong indication that we are converging on a distinguished generalized logarithm, for example, the one with the most rapidly convergent Maclaurin expansion. However, it is difficult to confirm this by computation. The infinite system is singular (there is a multiplicity of solutions) and this is reflected in increasing ill-conditioning of the finite system as N increases. In an attempt to improve convergence properties, Clenshaw modified Walker's procedure by assuming an expansion for $\psi(x)$ in series of Chebyshev polynomials of the form

$$\psi(x) = \tfrac{1}{2}b_0 + b_1T_1^*(x) + b_2T_2^*(x) + \ldots, \quad 0 \leqslant x \leqslant 1,$$

in place of (7.1.1). Here $T_n^*(x) = \cos\{n\cos^{-1}(2x-1)\}$. His procedure is described in [6]. Again, unfortunately, there is no indication of convergence as more and more terms in the series are taken into account.

It would appear that we have a profound open analytical problem, not unlike the classical problem of interpolating the factorial function. It will be recalled that the defining equations for this problem are given by

$$f(x+1) = xf(x), \qquad f(1) = 1. \tag{7.1.3}$$

Table 7.1.1. Maclaurin coefficients for an analytic generalized logarithm

n	c_n	n	c_n
1	0.91594 47685 85476	21	0.00001 18526 17128
2	0.24934 81945 92665	22	−0.00004 89246 56622
3	−0.11046 07876 84269	23	−0.00001 57289 01181
4	−0.09391 25936 15892	24	0.00001 97005 35453
5	0.01001 02605 99740	25	0.00001 04962 58214
6	0.03585 93980 77603	26	−0.00000 77867 65775
7	0.00653 74106 97402	27	−0.00000 56204 47841
8	−0.01227 77555 14280	28	0.00000 33701 57518
9	−0.00632 86813 84804	29	0.00000 25680 67807
10	0.00327 68514 84459	30	−0.00000 16714 69443
11	0.00370 76840 29147	31	−0.00000 10528 55835
12	−0.00031 68461 28022	32	0.00000 11049 81374
13	−0.00173 70955 69492	33	0.00000 00829 58629
14	−0.00037 58262 49465	34	−0.00000 05932 50588
15	0.00067 28599 17915	35	0.00000 04255 76148
16	0.00036 59520 80618	36	−0.00000 01678 16704
17	−0.00020 29937 20169	37	0.00000 00418 95426
18	−0.00022 39036 47862	38	−0.00000 00066 50693
19	0.00003 33084 23216	39	0.00000 00006 18375
20	0.00011 17042 00426	40	−0.00000 00000 25802

Table 7.1.2. Comparison of results from Maclaurin expansions

x	$\psi(x) + 1$	$\psi(e^x)$
−0.7	0.49793 18317 67538	0.49793 18317 68673
−0.6	0.55246 36057 13683	0.55246 36057 13729
−0.5	0.61246 50225 53202	0.61246 50225 53202
−0.4	0.67821 06014 02583	0.67821 06014 02586
−0.3	0.74987 93749 21122	0.74987 93749 21111
−0.2	0.82751 73803 26420	0.82751 73803 26420
−0.1	0.91100 00096 02596	0.91100 00096 02592
0.0	1.00000 00000 00000	0.99999 99999 99979
0.1	1.09396 82432 4443	1.09396 82432 4444
0.2	1.19213 44827 5065	1.19213 44827 5065
0.3	1.29353 26263 3353	1.29353 26263 3353
0.4	1.39705 08384 0543	1.39705 08384 0541
0.5	1.50150 08959 3100	1.50150 08959 3080
0.6	1.60569 64513 3134	1.60569 64514 7089

A solution that is analytic on a complex domain that includes the open interval $(0,\infty)$ was known to Euler:

$$f(x) = \int_0^\infty e^{-t} t^{x-1}\, dt, \qquad \text{Re } x > 0. \tag{7.1.4}$$

But this is only one member of an infinite set of solutions of equations (7.1.3), and it is not the only solution that is analytic on $(0,\infty)$. Almost 200 years elapsed before analysts arrived at an acceptable distinguishing property, namely that $f(x) > 0$ when $x > 0$ and the graph of $f(x)$ be logarithmically convex [14]. Quite possibly many more years may pass before acceptable distinguishing properties of the generalized logarithmic and exponential functions emerge. Indeed, in an important way we are worse off than in the interpolation problem for the factorial function. We do not have an explicit formula such as Euler's integral representation (7.1.4) for a prime candidate, nor are we ever likely to be able to find one. Again, it is fortunate that although this problem is pertinent to level-index arithmetic, its solution is not crucial for a successful prosecution of the whole project.

7.2 Surface fitting

In Lecture 3 we described algorithms for performing addition, subtraction, multiplication and division in ℓi and $s\ell i$ arithmetics. In Lecture 5 we gave an account of efforts we have been making to achieve fast implementations of these algorithms in hardware, by use of devices such as CORDIC, Carry Save Adders, Carry Propagate Adders and double numbers. Even with the use of these artifices and the maximum amount of parallelism, however, it is unlikely that we shall be able to achieve execution speeds better than 1/10th of those of the corresponding floating-point operations. This would be quite satisfactory for many purposes, especially exploratory scientific work in which ease of programming would usually outweigh a loss of execution speed. Moreover, because only a fraction of processing time is actually spent on arithmetic operations, the overall loss of speed might amount to a factor of 3 or 4 rather than 10. (Again, see Lecture 5.)

There is an alternative. The ℓi operation of addition, for example, consists in finding the value of z that satisfies the equation

$$\phi(z) = \phi(x) + \phi(y)$$

for given values of x and y such that $0 \leqslant y \leqslant x$. This equation defines a surface $z = z(x,y)$, and it is possible - at least in theory - to fit this surface by a double Chebyshev-series expansion of the form

$$z = \sum_{m,n} a_{mn}\, T_m^*(\rho x)\, T_n^*(y/x),$$

where ρ is a suitable scaling factor. The smoother the surface, the fewer the number of terms that will be needed. This is where the C^∞ generalized exponential functions described in Section 7.1 assume importance.

Initial attempts on these lines were made in 1984 and reported briefly in [6]. These attempts were unsuccessful, however, owing to the rapidly changing nature of the surface near the plane $y = x$, both for addition and subtraction. At present other ways of solving the surface-fitting problem are being explored.

In one method, the surface $z = z(x,y)$ is divided into several parts. One part, for example, comprises those points for which $z = x$ to the working precision. Another part is the "boundary layer" near the plane $y = x$. Each part is fitted with a double Chebyshev-series expansion that does not require a prohibitively large number of terms. To compute the sum of such an expansion we may either use a two-dimensional version of the Clenshaw algorithm [5], or we may rearrange the expansion in polynomial form and use two-dimensional nested multiplication. Either way admits extensive use of parallelism, including the use of double numbers and Carry Save Adders as in Lecture 5.

A second method is to fit the surface $z = z(x,y)$ using B-splines with curved knot lines. The use of curved knot lines has been developed by M.G.Cox and J.G.Hayes at the National Physical Laboratory; see, for example, [13,29], and appears to be tailor-made for the present problem. If the fitting by B-splines proves to be successful, then the splines can be computed rapidly with the aid, again, of double numbers and Carry Save Adders.

Quite possibly, with either of these proposed methods only one Carry Propagate Adder will be needed, in which event the speed ratio to floating-point arithmetic may approach 1/2. The ability to use simpler programs would offset much - or probably all - of this loss, thereby rendering overall running times for algorithms executed in ℓi or $s\ell i$ arithmetic competitive with those for the same algorithms executed in floating-point arithmetic.

7.3 Conclusions

Let us suppose that ongoing research on the ℓi and $s\ell i$ systems eventually proves to be fully successful, and at some future time computers become available equipped with hardware facilities for carrying out arithmetic operations in these systems. How would the average user be affected? In the first place the average user need not be familiar with the new systems, any more than he or she needs to be familiar with CORDIC and other devices in daily use for generating elementary mathematical functions on hand-held calculators. The new arithmetics are designed primarily for internal use in computers. Input and output in the more familiar fixed-point and floating-point formats would normally be used, unless of course the numbers involved are too large to be represented this way.

Next, compared with floating-point hardware there might be a small gain or loss of speed, and a small gain or loss of precision. Error analysis would also be changed, but this is not a topic that concerns the average user. (Readers who are interested in error analysis in ℓi arithmetic may consult [51]).

The main changes will be the simplification of the tasks of writing and debugging programs, especially the construction of robust software. Portability of programs will also be eased, because of the disappearance of "machine constants" associated with bounds on representation ranges in the floating-point system. Some examples of simplified programs have been provided in earlier lectures. Here is another example, which is particularly interesting in view of the efforts that have been made to produce robust floating-point software.

The Euclidean norm of a vector is defined by

$$\|X\| = (X_1^2 + X_2^2 + \ldots + X_n^2)^{1/2} .$$

The direct algorithm for its evaluation occupies four lines:

$$\begin{array}{l} Y = 0 \\ \text{for } i = 1 \text{ through } n \\ \quad Y \leftarrow Y + X_i^2 \\ \|X\| = Y^{1/2} . \end{array}$$

However, when executed in floating-point arithmetic, this algorithm fails when any of the squares X_i^2 or their partial sums, overflows, even though the final answer may be well within the range of machine-representable numbers. Failure can also stem from underflow, including gradual underflow.

To circumvent these problems careful rescalings are needed. In 1978 J. L.Blue devoted a nine-page paper to this topic [4]. In the introduction he observed:

"A set of Fortran subprograms for performing the basic operations of linear algebra . . . should include a subprogram to find the Euclidean norm of an n-vector . . . . Such a subprogram should be accurate and efficient, and should avoid all overflows and underflows.

The problem appears to be much easier than it is. Preliminary versions of the subprogram, by several authors, failed at least two of these requirements."

```
if n = 0, set ||x|| = 0 and return.
if n < 0, set an error flag and stop.
if n > N, set an error flag and stop.
a_sml = 0; a_med = 0; a_big = 0
for i = 1 through n
    if |x_i| > B, a_big ← a_big + (x_i/S)^2
    else if |x_i| < b, a_sml ← a_sml + (x_i/s)^2
    else a_med ← a_med + x_i^2
if a_big is nonzero
    if a_big^½ > R/S, ||x|| > R and overflow would occur.
                      Set ||x|| = R, set an error flag, and return.
    if a_med is nonzero
        y_min = min(a_med^½, S a_big^½)
        y_max = max(a_med^½, S a_big^½)
    else set ||x|| = S a_big^½ and return
else if a_sml is nonzero
    if a_med is nonzero
        y_min = min(a_med^½, s a_sml^½)
        y_max = max(a_med^½, s a_sml^½)
    else set ||x|| = s a_sml^½ and return
else set ||x|| = a_med^½ and return.
if y_min < ε^½ y_max, set ||x|| = y_max.
else set ||x|| = y_max (1 + (y_min/y_max)^2)^½
```

Table 7.3.1 Blue's Euclidean norm program

Blue's algorithm is reprinted in Table 7.3.1 (with the kind permission of ACM and the author). The part that corresponds to the direct algorithm given above occupies the last 20 lines and requires the setting of six machine-dependent constants. In contrast, with the sℓi system we can use the four line algorithm, no machine-dependent constants need be present and fewer arithmetic operations (especially divisions) are required. (We note that the ACRITH floating-point package [36], [43] could also be used in conjunction with the direct algorithm. However, this package would not succeed with the problem of the p-norm described below.)

Given the choice, is there any reasonable doubt which route a programmer would prefer?

An even more formidable problem for the floating-point system would be the construction of a robust program for evaluating the p-norm

$$\|X\|_p = (|X_1|^p + |X_2|^p + \ldots + |X_n|^p)^{1/p}$$

for high values of p. A generalization of Blue's program would lead to further complications, whereas the obvious modification of the direct algorithm for the 2-norm given above would present no problems for the sli system.

Is it likely that one form of computer arithmetic will eventually supersede all others? Probably not, nor should any attempt be made to force the issue in this way. We envision that eventually a variety of computer arithmetics will become available and programmers will be able to declare at the outset of their programs which arithmetic is to be used, just as declarations are now made concerning the choice of single or double precision and real or complex numbers.

For many purposes, floating-point may remain the preferred system. These include vast number-crunching operations in the solution of partial differential equations for which even a slight reduction in overall running time will save considerable expense. (Though here, again, the fixed-point system with appropriate programmed scalings, might prove slightly faster still). Or we might wish to take advantage of the fact that with double-length accumulators many floating-point subtractions can be carried out exactly. This might be useful, for example, in the iterative refinement of solutions of systems of linear algebraic equations having exact coefficients. Nevertheless, it is conceivable that the majority of scientific computations in the 21st Century will be performed in arithmetics other than the floating-point system.

REFERENCES

[1] E.H.Bareiss, *Resultant procedure and the mechanization of the Graeffe process*, J.Assoc. Comput. Mach., 7 (1960), pp. 346-386.

[2] J.L.Barlow and E.H.Bareiss, *On roundoff error distributions in floating-point and logarithmic arithmetic*, Computing, 34 (1985), pp. 325-341.

[3] F.Benford, *The law of anomalous numbers*, Proc. Am. Phil. Soc., 78 (1938), pp. 551-572.

[4] J.L.Blue, *A portable Fortran program to find the Euclidean norm of a vector*, ACM Trans. Math. Software, 4 (1978), pp. 15-23.

[5] C.W.Clenshaw, *A note on the summation of Chebyshev series*, Math. Comp., 9 (1955), pp. 118-120.

[6] C.W.Clenshaw, D.W.Lozier, F.W.J.Olver and P.R.Turner, *Generalized exponential and logarithmic functions*, Comput. Math. Appl., 12B (1986), pp. 1091-1101.

[7] C.W.Clenshaw and F.W.J.Olver, *An unrestricted algorithm for the exponential function*, SIAM J. Numer. Anal., 17 (1980), pp. 310-331.

[8] C.W.Clenshaw and F.W.J.Olver, *Beyond floating-point*, J. Assoc. Comput. Mach., 31 (1984), pp. 319-328.

[9] C.W.Clenshaw and F.W.J.Olver, *Level-index arithmetic operations*, SIAM J. Numer. Anal., 24 (1987) pp. 470-485.

[10] C.W.Clenshaw and P.R.Turner, *The symmetric level-index system*, IMA J. Numer. Anal. [In press].

[11] C.W.Clenshaw and P.R.Turner, *Root-squaring using level-index arithmetic*, [Manuscript].

[12] W.J.Cody Jr., and W.Waite, *Software Manual for the Elementary Functions*, Prentice-Hall, Englewood Cliffs, N.J., 1980.

[13] M.G.Cox, *The numerical evaluation of B-splines*, J. Inst. Maths Applics., 10 (1972), pp. 134-149.

[14] P.J.Davis, *Leonhard Euler's integral: a historical profile of the Gamma function* , Amer. Math. Monthly, 66 (1959), pp. 849-869.

[15] J.W.Demmel, *Underflow and the reliability of numerical software*, SIAM J. Sci. Statist. Comp., 5 (1984), pp. 887-919.

[16] J.W.Demmel, *On error analysis in arithmetic with varying relative precision*, Proc. 8th Symposium on Computer Arithmetic, M.J. Irwin and R. Stefanelli, eds., IEEE Computer Society Press, Washington, D.C., 1987, pp. 148-152.

[17] A.Feldstein and R.Goodman, *Loss of significance in floating-point subtraction and addition*, IEEE Trans. Comp., 31 (1982), pp. 328-335.

[18] A.Feldstein and J.F.Traub, *Asymptotic behavior of vector recurrences with applications*, Math. Comp., 31 (1977), pp. 180-192.

[19] A.Feldstein and P.R.Turner, *Overflow, underflow, and severe loss of significance in floating-point addition and subtraction*, IMA J. Numer. Anal., 6 (1986), pp. 241-251.

[20] B.J.Flehinger, *On the probability that a random number has leading digit A,* Amer. Math. Monthly, 73 (1966), pp. 1056-1061.

[21] R.Goodman, *Some models of error in floating-point multiplication.* Computing, 27 (1981), pp. 227-236.

[22] R.Goodman and A.Feldstein, *Effect of guard digits and normalisation options on floating-point multiplication*, Computing, 18 (1977), pp. 93-106.

[23] R.Goodman, A.Feldstein and J.Bustoz, *Relative error in floating-point multiplication*, Computing, 35 (1985), pp. 137-139.

[24] J.B.Gosling, *Design of arithmetic units for digital computers*, MacMillan, London, 1980.

[25] H.Hamada, *URR: Universal representation of real numbers*, New Generation Computing, OHM-Sha, Springer-Verlag 1 (1983), pp. 205-209.

[26] H.Hamada, *A new real number representation and its operation*, Proc. 8th Symposium on Computer Arithmetic, M.J.Irwin and R.Stefanelli, eds., IEEE Computer Society Press, Washington D.C., 1987, pp. 153-157.

[27] J.M.Hammersley, *Probability and arithmetic in science*, IMA Bulletin, 21 (1985), pp. 114-120.

[28] R.W.Hamming, *On the distribution of numbers*, Bell Systems Tech. J., 49 (1970), pp. 1609-1625.

[29] J.G.Hayes, *Curved knot lines and surfaces with ruled segments*, Numerical Analysis: Proceedings of the 9th Biennial Conference, Dundee 1981, G.A.Watson, ed., Lecture Notes in Mathematics No. 912, Springer-Verlag, Berlin, 1982, pp. 140-156.

[30] P.Henrici, *Applied and Computational Complex Analysis*, Vol. 3, J. Wiley and Sons, New York, 1986.

[31] F.B.Hildebrand, *Introduction to Numerical Analysis*, McGraw-Hill, New York, 1956.

[32] T.E.Hull, *Precision control, exception handling and a choice of numerical algorithms*, Lecture Notes in Mathematics No. 912, Numerical Analysis, G.A.Watson ed., Springer-Verlag, Berlin, pp. 169-178, 1982.

[33] T.E.Hull and M.S.Cohen, *Toward an ideal computer arithmetic*, Proc. 8th Symposium on Computer Arithmetic, M.J.Irwin and R.Stefanelli, eds. IEEE Computer Society Press, Washington, D.C., 1987, pp. 131-138.

[34] T.E.Hull et al, *Numerical Turing*, ACM SIGNUM Newsletters, 20, No. 3 (1985), pp. 26-32.

[35] K.Hwang and F.A.Briggs, *Computer Architecture and Parallel Processing*, McGraw-Hill, New York, 1984.

[36] IBM, *High-accuracy arithmetic*, General Information Manual, GC33-6163-1, second edition, IBM Corporation, Mechanicsburg, PA, 1984.

[37] IEEE Standard 754, *Binary floating-point arithmetic*, The Institute of Electrical and Electronic Engineers, New York, 1985.

[38] M.J.Irwin and R.Stefanelli, eds., Proc. 8th Symposium on Computer Arithmetic, IEEE Computer Society, Washington D.C., 1987.

[39] H.Kneser, *Reelle analytische Lösungen der Gleichung* $\phi(\phi(x)) = e^x$ *und verwandte Funktionalgleichungen*, J.Reine Angew. Math., 187 (1950) pp. 56-67.

[40] D.E.Knuth, *The art of computer programming; Volume 2, Semi numerical algorithms*, Addison-Wesley, Reading, Mass., 1969.

[41] P.Kornerup and D.W.Matula, *Finite precision lexicographic continued fraction number systems*, Proc. 7th Symposium on Computer Arithmetic, K.Hwang, ed., IEEE Computer Society Press, Washington, D.C., 1985, pp. 207-214.

[42] P.Kornerup and D.W.Matula, *A bit-serial arithmetic unit for rational arithmetic*, Proc. 8th Symposium on Computer Arithmetic., M.J.Irwin and R.Stenanelli, eds., IEEE Computer Society Press, Washington, D.C., 1987, pp. 204-211.

[43] U.W.Kulisch and W.L.Miranker, *The arithmetic of the digital computer: a new approach*, SIAM Review, 28 (1986), pp. 1-40.

[44] D.W.Lozier and F.W.J.Olver, *Closure and precision in level-index arithmetic*, [Manuscript].

[45] S.Matsui and M.Iri, *An overflow/underflow-free floating-point representation of numbers*, J.Inform. Process., 4 (1981), pp. 123-133.

[46] D.W.Matula and P.Kornerup, *Foundations of finite precision rational arithmetic*, Foundations of Numerical Computation (Computer-Orientated Numerical Analysis), G.Alefeld and R.D.Grigorieff, eds., Computing, Suppl., 2 (1980), pp. 85-111.

[47] F.W.J.Olver, *The evaluation of zeros of high-degree polynomials*, Phil. Trans. Royal Soc. A, 244 (1952), pp. 385-415.

[48] F.W.J.Olver, *A new approach to error arithmetic*, SIAM J. Numer. Anal. 15 (1978), pp. 368-393.

[49] F.W.J.Olver, *Further developments of rp and ap error analysis*, IMA J. Numer. Anal., 2 (1982), pp. 249-274.

[50] F.W.J.Olver, *A closed computer arithmetic*, Proc. 8th Symposium on Computer Arithmetic, M.J.Irwin and R.Stefanelli, eds., IEEE Computer Society Press, Washington, D.C., 1987, pp. 139-143.

[51] F.W.J.Olver, *Roundings errors in algebraic processes - in level-index arithmetic*, Proceedings of a Conference on Reliable Numerical Computation (in memoriam of J.H.Wilkinson), M.G.Cox and S.Hammarling, eds., Oxford University Press. [In press].

[52] F.W.J.Olver and P.R.Turner, *Implementation of level-index arithmetic using partial table look-up*, Proc. 8th Symposium on Computer Arithmetic, M.J.Irwin and R.Stefanelli, eds., IEEE Computer Society Press, Washington, D.C., 1987, pp. 144-147.

[53] R.S.Pinkham, *On the distribution of first significant digits*, Ann. Math. Stat., 32 (1981), pp. 1223-1230.

[54] R.A.Raimi, *On the distribution of first significant figures*, Amer. Math. Monthly, 76 (1969), pp. 342-348.

[55] A.Ralston, *A first course in numerical analysis*, McGraw-Hill, New York, 1965.

[56] C.W.Schelin, *Calculator function approximation*, Amer. Math. Monthly 90 (1983), pp. 317-325.

[57] P.H.Sterbenz, *Floating-point Computation*, Prentice-Hall, Englewood Cliffs, N.J., 1974.

[58] D.W.Sweeney, *An analysis of floating-point addition*, IBM Systems J., 4(1965), pp. 31-42.

[59] P.R.Turner, *The distribution of leading significant digits*, IMA J. Numer. Anal. 2 (1982), pp. 407-412.

[60] P.R.Turner, *Further revelations on l.s.d.*, IMA J. Numer. Anal. 4 (1984), pp. 225-231.

[61] P.R.Turner, *Towards a fast implementation of level-index arithmetic*, Bull. Inst. Math. Appl., 22 (1986), pp. 188-191.

[62] P.R.Turner, *The use of level-index arithmetic to avoid overflow/underflow in floating-point computation*, University of Lancaster, Tech. Rep., 1986.

[63] J.Volder, *The CORDIC computing technique*, IRE Trans. Computers EC8 (1959), pp. 330-334.

[64] J.S.Walther, *A unified algorithm for elementary functions*, AFIPS Conference Proc., 38 (1971), pp. 379-395.

[65] S.Waser and M.J.Flynn, *Introduction to Arithmetic for Digital Systems Designers*, Holt, Rinehart and Winston, New York, 1982.

SOME ASPECTS OF FLOATING POINT COMPUTATION

Alan Feldstein
Department of Mathematics
Arizona State University
Tempe, AZ 85287, USA

and

Richard H. Goodman
Department of Mathematics and Computer Science
University of Miami
Coral Gables, FL 33124, USA

ABSTRACT: The floating point format is widely used in computing because it allows for the representation of numbers with greatly varying magnitude. This paper introduces some general aspects of floating point representation and includes a discussion of the implications of balanced and unbalanced exponent ranges. A statistical approach is presented in order to analyze the error in floating point computations. The shortening of computer words is discussed and is applied to study aspects of the fraction error and the relative error.

1. FLOATING POINT NUMBERS

A. INTRODUCTION

In base β, a *normalized floating point number* X is represented as $X = \beta^E \cdot F$. The *exponent* E and the *fraction (mantissa)* F are given by

$$E = \left[\log_\beta X\right] + 1 \quad \text{and} \quad \beta F = \beta^{((\log_\beta X))}, \tag{1.1}$$

where any number t is represented as

$$t = [t] + ((t)),$$

the sum of its integer part plus its fraction part, respectively. The *normalization* requirement is $\beta^{-1} \leq F < 1$.

Let βit denote a base $\underline{\beta}$ dig$\underline{it}$. The fraction, F, of a mathematically real, floating point number may be represented by infinitely many βits. On the other hand, a computer has only a finite number (say, M) of βits for representing the fraction of a floating point number. Sterbenz [74] gives a definitive discussion of floating point computation. This manuscript will, where possible, follow (or sometimes extend) Sterbenz's notation.

The set of computational floating point numbers in base β with M βits of precision is denoted by $S(\beta, M)$ and is defined to be the union of the number zero together with all normalized floating point numbers of the form $X = \pm\beta^E \cdot F$, where the fraction F can be expressed in base β using at most M βits. Usually the M βits of the finite fraction F are denoted as $f_1, f_2, \cdots, f_M$. Thus

$$\begin{cases} F \equiv f_1\beta^{-1} + f_2\beta^{-2} + \cdots + f_M\beta^{-M}, \text{ where} \\ 1 \leq f_1 \leq \beta - 1, \text{ and } 0 \leq f_i \leq \beta - 1 \text{ for } 2 \leq i \leq M. \end{cases} \tag{1.2}$$

Note from Eq. (1.2) that if $\beta = 2$, then $f_1 = 1$. So, in binary arithmetic it is possible (and typical) to obtain an extra binary bit by suppressing the storage of f_1. This procedure, called using a *hidden (or implicit) leading bit*, means that a bit is not needed and not used for storing f_1, because f_1 always equals one. The leading hidden bit situation in binary is denoted by $\beta = 2'$.

B. FUNDAMENTAL OPERATIONS

The four fundamental floating point arithmetic operations are denoted by $\oplus$, $\ominus$, $*$, and $\div$. When a floating point number (or fraction) is shortened to M βits by one of the two commonly used rounding rules (*chopping or symmetric rounding*), this is denoted by "c" or by "R", respectively. Floating point arithmetic with numbers in $S(\beta, M)$ is then denoted either by $FP(\beta, M, c)$ or by $FP(\beta, M, R)$.

The arithmetic registers on some computers allot a full double length word ($2M$ βits) for holding intermediate results. Other computers allot a single length word (M βits) plus some (say, q) βits for holding intermediate results. These q βits are called ***guard βits***. The number q of guard βits may vary by computer, by arithmetic chip, by floating point operation, or by precision (single precision versus double precision). If a computer has no guard βits at all (for example, double precision on early versions of the IBM System/360), then very unpleasant things can happen. For example, one such unpleasant result is that multiplication by one causes the trailing βit of the normalized product to become zero; so, $1 * X \neq X$ when there are no guard βits.

The computational version of the fundamental laws of mathematical algebra are subject to possible failure. Some of these laws of computational algebra hold exactly. Some of them hold approximately. Yet others (typically, those computational laws which involve a subtract magnitude operation) do not even hold approximately. See Chapters 1 and 3 of Sterbenz [74] for details.

C. EXPONENT SPILL

Let E^* and E_* denote, respectively, the *largest and smallest exponents* in a computer representation; that is, the exponent E satisfies $E_* \leq E \leq E^*$. Let Ω and ω denote, respectively, the *largest and smallest floating point numbers*. Then

$$\Omega = (1 - \beta^{-M})\beta^{E^*} \quad \text{and} \quad \omega = \beta^{E_* - 1}. \tag{1.3}$$

Exponent spill means that a result occurs which is greater than Ω (called *overflow*) or is less than ω (called *underflow*).

On many computers a *characteristic* is stored instead of an exponent E. The *characteristic* is defined as $E + \gamma$, where γ is a constant called the *bias*. Define the computer dependent constant c:

$$c = E^* - E_* + 1. \tag{1.4}$$

Then the characteristic takes on exactly c distinct values ranging from 0 (which corresponds to the smallest exponent E_*) through $c - 1$ (which corresponds to E^*). Furthermore,

$$E^* = c - 1 - \gamma \quad \text{and} \quad E_* = -\gamma. \tag{1.5}$$

On most computers $E^* + E_* = -1$. For example, the IBM hexadecimal computers, beginning with the System/360, have $\beta = 16$, $E^* = 63$, $E_* = -64$ (thus $E^* + E_* = -1$), and $\gamma = 64$. So the characteristic of these computers ranges from 0 through 127, and it is stored in a 7 bit field, since $128 = 2^7$.

D. BALANCED EXPONENT RANGE

While $E^* + E_* = -1$ is a very convenient choice for computer design engineers and thus is very commonly used, it has significant computational disadvantages, because many perfectly valid floating point numbers have reciprocals that overflow. Indeed, $E^* + E_* = -1$ and $\omega \le X \le \beta^2\omega$ together imply that $X^{-1} > \Omega$; that is, the arithmetic done to form $1 \div X$ must cause overflow to occur for this combination of X and design parameters. This disadvantage can be virtually eliminated by requiring that the exponent range be *balanced*, which means that

$$\Omega\omega \approx 1.$$

It is shown below in Theorem 1.1 that if $E^* + E_* = +1$, then $\Omega\omega \approx 1$. However, if $E^* + E_* = -1$, as on IBM hexadecimal machines, then

$$\Omega\omega \approx \beta^{-2} \not\approx 1,$$

a design situation which is called the *usual unbalanced exponent* case.

The important point is that if the exponent range is balanced, then reciprocation will never cause an underflow and will cause an overflow only for one extreme case—taking the reciprocal of ω, the smallest positive floating point number. The following theorem summarizes the relationship between a balanced or an unbalanced exponent range and the various computer design parameters.

THEOREM 1.1.

A. *The Balanced Exponent Case* ($\Omega\omega \approx 1$). *The following are equivalent:*

(1.6) $c = 2\gamma + 2$

(1.7) $E^* + E_* = +1$

(1.8) $E^* = \gamma + 1$ *and* $E_* = -\gamma$

(1.9) $\Omega\omega = 1 - \beta^{-M} \approx 1$

B. *The Usual Unbalanced Exponent Case* ($\Omega\omega \approx \beta^{-2}$). *The following are equivalent:*

(1.10) $c = 2\gamma$

(1.11) $E^* + E_* = -1$

(1.12) $E^* = \gamma - 1$ *and* $E_* = -\gamma$

(1.13) $\Omega\omega = \beta^{-2}(1 - \beta^{-M}) \approx \beta^{-2}$

PROOF: All of the equivalences, except those involving Eq. (1.9) or (1.13), follow easily from Eqs. (1.4) and (1.5). To show that Eq. (1.7) is equivalent to (1.9), multiply the two equations of (1.3) together to get

$$\Omega\omega = (1 - \beta^{-M})\beta^{E^* + E_* - 1}. \tag{1.14}$$

The equivalence of Eqs. (1.7) and (1.9) follows from Eq. (1.14). The equivalence of Eqs. (1.11) and (1.13) follows similarly from Eq. (1.14). ■

While a balanced exponent range combats one overflow problem caused by the customary lack of symmetry, there are some other overflow problems which can arise on high speed, floating point computers. For logarithmically distributed numbers, even floating point addition and subtraction can cause overflow or catastrophic cancellation to occur with alarming frequency. If numbers are assumed to be distributed in accordance with Sweeney's [65] experiments, then severe loss of significance occurs with large probability in floating point subtraction. These problems are pointed out in Turner [89], the first lecture in this volume, and in Feldstein and Turner [86]. In the latter paper, suggestions are given to combat these problems by dramatically increasing the word-length to 128 binary bits (100 bits in the fraction and 27 bits in the exponent). An alternative for coping with overflow is to replace the conventional floating point format with another arithmetic system, such as level index arithmetic. See Clenshaw and Olver [84] and Turner [86 and 89] for details.

2. SHORTENING COMPUTER NUMBERS: BASIC CONCEPTS

This section studies the error that arises from *shortening (or abbreviating)* a normalized floating point number X. Additional details may be found in BFGL [79] and GFB [85]. For simplicity, let $X > 0$. Recall that $X = \beta^E \cdot F$ where $\beta^{-1} \leq F < 1$ and Eq. (1.1) holds. Let $\widetilde{X} \approx X$ be an approximate representation of X. Assume that the fraction of $\widetilde{X}$ has finitely many βits. Define

$$
\begin{aligned}
R_0(X) &= \widetilde{X} - X && \textit{(actual error)}\\
R_2(X) &= R_0(X)/\beta^E && \textit{(fraction error or precision error)}\\
R_3(X) &= R_0(X)/X && \textit{(relative error)}
\end{aligned}
$$

Assume $\widetilde{\beta^E \cdot F} = \beta^E \cdot \widetilde{\beta^0 \cdot F}$. Then $R_i(F) = R_i(X)$ for $i = 2, 3$. Since $R_2(X) < R_3(X) \leq \beta R_2(X)$, then relative error and fraction error are of the same order of magnitude.

Consider a fraction F of infinite or finite ($M + K$ βit) length, which will be shortened to M βits either by chopping or by symmetric rounding. Denote

CH = *chopping*

SE = *symmetric rounding with even base* β

SO = *symmetric rounding with odd base* β

DEFINITION 2.1. Fix the integers $K \geq 1$, and $M \geq 1$. Let N denote the number of βits in the fraction of a floating point computer word. Denote

$$u = \beta^{-N}, \quad v = \beta^{-K}, \quad w = \beta^{-M}, \qquad \text{and} \qquad F = f_1\beta^{-1} + f_2\beta^{-2} + \dots, \tag{2.1}$$

where the f_i are integers (βits) satisfying Eq. (1.2). The collection of βits that will be lost in the shortening process (namely $f_{M+1}, f_{M+2}, \dots$) are called the *trailing βits* of F. Denote

$$h(F) = (f_{M+1}\beta^{-(M+1)} + f_{M+2}\beta^{-(M+2)} + \dots) + \begin{cases} \beta^{-M}, & \text{for rounding up} \\ 0, & \text{for rounding down.} \end{cases} \tag{2.2}$$

If F has finite length ($M + K$ βits), denote

$$G = v^{-1}((Fw^{-1})) \equiv \beta^K(f_{M+1}\beta^{-1} + f_{M+2}\beta^{-2} + \dots) \tag{2.3}$$

It is important to note that the error R_2 or R_3 when shortening F to M βits is a function of the trailing βits of F. The following is proved in GFB [85].

LEMMA 2.1. *Let F be a positive normalized fraction of finite or infinite length. If F is rounded (either up or down) to M βits, then the trailing βits of F are $(f_{M+1}, f_{M+2}, \ldots)$ if and only if $R_2(F) = h(F)$. Furthermore*

$$h(F) = -w((Fw^{-1})) + \begin{cases} w, & \text{for rounding up} \\ 0, & \text{for rounding down} \end{cases} \tag{2.4}$$

$$h(F) = -\beta^{-M}((F\beta^{M})) + \begin{cases} \beta^{-M}, & \text{for rounding up} \\ 0, & \text{for rounding down.} \end{cases} \tag{2.5}$$

If, in addition, $f_{M+K+1} = f_{M+K+2} = \ldots = 0$ (that is, if F has finite length $= M + K$ βits), then

$$h(F) = -Gwv + \begin{cases} w, & \text{for rounding up} \\ 0, & \text{for rounding down} \end{cases} \tag{2.6}$$

$$G = \sum_{i=0}^{K-1} f_{M+1+i}\beta^{K-1-i} \qquad \text{and} \qquad R_2(F) = h(F) = mwv \tag{2.7}$$

where the integer m is given by

$$m = -G + \begin{cases} v^{-1}, & \text{for rounding up} \\ 0, & \text{for rounding down.} \end{cases} \tag{2.8}$$

Furthermore, m satisfies

$$-v^{-1} + 1 \le m \le 0 \qquad \text{for CH} \tag{2.9a}$$

$$-\frac{1}{2}v^{-1} + 1 \le m \le \frac{1}{2}v^{-1} \qquad \text{for SE} \tag{2.9b}$$

$$-\frac{1}{2}v^{-1} + \frac{1}{2} \le m \le \frac{1}{2}v^{-1} - \frac{1}{2} \qquad \text{for SO.} \tag{2.9c}$$

Many authors have adopted the logarithmic distribution as the realistic one for numbers which occur in real life. (See, for example, p. 239 ff. of Knuth [81], Turner [89] and the references in each of these.)

CONVENTION: For the rest of this manuscript, assume that F (or X) is a positive, base β, normalized, floating point number which is selected at random from the logarithmic distribution. This means that $\log_\beta F \pmod 1$ is uniformly distributed on $[0,1)$.

The following four results are taken from Feldstein and Goodman [76]. They concern the distribution of βits of a logarithmically distributed fraction.

THEOREM 2.1. *The probability that the first K βits of F are $f_1, f_2, \ldots, f_K$ (respectively) equals*

$$\log_\beta \frac{\beta^{K-1}f_1 + \beta^{K-2}f_2 + \cdots + f_K + 1}{\beta^{K-1}f_1 + \beta^{K-2}f_2 + \cdots + f_K} = \log_\beta\left(1 + 1/(F\beta^K)\right).$$

DEFINITION 2.2. $p_{M,\beta}(f)$ denotes the probability that the Mth βit of F equals f.

THEOREM 2.2. *Let* $0 \leq f \leq \beta - 1$. *Then*

$$p_{1,\beta}(f) = \log_\beta \left(\frac{f+1}{f}\right), \qquad \text{for} \quad f \geq 1$$

$$p_{M+1,\beta}(f) = \log_\beta \prod_{m=\beta^{M-1}}^{\beta^M - 1} \frac{m\beta + f + 1}{m\beta + f}, \qquad \text{for} \quad M \geq 1.$$

Also, $p_{M,\beta}(f) = \beta^{-1} + O(\beta^{-M})$ *as* $M \to \infty$.

DEFINITION 2.3. Suppose that F is a fraction of finite length; that is, suppose that

$$F = f_1\beta^{-1} + f_2\beta^{-2} + \cdots + f_{M+K}\beta^{-(M+K)}.$$

Let P denote the probability that the K trailing βits of F (that is, βits number $M+1, \ldots, M+K$) are equal to, respectively, the values $f_{M+1}, f_{M+2}, \ldots, f_{M+K}$ where $0 \leq f_j \leq \beta - 1$.

COROLLARY 2.2. *By Lemma 2.1, P also is equal to the probability that the relative error $R_2(F)$ in shortening F from $(M+K)$ βits to M βits satisfies $R_2(F) = h(F)$.*

THEOREM 2.3. *Let* $M \geq 1$ *and* $K \geq 1$. *Then*

$$P = \log_\beta \prod_{\ell=\beta^{M-1}}^{\beta^M - 1} \frac{\ell\beta^K + G + 1}{\ell\beta^K + G}.$$

It is shown in Theorem 3.1 of BFGL [79] that

THEOREM 2.4.

$$P - \beta^{-K} = \frac{\beta(\beta-1)}{2\ell n\beta} \cdot \; \beta^{-K}\beta^{-(M+1)} + O(\beta^{-2K}\beta^{-M})$$

Note that if F were uniformly distributed instead of logarithmically distributed, then $P = \beta^{-K}$ would hold. So $P - \beta^{-K}$ measures the difference between the resulting probabilities for the logarithmic versus the uniform distributions.

3. FRACTION ERROR IN MULTIPLICATION

Consider a computer with an N βit fraction and a machine unit $u = \beta^{-N}$. To study the error in floating point multiplication, let $X_i = \beta^{E_i} \cdot F_i$ for $i = 1, 2, 3$ be three normalized floating point numbers, where $X_3 = X_1 * X_2$. Assume F_1 and F_2 are positive, normalized, base β numbers, which are random variables selected from the logarithmic distribution on $[\beta^{-1}, 1)$. Then it is well-known that the product F_1F_2 is logarithmically distributed on $[\beta^{-2}, 1)$. There are two cases for F_1F_2:

Case 0. $\quad \beta^{-1} \leq F_1F_2 < 1 \qquad$ *(no post-arithmetic normalization shift is needed)*

Case 1. $\quad \beta^{-2} \leq F_1F_2 < \beta^{-1} \qquad$ *(post-arithmetic normalization shift of 1 βit is needed)*

In case j (for $j = 0, 1$), the quantity $F_1F_2\beta^j$ is logarithmically distributed on $[\beta^{-1}, 1)$.

To study the error in the product, consider both Cases 0 and 1. Apply to each case the results presented in Section 2 about the error in shortening a normalized floating point fraction F from $(M+K)$ βits to M βits. For Case 0, $M = N$ and $K = N$ (so $v = w = u$). For Case 1, the values of M and K (and so by Eq. (2.1) also the values of w and v) depend on the shortening process used, upon the normalization option, and upon the number of guard βits. Since Case 1 means that a post-arithmetic normalization shift is needed, a computer can perform the shift either before or after it does the shortening process. To be precise, for chopping there are two cases: let NCH denote Normalize before CHop and let CHN denote CHop before Normalize. Similarly for symmetric rounding, there are four cases: NSE, SEN, NSO, SON. The number of guard βits is written in parentheses. For example NSE(≥ 2) means normalize before doing symmetric rounding on an even base computer that has two or more guard βits. There are eight different rounding rules to consider for multiplication: NCH(0), NCH(≥ 1), CHN(≥ 0), NSE(1), NSE(≥ 2), SEN(≥ 1), NSO(N), SON(N). The values of v and w (see Definition 2.1) for these eight rounding rules are given below in Table 3.1. See p. 725 of BFGL [79] for an explanation of the reasons for each value.

TABLE 3.1

Values of v and w

$u = \beta^{-N}$ = **machine unit**

Rounding Rule	**Case 0**		**Case 1**	
	v	w	v	w
NCH(0), CHN(≥ 0), SEN(≥ 1), SON(N)	u	u	u	βu
NCH(≥ 1), NSE(1)†, NSE(≥ 2), NSO(N)	u	u	$\beta u^\dagger$	$u^\dagger$

† Chopping is done for NSE(1) in case 1, so the CH formula must be used for NSE(1) in case 1.

REMARK 3.1 Symmetric rounding with 0 guard βits need not be considered separately. In such a case there would be no guard βit upon which to base a symmetric rounding decision, so chopping actually would occur.

REMARK 3.2 NSE(1) is a chop-like rule. The reason is that when Case 1 occurs (which happens half the time), there is no post-normalization βit upon which to base a symmetric rounding decision. So chopping would be done in Case 1.

THEOREM 3.1. *Let X_1 and X_2 be two N-βit numbers selected at random from the logarithmic distribution. Then the fraction error $R_2(X_1 * X_2)$ has mean μ_2 and variance V_2 as follows:*

	$\mu_2 + O(u^3)$	$V_2 + O(u^3 \text{ or } u^4)$
NCH(≥ 1)	$-u/2+u^2\{3+3\beta+(\beta-1)/\ell n\beta\}/12$	$u^2/12$
CHN(≥ 0), *NCH*(0)	$-u(1+\beta)/4+u^2\{6+6\beta+(\beta-1)(\beta^2+1)/\ell n\beta\}/24$	$u^2(5\beta^2-6\beta+5)/48$
NSE(1)	$-u/4+u^2\{12+12\beta+(\beta-1)/\ell n\beta\}/48$	$7u^2/48$
NSE(≥ 2)	$u^2\{6+6\beta-(\beta-1)/\ell n\beta\}/24$	$u^2/12$
SEN(≥ 1)	$u^2\{12+12\beta-(\beta-1)(\beta^2+1)/\ell n\beta\}/48$	$u^2(1+\beta^2)/24$
NSO(N)	$-u^2(\beta-1)/(24\ell n\beta)$	$u^2/12$
SON(N)	$-u^2(\beta-1)(\beta^2+1)/(48\ell n\beta)$	$u^2(\beta^2+1)/24$

PROOF: For $j = 0,1$, define $P^{(j)}(m) \equiv$ the conditional probability that $R_2 = mwv$, given that case j occurs. The assumption of logarithmic distribution implies that Cases 0 and 1 each occur with probability $\frac{1}{2}$. Then the mean μ_2 of fraction error R_2 is given by

$$\mu_2 = \frac{1}{2}\sum\nolimits^{(0)} R_2 P^{(0)}(m) + \frac{1}{2}\sum\nolimits^{(1)} R_2 P^{(1)}(m) \tag{3.1}$$

where $\sum^{(j)}$ refers to case j. Recall that $R_2 = mwv$ where m is given by the appropriate range of summation in Eq. (2.9). So

$$\mu_2 = \frac{1}{2}\sum\nolimits^{(0)} mwvP^{(0)}(m) + \frac{1}{2}\sum\nolimits^{(1)} mwvP^{(1)}(m). \tag{3.2}$$

Refer to Table 3.1. For all the rounding rules, note that $wv = u^2$ for Case 0 and that $wv = \beta u^2$ for Case 1. Substitute these values for wv into Eq. (3.2) to obtain

$$\mu_2 = \frac{1}{2}u^2\left\{\sum\nolimits^{(0)} mP^{(0)}(m) + \beta\sum\nolimits^{(1)} mP^{(1)}(m)\right\}. \tag{3.3}$$

To obtain $P^{(j)}(m)$, apply Theorem 4.3 of BFGL [79]. This yields

$$P^{(j)}(m) = v + wv\frac{\beta-1}{\ell n\beta}(mv + \delta_m) + O(u^3), \quad \text{where} \tag{3.4}$$

$$\delta_m = \begin{cases} -\dfrac{1}{2}, & \text{rounding up} \\ +\dfrac{1}{2}, & \text{rounding down} \end{cases}$$

$$m = -G + (\frac{1}{2} - \delta_m)v^{-1} \qquad \text{(by Eq. (2.8))}.$$

To evaluate $\sum^{(j)} mP^{(j)}(m)$, substitute $P^{(j)}(m)$ from Eq. (3.4) into Eq. (3.3). After some algebra (see p. 727 of BFGL [79]), one obtains:

$$\sum\nolimits^{(j)} mP^{(j)}(m) = \begin{cases} -\dfrac{1}{2}v^{-1} + \dfrac{1}{2} + \dfrac{1}{12}\dfrac{\beta-1}{\ell n\beta}wv^{-1} + O(u) & \text{for CH} \\ \dfrac{1}{2} - \dfrac{1}{24}\dfrac{\beta-1}{\ell n\beta}wv^{-1} + O(u) & \text{for SE} \\ -\dfrac{1}{24}\dfrac{\beta-1}{\ell n\beta}wv^{-1} + O(u) & \text{for SO} \end{cases} \tag{3.5}$$

To evaluate the mean μ_2 of the fraction error for each rounding rule, substitute Eq. (3.5) into (3.3) and substitute the appropriate values of v and w from Table 3.1. This yields the means in Theorem 3.1. The variances in Theorem 3.1 are derived similarly in Goodman and Feldstein [77]. ■

4. RELATIVE ERROR IN MULTIPLICATION

A similar computation for relative error in multiplication yields the means μ_3 and variances V_3 of relative error R_3. They are given below in Theorem 4.1, which is proved in Sections 4 and 5 of GFB [85]. The corrected coefficient for μ_3 for NSE(≥ 2) is given here.

THEOREM 4.1. *Let X_1 and X_2 be two N-βit numbers selected at random from the logarithmic distribution. Then the relative error $R_3(X_1 * X_2)$ has mean and variance shown in Table 4.1.*

TABLE 4.1

Means and Variances of Relative Error

	μ_3	V_3/u^2
NCH(≥ 1)	$-(\beta-1)u/(2L)$	$(\beta^2-1)/(6L)-(\beta-1)^2/(4L^2)$
CHN(≥ 0), NCH(0)	$-(\beta^2-1)u/(4L)$	$(\beta^4-1)/(12L)-(\beta^2-1)^2/(16L^2)$
NSE(1)	$-(\beta-1)u/(4L)$	$5(\beta^2-1)/(48L)-(\beta-1)^2/(16L^2)$
NSE(≥ 2)	$5(\beta^2-1)u^2/(24L)$	$(\beta^2-1)/(24L)$
SEN(≥ 1)	$(\beta^2-1)(11-\beta^2)u^2/(48L)$	$(\beta^4-1)/(48L)$
NSO(N)	$-(\beta^2-1)u^2/(24L)$	$(\beta^2-1)/(24L)$
SON(N)	$-(\beta^4-1)u^2/(48L)$	$(\beta^4-1)/(48L)$
Plus higher order terms in $u \equiv \beta^{-N}$, $L \equiv \ell n\beta$.		

Chop-like rules—the first three—always produce negative errors. It is interesting to compare the statistical estimates in Theorem 4.1 with the error bounds that Sterbenz derived. For chopping, Eq. (3.2.6) on page 74 of Sterbenz [74] is

$$0 < -(\textit{relative error}) < (\textit{base}) \cdot (\textit{machine unit}) \equiv B. \tag{4.1}$$

In the notation of this paper, the upper bound in Eq. (4.1) becomes βu both for NCH(≥ 1) and for NSE(1). But the situation changes when chopping precedes the post-multiplicative normalization; that is, for CHN(≥ 0). When a post-multiplicative shift occurs in CHN(≥ 0), the needed βit has already been chopped off and therefore a zero is shifted into the last βit position. The effect in this case is as though the machine were to have one fewer βit than it actually has; that is, the effect is as though the machine unit were $= \beta u$. Thus, the Sterbenz upper bound in Eq. (4.1) for CHN(≥ 0) is

$$B \equiv (\textit{base}) \cdot (\textit{effective machine unit}) = \beta \cdot \beta u = \beta^2 u.$$

For chop-like rules, the mean is of order u; so define the *magnified mean* as μ/u. For symmetric rounding rules, the mean is of order u^2, so define the *magnified mean* as μ/u^2. For all rules, the standard deviation σ is of order u, so define the *magnified standard deviation* as σ/u.

Table 4.2 (taken from Table 6.1 of GFB [85]) presents the magnified means and the magnified standard deviations, along with certain ratios for chop-like rules. Columns 3 and 5 in Table 4.2 present the numerical values of the ratios μ_3 and σ_3, each divided by the appropriate Sterbenz bound (either βu or $\beta^2 u$), as discussed above. The values in columns 3 and 5 can be used to compare the results from the stochastic model with those from the Sterbenz bounds. To do so, let $B \equiv$ *the Sterbenz bound*. Define

$$P = min\left(\left|\frac{B-\mu}{\sigma}\right|, \left|\frac{B+\mu}{\sigma}\right|\right). \tag{4.2}$$

Then P is a measure of how pessimistic the Sterbenz bound is compared with the stochastic estimates. The values of P, computed from columns 3 and 5 of Table 4.2, are presented in column 6.

TABLE 4.2

Magnified mean, magnified standard deviation, and ratios for relative error in floating point multiplication.

Chop-like Rules

($\beta = 2'$ *means binary machine with a hidden bit*)

NCH(≥ 1)					
β	$-\mu/u$	$-\mu/(\beta u)$	σ/u	$\sigma/(\beta u)$	P
2'	.361	.361	.224	.224	2.9
2	.721	.361	.448	.224	2.9
4	1.082	.271	.795	.199	3.7
8	1.683	.210	1.489	.186	4.2
10	1.954	.195	1.829	.183	4.4
16	2.705	.169	2.830	.177	4.7
NSE(1)					
β	$-\mu/u$	$-\mu/(\beta u)$	σ/u	$\sigma/(\beta u)$	P
2'	.180	.180	.283	.283	2.9
2	.361	.180	.566	.283	2.9
4	.541	.135	.914	.228	3.8
8	.842	.105	1.565	.196	4.6
10	.977	.098	1.877	.188	4.8
16	1.353	.085	2.784	.174	5.3
CHN(≥ 0) and NCH(0)					
β	$-\mu/u$	$-\mu/(\beta^2 u)$	σ/u	$\sigma/(\beta^2 u)$	P
2'	.541	.271	.397	.199	3.7
2	1.082	.271	.795	.199	3.7
4	2.705	.169	2.830	.177	4.7
8	7.574	.118	10.331	.161	5.5
10	10.748	.107	15.695	.157	5.7
16	22.993	.090	37.961	.148	6.1

Similarly for SE rules, the statistical estimates for relative error are compared with the Sterbenz bounds. Columns 2 and 3 of Table 4.3 present the magnified mean and the magnified standard deviations for SE rules (taken from Table 6.2 of GFB [85]; corrected values are given here for μ/u^2 for NSE(≥ 2)). For symmetric rounding, Eq. (3.2.9) on page 74 of Sterbenz is

(4.3) $$|\textit{relative error}| \leq \frac{1}{2} \cdot (\textit{base}) \cdot (\textit{machine unit}) \equiv B.$$

For NSE(≥ 2), the bound B in (4.3) becomes $\frac{1}{2}\beta u$. But for SEN(≥ 1), the effective machine unit becomes βu; so the Sterbenz bound becomes $B = \frac{1}{2}\beta(\beta u) = \frac{1}{2}\beta^2 u$. For SE rules it is very important to note that the mean μ is of order u^2, while the Sterbenz bound, B, and the standard deviation, σ, are of order u. So σ and B dominate μ. Thus one may neglect μ in (4.2) and obtain $P \approx B/\sigma$. These ratios are in the last column of Table 4.3.

TABLE 4.3			
Magnified mean, magnified standard deviation, and ratio for relative error in floating point multiplication. Symmetric Rounding Rules with Even Base ($\beta = 2'$ *means binary machine with a hidden bit*)			
NSE(≥ 2)			
β	μ/u^2	σ/u	$P \approx (\beta u/2)/\sigma$
$2'$	.225	.212	2.4
2	.902	.424	2.4
4	2.254	.671	3.0
8	6.312	1.124	3.6
10	8.956	1.338	3.7
16	19.161	1.958	4.1
SEN(≥ 1)			
β	μ/u^2	σ/u	$P \approx (\beta^2 u/2)/\sigma$
$2'$	0.158	.336	3.0
2	0.631	.671	3.0
4	-1.127	1.958	4.1
8	-33.452	6.405	5.0
10	-79.712	9.511	5.3
16	-469.439	22.191	5.8

CONCLUSIONS ABOUT RELATIVE ERROR

1. Table 3.2 of fraction error and Table 4.1 of relative error confirm the superiority of symmetric rounding versus rounding by chopping. Indeed, the means for symmetric rounding rules are $O(u^2)$, whereas the means for chopping rules are an order of magnitude larger, $O(u)$.
2. Tables 4.2 and 4.3 show the desirability of a small base. So binary has an advantage over the widely used base 16. (Similar studies show the desirability of a small base also for odd base with symmetric rounding.)
3. NSE(1) is a chop-like rule with chop-like error behavior. The results show that NSE(≥ 2) is an order of magnitude better than NSE(1).
4. The pessimism of using the Sterbenz bounds instead of our statistical estimates for relative error, becomes worse as β increases. So as β increases, the statistical study becomes more important. See P in Tables 4.2 and 4.3.

A similar study for products has been done for the magnitude relative error R_6, which is defined to be the absolute value of relative error. See Goodman [89]. Using magnitude relative error implies that the mean error will not accidentally be small only because of the cancellation of very positive and very negative errors. The means μ_6 and variances V_6 of magnitude relative error R_6 in multiplication are given in Theorem 4.2.

THEOREM 4.2. *Let X_1 and X_2 be two N βit numbers selected at random from the logarithmic distribution. Then the magnitude relative error $R_6(X_1 * X_2)$ has the mean and variance shown in Table 4.4.*

TABLE 4.4

Means and Variances of Magnitude Relative Error

	μ_6/u	V_6/u^2
NCH(≥ 1)	$(\beta-1)/2L$	$(\beta^2-1)/6L-(\beta-1)^2/4L^2$
CHN(≥ 0), NCH(0)	$(\beta^2-1)/4L$	$(\beta^4-1)/12L-(\beta^2-1)^2/16L^2$
NSE(1)	$3(\beta-1)/8L$	$5(\beta^2-1)/48L-9(\beta-1)^2/64L^2$
NSE(≥ 2), NSO(N)	$(\beta-1)/4L$	$(\beta^2-1)/24L-(\beta-1)^2/16L^2$
SEN(≥ 1), SON(N)	$(\beta^2-1)/8L$	$(\beta^4-1)/48L-(\beta^2-1)^2/64L^2$
Plus higher order terms in $u \equiv \beta^{-N}$. $L \equiv \ell n\beta$.		

OBSERVATION 4.1. The means and standard deviations of R_6 for all the rounding rules in Table 4.4 are of order u.

OBSERVATION 4.2. To make the mean μ_6 small for any of the rounding rules in Table 4.4, take the base β to be small.

ACKNOWLEDGEMENT:

The work of Richard H. Goodman was supported in part by NSF Grant #CCR8602064 and in part by a sabbatical leave from the University of Miami.

REFERENCES

BFGL [79]. Bustoz, J., Feldstein, A., Goodman, R., Linnainmaa, S., Improved Trailing Digits Estimates Applied to Optimal Computer Arithmetic. *J.ACM* **26**, 716–730 (1979).

Clenshaw and Olver [84]. Clenshaw, C. W. and Olver, F. W. J., Beyond Floating Point. *J.ACM* **31**, 319–328 (1984).

Feldstein and Goodman [76]. Feldstein, A. and Goodman, R., Convergence Estimates for the Distribution of Trailing Digits. *J.ACM* **23**, 287–297 (1976).

Feldstein and Turner [86]. Feldstein, A. and Turner, P., Overflow, Underflow, and Severe Loss of Significance in Floating-Point Addition and Subtraction. *IMA J. of Numerical Analysis* **6**, 241–251 (1986).

Goodman [89]. Goodman, R. H., Some Models of Relative Error in Products. *Applied Numerical Mathematics*, to appear (1989 or 1990).

Goodman and Feldstein [77]. Goodman, R. and Feldstein, A., Effect of Guard Digits and Normalization Options on Floating Point Multiplication. *Computing* **18**, 93–106 (1977).

GFB [85]. Goodman, R. H., Feldstein, A., and Bustoz, J., Relative Error in Floating Point Multiplication. *Computing* **35**, 127–139 (1985).

Knuth [81]. Knuth, D. E., *The Art of Computer Programming*, Vol. 2, *Semi-numerical Algorithms*, Second Edition, Addison-Wesley, Reading, Mass., 1981.

Sterbenz [74]. Sterbenz, P. H., *Floating Point Computation*, Prentice-Hall, Inc., Englewood Cliffs, N.J., 1974.

Sweeney [65]. Sweeney, D. W., An Analysis of Floating Point Addition, *IBM Systems J.* **4**, 31–42 (1965).

Turner [86]. Turner, P. R., Towards a Fast Implementation of Level Index Arithmetic. *Bull. IMA* **22**, 188–191 (1986).

Turner [89]. Turner, P. R., Alternatives to Floating Point: I. The Need, *These Proceedings*, Springer-Verlag, Heidelberg, 1989.

SOME GRADIENT SUPERCONVERGENCE RESULTS IN THE FINITE ELEMENT METHOD

J.R. WHITEMAN and G. GOODSELL
Institute of Computational Mathematics
Department of Mathematics and Statistics
Brunel University, Uxbridge, UB8 3PH, England.

1. Introduction

1.1 Review of Gradient Superconvergence

This paper is concerned with the phenomenon of *gradient superconvergence* of finite element approximations to the solutions of two-dimensional second order elliptic boundary value problems. The original concept of gradient superconvergence is that there exist certain points in each element at which the gradient of the finite element approximation has a higher rate of convergence to the gradient of the true solution in a discrete analogue of a Lebesgue norm than that found globally in the Lebesgue norm itself.

The first observations of the existence of points of high accuracy of gradients were made by Veryard [36], Strang and Fix [35], Barlow [3] and Zlamal [45]. Such points became known as *stress points*. Subsequently many systematic studies of gradient superconvergence have been undertaken. In particular Strang and Fix stated in [35] that for a specific piecewise polynomial finite element approximation the stress points occur at points where the gradient of the corresponding interpolant is exact for true solutions of one order higher than the polynomials in the finite element space. Levine [23] highlighted this effect as being at the heart of any superconvergence result. This is because the gradient of the (unknown) interpolant is at *all* points a superconvergent approximation to the gradient of the finite element approximation and thus the stress points of these two functions are identical.

The first precise description of and full proof of gradient superconvergence at element second order Gauss-points in the discrete ℓ_1 sense was given by Zlamal [45] for the eight-node isoparametric plane element. In [46] Zlamal improved his results to the discrete ℓ_2 setting.

In the context of finite element methods on *triangular* partitions the study of superconvergence has been limited almost entirely to piecewise linear approximations, for which there are no stress points. It was conjectured by Strang and Fix [35] that (triangle) element edge mid-points are points of exceptional accuracy for derivatives tangential to the edge. This was considered by Levine [22], who proved the superconvergence in the discrete ℓ_2 sense of these tangential derivatives.

One can obtain recovered derivatives of the Galerkin solution through post-processing the derivatives themselves and this can produce an acceleration of convergence. This acceleration is also called superconvergence. Such post-processing can produce an increase of accuracy not only at specific points but also throughout a subdomain or even the complete domain of a problem. This leads on to the investigation of *recovered gradient superconvergence*.

For piecewise linear finite element approximations on triangular meshes Levine [23] defined the recovered gradient at an element edge mid-point and proved the superconvergence of the discrete ℓ_2 average error. These results were extended in Lin, Lü and Shen [25], Neittaanmäki and Krizek [27] and Levine [24]. Recovered gradients at the element vertices were then considered by Krizek and Neittaanmäki [19], [20], using an averaging scheme of Zienkiewicz and Cheung [44]. All the above results demand high regularity of the solutions of the boundary value problems, which is not practical. Wheeler and Whiteman [38] addressed this problem and derived superconvergent ℓ_2 average error estimates on subdomains. Hlavacek and Krizek [17] considered elliptic systems and also produced equivalent results on subdomains. An extensive survey of superconvergence techniques, including the above, is given by Krizek and Neittaanmäki [21].

Following on from all the above work we now proceed in this paper to give a unified treatment of recovered gradient superconvergence, for Poisson problems and for problems of planar linear elasticity in polygonal domains, for piecewise linear finite element approximations on triangular partitions.

In Section 1.2 we set out the Sobolev space theory necessary for deriving the error estimates. Section 2 contains the Poisson problem, the descriptions of the meshes and the definitions of the various recovered gradients. Superconvergence results on general polygonal domains with mixed boundary conditions are produced in Section 3. In Section 4 we extend these to problems of planar elasticity, and in Section 5 to the particular case of linear elastic fracture. The aim in the fracture context is to estimate the errors in the approximations to the fracture criteria. In the work of Section 5 it was found that superconvergence results in the L_2 norm were inadequate to analyse the problems, so that recourse to L_∞ results was needed. Thus in Section 6 pointwise superconvergence of recovered gradients is considered firstly for Poisson problems and then for problems of planar elasticity. Some final comments are made at the end.

1.2 Some Sobolev Space Theory

1.2.1 Sobolev Spaces

The analysis of this paper is given in a Sobolev space setting, see Adams [1], and so we first state some results for Sobolev spaces. We work only with open bounded regions in $\mathbb{R}^2$ satisfying the strong cone property, Bramble and

Hilbert [4]. This includes all polygonal regions but excludes regions with outward facing cusps.

Let Ω be such a region and denote by $W_p^m(\Omega)$, $1 \leqq p \leqq \infty$ the Sobolev space of functions which together with their generalised derivatives up to and including order m are in $L_p(\Omega)$. In the usual way we denote by $\|v\|_{m,p,\Omega}$ and $|v|_{m,p,\Omega}$ respectively the norm and seminorm defined for $1 \leqq p < \infty$ by

$$\|v\|_{m,p,\Omega} \equiv \left\{ \sum_{|\alpha| \leqq m} \int_\Omega |D^\alpha v|^p d\mathbf{x} \right\}^{1/p} \tag{1.1}$$

$$|v|_{m,p,\Omega} \equiv \left\{ \sum_{|\alpha| = m} \int_\Omega |D^\alpha v|^p d\mathbf{x} \right\}^{1/p} \tag{1.2}$$

and for $p = \infty$,

$$\|v\|_{m,\infty,\Omega} \equiv \max_{|\alpha| \leqq m} \ \operatorname{ess.sup}_{\Omega} |D^\alpha v| \tag{1.3}$$

$$|v|_{m,\infty,\Omega} \equiv \max_{|\alpha| = m} \ \operatorname{ess.sup}_{\Omega} |D^\alpha v| \ . \tag{1.4}$$

For $p = 2$, we have the *Hilbert spaces* $H^m(\Omega) \equiv W_2^m(\Omega)$, the norm and seminorm on which we denote by $\|v\|_{m,\Omega} \equiv \|v\|_{m,2,\Omega}$ and $|v|_{m,\Omega} \equiv |v|_{m,2,\Omega}$. We denote by $\overset{\circ}{W}{}_p^m(\Omega)$ that subspace of $W_p^m(\Omega)$ such that

$$\overset{\circ}{W}{}_p^m(\Omega) \equiv \left\{ v : v \in W_p^m(\Omega) \ , \ D^\alpha v = 0 \text{ on } \partial\Omega : |\alpha| \leqq m-1 \right\} .$$

We note that $\overset{\circ}{W}{}_2^m(\Omega)$ is the space $\overset{\circ}{H}{}^m(\Omega)$. The seminorm, $|v|_{m,p,\Omega}$ is a norm on $\overset{\circ}{W}{}_p^m(\Omega)$ *equivalent* to the norm $\|v\|_{m,p,\Omega}$. Thus there exists a constant, $c > 0$, such that

$$|v|_{m,p,\Omega} \leqq \|v\|_{m,p,\Omega} \leqq c|v|_{m,p,\Omega} \ , \quad \forall \ v \in \overset{\circ}{W}{}_p^m(\Omega) \ ; \tag{1.5}$$

this follows from the Poincaré inequality, see[1,pp.158-159].

The Sobolev space, $\left[W_p^m(\Omega)\right]^2$, is that space of vector valued functions, each component of which belongs to $W_p^m(\Omega)$. For $1 \leqq p < \infty$, the norm is defined by

$$\|\mathbf{v}\|_{m,p,\Omega} \equiv \left\{ \sum_{j=1}^{2} \|v_j\|_{m,p,\Omega}^p \right\}^{1/p} \tag{1.6}$$

and for $p = \infty$

$$\|\mathbf{v}\|_{m,\infty,\Omega} \equiv \max_{1\leq j\leq 2} \|v_j\|_{m,\infty,\Omega} . \tag{1.7}$$

The seminorm is defined similarly.

In Section 6, we shall need the norm for Ω in $\mathbb{R}^2$ of the Sobolev space $W_p^s(\Omega)$, $s = m + \sigma : 0 < \sigma < 1$, of fractional order derivatives. This is defined for $1 \leq p < \infty$ by

$$\|v\|_{s,p,\Omega} \equiv \left\{ \|v\|^p_{m,p,\Omega} + \sum_{|\alpha|=m} \int_\Omega\int_\Omega \frac{\left|D^\alpha\{v(\mathbf{x})-v(\chi)\}\right|^p}{|\mathbf{x}-\chi|^{2+\sigma p}} \, d\mathbf{x}\, d\chi \right\}^{1/p} \tag{1.8}$$

and for $p = \infty$, see [1,p.214],

$$\|v\|_{s,\infty,\Omega} \equiv \max\left\{ \|v\|_{m,\infty,\Omega} , \max_{|\alpha|=m} \operatorname*{ess.sup}_{\substack{\mathbf{x},\chi\in\Omega\\ \mathbf{x}\neq\chi}} \frac{\left|D^\alpha\{v(\mathbf{x})-v(\chi)\}\right|}{|\mathbf{x}-\chi|^\sigma} \right\} . \tag{1.9}$$

1.2.2 Sobolev Embeddings

In consequence of the *Sobolev embedding*

$$W_p^m(\Omega) \longrightarrow X , \tag{1.10}$$

where X is a Banach space of functions defined over Ω, we can infer that

$$W_p^m(\Omega) \subseteq X \tag{1.11}$$

and that there exists an *embedding constant*, $C > 0$, dependent only upon m, p, the function space X, and the region Ω, such that

$$\|v\|_X \leq C\|v\|_{m,p,\Omega} , \quad \forall \ v \in W_p^m(\Omega) . \tag{1.12}$$

The embeddings, corresponding to (1.10) above, are given by the Sobolev Embedding Theorem, see Adams [1,pp.95-99]. A particular consequence of this theorem is the Sobolev Lemma, see Section 1.2.3.

There also exist Sobolev mappings of the form

$$W_p^m(\Omega) \longrightarrow L_q(\partial\Omega) . \tag{1.13}$$

From this we can infer that

$$v \in W_p^m(\Omega) \Rightarrow \text{trace}(v) \in L_q(\partial\Omega) \tag{1.14}$$

and, as before, that there exists an embedding constant, $C > 0$, dependent only upon m, p, q, and the region Ω, such that

$$|v|_{0,q,\partial\Omega} \leq C\|v\|_{m,p,\Omega} , \quad \forall \; v \in W_p^m(\Omega) . \tag{1.15}$$

The embeddings, corresponding to (1.13), are given by the Sobolev Embedding Theorem for traces of functions, see Adams [1,pp.113-115]. One that we shall require later is

$$W_p^1(\Omega) \longrightarrow L_2(\partial\Omega) , \quad \forall \; p \geq \frac{4}{3} . \tag{1.16}$$

It is convenient to introduce here a final piece of notation. Let the region Ω be an (n-sided) polygon, with boundary $\partial\Omega \equiv \bigcup_{i=1}^{n} \partial\Omega_i$, and consider the function

$$g \in W_p^m(\partial\Omega_i) , \quad i = 1,2,\ldots,n , \tag{1.17}$$

which, however, because of the discontinuity in the direction of the derivative at the corners of $\partial\Omega$, is such that

$$g \notin W_p^m(\partial\Omega) . \tag{1.18}$$

For such functions, we define the following norm for $1 \leq p < \infty$

$$\|g\|'_{m,p,\partial\Omega} \equiv \left\{ \sum_{i=1}^{n} \|g\|^p_{m,p,\partial\Omega_i} \right\}^{1/p} \tag{1.19}$$

and, for $p = \infty$,

$$\|g\|'_{m,\infty,\partial\Omega} \equiv \max_{1\leq i\leq n} \|g\|_{m,\infty,\partial\Omega_i} \tag{1.20}$$

(with the seminorm defined similarly). We note, of course, that on $W_p^m(\partial\Omega)$, the norms $\|\cdot\|_{m,p,\partial\Omega}$ and $\|\cdot\|'_{m,p,\partial\Omega}$ are identical.

1.2.3 Sobolev Lemma

As a result of the Sobolev embedding

$$W_p^m(\Omega) \longrightarrow C(\bar{\Omega}) , \quad mp > 2 , \tag{1.21}$$

see [1,p.97], we have the following results:

$$\sup_{\Omega} |v| \leq C\|v\|_{m,p,\Omega} , \quad \forall \; v \in W_p^m(\Omega) : mp > 2 \tag{1.22}$$

and

$$\sup_{\Omega} |\nabla v| \leq C\|v\|_{m+1,p,\Omega} , \quad \forall \; v \in W_p^{m+1}(\Omega) : mp > 2 , \tag{1.23}$$

with the constants C dependent only upon m, p, and the region Ω. See also Oden and Reddy [29,p.79].

1.2.4 Bramble-Hilbert Lemma

Let F be a *linear* functional on $W_p^m(\Omega)$ such that

$$|F(v)| \leq c_1 \|v\|_{m,p,\Omega} , \quad \forall \; v \in W_p^m(\Omega) \tag{1.24}$$

and

$$F(p) = 0 , \quad \forall \; p \in P_{m-1}(\Omega) , \tag{1.25}$$

where c_1 is a constant dependent only upon m, p, the functional F, and the region Ω, and $P_k(\Omega)$ denotes the space of all polynomials of degree not exceeding k in Ω. Then we have that

$$|F(v)| \leq c_2 |v|_{m,p,\Omega} , \quad \forall \; v \in W_p^m(\Omega) , \tag{1.26}$$

with the constant c_2 again dependent only upon m, p, F, and Ω.

The lemma and proof are due to Bramble and Hilbert [4], and are given for scalar functionals of scalar functions. The extension of the result to vector valued functionals of vector valued functions is straightforward.

2. Poisson Problems: Superconvergence of Recovered Gradients for Piecewise Linear Finite Element Approximations

2.1 Boundary Value Problem and Finite Element Discretisation

For $\Omega \subset \mathbb{R}^2$ a simply connected rectangular domain with boundary $\partial\Omega$ consisting of sides parallel to the x- and y- axes, we consider the Poisson problem in which $u \equiv u(\mathbf{x})$ satisfies

$$\left.\begin{aligned} -\Delta[u(\mathbf{x})] &= f(\mathbf{x}) , \quad \mathbf{x} \in \Omega , \\ u(\mathbf{x}) &= g(\mathbf{x}) , \quad \mathbf{x} \in \partial\Omega , \end{aligned}\right\} \tag{2.1}$$

where it is assumed that f, g are such that $u \in H^3(\Omega)$. A weak formulation of (2.1) is found by multiplying the differential equation by $v \in \overset{\circ}{H}{}^1(\Omega)$ and integrating by parts, so that we seek

$$u \in H_E^1 \cap H^3(\Omega) \quad \text{such that}$$

$$a(u,v) = (f,v) , \quad \forall \; v \in \overset{\circ}{H}{}^1(\Omega) , \tag{2.2}$$

where $\overset{\circ}{H}{}^1(\Omega)$ is the usual Sobolev space and

$$H_E^1 \equiv \left\{ v : v \in H^1(\Omega) \;,\; v = g \text{ on } \partial\Omega \right\} ,$$

$$a(u,v) \equiv \int_\Omega \nabla u . \nabla v \, d\mathbf{x} \;, \tag{2.3}$$

$$(f,v) \equiv \int_\Omega fv \, d\mathbf{x} \;. \tag{2.4}$$

For the application of the finite element method here the region Ω is first partitioned into squares, each of which is then subdivided into two isosceles right-angled triangles using the diagonal with slope unity. For this fully uniform mesh of triangles T_k, $k = 1,2,\ldots,K$, we define

$K_1 \equiv$ number of element edge mid-points internal to Ω. Any such mid-point is associated with two triangles T_k and T_k'.

$K_2 \equiv$ number of element vertices internal to Ω.

$K_3 \equiv$ number of element edge mid-points in $\overline{\Omega}$.

$K_4 \equiv$ number of element vertices in $\overline{\Omega}$.

The finite dimensional subspace $S^h(\Omega) \subset H^1(\Omega)$ is defined as the space of piecewise linear functions over the partition of triangles. The problem of approximating (2.2) is thus to find $u_h \in S_E^h$ such that

$$a(u_h,v_h) = (f,v_h) \;, \quad \forall \; v_h \in S_0^h(\Omega) \;, \tag{2.5}$$

where

$$S_0^h(\Omega) \equiv \left\{ v_h : v_h \in S^h(\Omega) \;,\; v_h = 0 \text{ on } \partial\Omega \right\}$$

and

$$S_E^h \equiv \left\{ v_h : v_h \in S^h(\Omega) \;,\; v_h = g_I \text{ on } \partial\Omega \right\} ,$$

with g_I the continuous piecewise linear interpolant to g on $\partial\Omega$.

For $v_h \in S^h(\Omega)$ the gradient of v_h is a piecewise constant vector which is in general discontinuous at the element edges. The *recovered gradient* of $v_h \in S^h(\Omega)$ at element edge mid-points internal to Ω is defined as the average of the two gradients in the triangles T_k and T_k' sharing the common edge. Thus

$$\mathbf{D}_k \, v_h \equiv \frac{1}{2} \left\{ \left[\nabla v_h\right]_{T_k} + \left[\nabla v_h\right]_{T_k'} \right\} . \tag{2.6}$$

In Levine [23] the following estimate for the ℓ_2 average error is given:

$$h \left\{ \sum_{k=1}^{K_1} \left| \mathbf{D}_k u_h - \left[\nabla u\right]_{M_k} \right|^2 \right\}^{\frac{1}{2}} \leq C \left\{ \left| u_I - u_h \right|_{1,\Omega} + h^2 \left| u \right|_{3,\Omega} \right\} \tag{2.7}$$

where C is a constant, (from now on all such constants will be designated C), the M_k are element edge mid-points internal to Ω and $u_I \in S_E^h$ is the continuous piecewise linear interpolant to u.

It has since been shown independently by Goodsell and Whiteman [13] and Levine [24] that (2.7) may be extended to element edge mid-points lying on $\partial\Omega$. This is done, for $v_h \in S^h(\Omega)$, by defining the recovered gradient at such a mid-point $M \in \partial\Omega$ as the weighted average of the three element gradients as shown in Fig. 2.1.

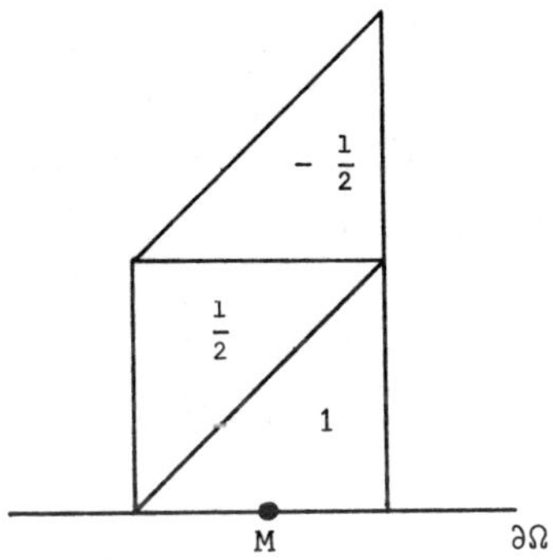

Fig. 2.1.

The point of crucial significance here is that for u, the solution of (2.2), quadratic on the union of triangles, the recovered gradient of the piecewise linear interpolant $u_I \in S_E^h$ to u is exact at the mid-point in question. The following estimate, extending (2.7), results

$$h\left\{\sum_{k=1}^{K}{}^{3}\left|D_k u_h - \left[\nabla u\right]_{M_k}\right|^2\right\}^{\frac{1}{2}} \leqq C\left\{\left|u_I - u_h\right|_{1,\Omega} + h^2\left|u\right|_{3,\Omega}\right\} . \qquad (2.8)$$

2.2 A Superconvergent $H^1(\Omega)$ Error Seminorm

For $v_h \in S^h(\Omega)$, the recovered gradient of the finite element approximation at the mid-points of element edges exists at all three edge mid-points of every triangle in Ω.

For $v_h \in S^h(\Omega)$ and the above recovered point gradients we define the *discontinuous* piecewise linear recovered gradient function

$$\nabla^* v_h \in \left[L_\infty(\Omega)\right]^2 \qquad (2.9)$$

in each triangle to be the linear interpolant to the three recovered gradient values at the edge mid-points. We now define the $H^1(\Omega)^*$ error seminorm

$$|u - u_h|^*_{1,\Omega} \equiv \left\{ \int_\Omega |\nabla u - \nabla^* u_h|^2 dx \right\}^{\frac{1}{2}} . \tag{2.10}$$

For this Goodsell and Whiteman [13] proved that:

Theorem 2.1 For u and u_h respectively the solutions of (2.2) and (2.5) and with $|u - u_h|^*_{1,\Omega}$ defined as in (2.10)

$$|u - u_h|^*_{1,\Omega} \leq C\left\{ |u_I - u_h|_{1,\Omega} + h^2 |u|_{3,\Omega} \right\} \tag{2.11}$$

where $u_I \in S^h_E$ is the piecewise linear interpolant to u.

The proof of Theorem 2.1 given in [13] depends on the fact that we have a regular mesh, thus making $\nabla^* u_I$ and ∇u identical for u quadratic. This latter property allows the Sobolev Lemma, see Section 1.2.3 and the Bramble-Hilbert Lemma, see Section 1.2.4, to be used, thus enabling an extra power of h (than would be expected if $\nabla^* u_I$ and ∇u were only identical for u linear) to be obtained in the estimate.

2.3 A Superconvergent Recovered Gradient at the Vertices

At each internal vertex z_k, $k = 1,2,\ldots,K_2$, we define the recovered gradient of the finite element approximation as the average of the gradients in the six triangles meeting at the vertex. For this we state a superconvergent estimate for the ℓ_2 average error.

For each z_k we consider the six triangles $T^1_k, T^2_k, \ldots, T^6_k$ meeting at the vertex

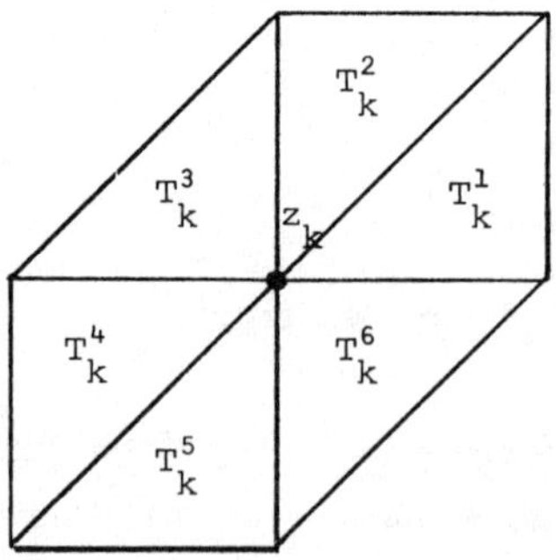

Fig. 2.2

as in Fig. 2.2. For $v_h \in S^h(\Omega)$ we define the recovered vertex gradient at z_k as

$$\mathbf{H}_k v_h \equiv \frac{1}{6} \sum_{i=1}^{6} [\nabla v_h]_{T^i_k} . \tag{2.12}$$

We have the theorem, proved in [13].

Theorem 2.2 If the recovered gradient $\mathbf{H}_k v_h$ is defined as in (2.12) for $v_h \in S^h(\Omega)$, and if u and u_h are respectively the solutions of (2.2) and (2.5), then

$$h\left\{\sum_{k=1}^{K_2} \left|\mathbf{H}_k u_h - \left[\nabla u\right]_{z_k}\right|^2\right\}^{\frac{1}{2}} \leq C\left\{\left|u_I - u_h\right|_{1,\Omega} + h^2\left|u\right|_{3,\Omega}\right\} . \tag{2.13}$$

In the proof of Theorem 2.2 the regular mesh property has again been exploited as it makes the recovered gradients of the interpolant $u_I \in S_E^h$ identical with the gradients of u at the nodal points if u is quadratic. As before the Sobolev Lemma and the Bramble-Hilbert Lemma can be used to produce the extra power of h in the estimate. As in the case of the recovered mid-point gradients of Section 2.1, the estimate (2.13) has been extended to include vertices on $\partial\Omega$.

Following Hlavacek and Krizek [17] for $v_h \in S^h(\Omega)$, we define the recovered gradient at such a vertex $z \in \partial\Omega$ as the weighted average of the element gradients as shown in Fig. 2.3.

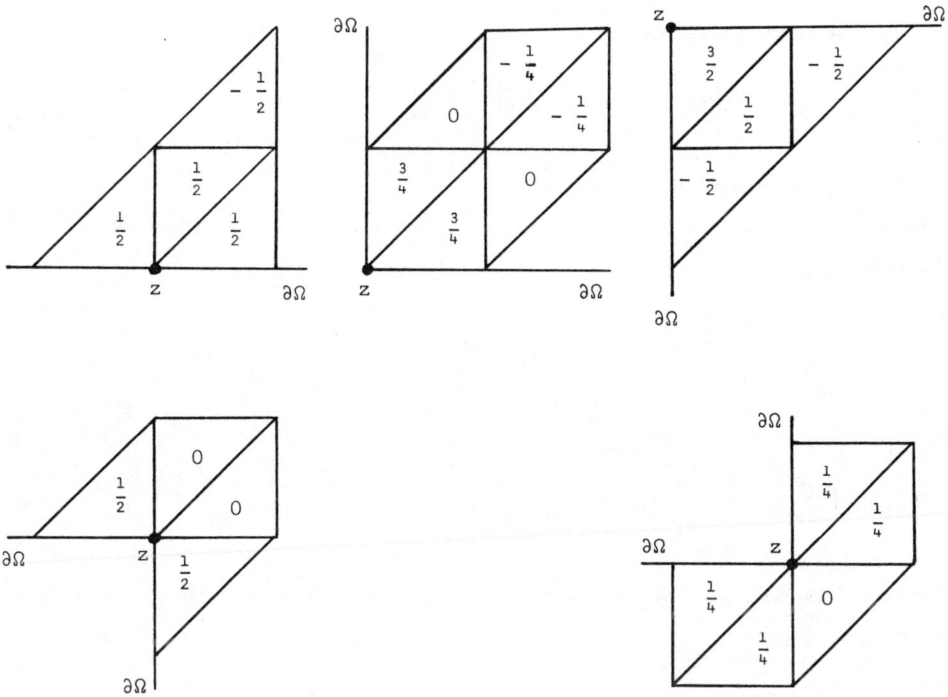

Fig. 2.3

Again the point of crucial significance in these definitions is that for u quadratic on a given union of triangles, the recovered gradient of the piecewise linear interpolant, $u_I \in S_E^h$, is exact at the vertex in question. We then have the following estimate which extends (2.13),

$$h\left\{ \sum_{k=1}^{K_4} \left|\mathbf{H}_k u_h - \left[\nabla u\right]_{z_k}\right|^2 \right\}^{\frac{1}{2}} \leq C\left\{\left|u_I - u_h\right|_{1,\Omega} + h^2\left|u\right|_{3,\Omega}\right\} . \qquad (2.14)$$

2.4 An Alternative Superconvergent $H^1(\Omega)$ Error Seminorm

For $v_h \in S^h(\Omega)$ we define the piecewise linear recovered gradient function

$$\nabla^+ v_h \in \left[W^1_\infty(\Omega)\right]^2 ,$$

as the linear interpolant to the three values of the recovered gradient of v_h at the vertices of each triangle. Unlike $\nabla^* v_h$ the function $\nabla^+ v_h$ is continuous. An analysis virtually the same as that in Section 2.2 shows that for the $H^1(\Omega)^+$ error seminorm defined as

$$\left|u - u_h\right|^+_{1,\Omega} \equiv \left\{ \int_\Omega \left|\nabla u - \nabla^+ u_h\right|^2 d\mathbf{x}\right\}^{\frac{1}{2}} , \qquad (2.15)$$

there is the superconvergent error estimate

$$\left|u - u_h\right|^+_{1,\Omega} \leq C\left\{\left|u_I - u_h\right|_{1,\Omega} + h^2\left|u\right|_{3,\Omega}\right\} . \qquad (2.16)$$

2.5 The Superconvergent Estimates

We now combine the results of this section and define, albeit somewhat artificially,

$$\left|u - u_h\right|^{RM}_{1,\Omega} \equiv h\left\{ \sum_{k=1}^{K_3} \left|\mathbf{D}_k u_h - \left[\nabla u\right]_{M_k}\right|^2 \right\}^{\frac{1}{2}} \qquad (2.17)$$

and

$$\left|u - u_h\right|^{RV}_{1,\Omega} \equiv h\left\{ \sum_{k=1}^{K_4} \left|\mathbf{H}_k u_h - \left[\nabla u\right]_{z_k}\right|^2 \right\}^{\frac{1}{2}} , \qquad (2.18)$$

which, as discrete ℓ_2 averages, are not really Sobolev norms. However we may informally think of

$$\left|u - u_h\right|^*_{1,\Omega} \quad \text{and} \quad \left|u - u_h\right|^+_{1,\Omega}$$

as like (2.17) and (2.18), but sampling at every point, rather than just at element edge mid-points and element vertices. From (2.8), (2.11), (2.14) and (2.16) we then have

$$|u - u_h|^{\circledast}_{1,\Omega} \leq C\{|u_I - u_h|_{1,\Omega} + h^2|u|_{3,\Omega}\} ,$$

$$\circledast \equiv RM, RV, *, + . \tag{2.19}$$

For these estimates to be superconvergent, as we have claimed, we must obtain an $O(h^2)$ estimate for the term

$$|u_I - u_h|_{1,\Omega} .$$

It is worth highlighting here that, throughout all this work, our estimates will involve both an interpolation result, which we can estimate fairly easily, and a norm of the difference between the gradients of the interpolant, u_I, and the finite element approximation, u_h, a global approximation; the major challenge of our work is the obtaining of sharp error estimates for this norm.

In the case of the L_2-norm of the gradient difference, we exploit the fact that ∇u_I is a superconvergent $(O(h^2))$ approximation to ∇u_h. This was first shown by Oganesjan and Ruchovec [30]; a simpler proof was given by Goodsell in [11,Ch.6]. Thus

$$|u_I - u_h|_{1,\Omega} \leq Ch^2|u|_{3,\Omega} \tag{2.20}$$

and hence from (2.19)

$$|u - u_h|^{\circledast}_{1,\Omega} \leq Ch^2|u|_{3,\Omega} . \tag{2.21}$$

By comparison with the standard $(O(h))$ L_2-error norm of the gradient, the estimate (2.21) has a full $O(h)$ faster convergence.

In the proof of (2.20) we consider the term $a(u_I-u,v_h)$ for arbitrary $v_h \in S_0^h(\Omega)$. Exploiting the fact that this term vanishes for quadratic u, using the Sobolev and Bramble-Hilbert Lemmas as before, we obtain the estimate

$$|a(u_I-u,v_h)| \leq Ch^2|u|_{3,\Omega}|v_h|_{1,\Omega} \quad \forall \quad v_h \in S_0^h(\Omega) . \tag{2.22}$$

An alternative form of this estimate, which will be required later in this paper is

$$|a(u_I-u,v_h)| \leq Ch^2|u|_{3,\infty,\Omega}|v_h|_{1,1,\Omega} \quad \forall \quad v_h \in S_0^h(\Omega) . \tag{2.23}$$

To complete the proof of (2.20) we note that $(u_I-u_h) \in S_0^h(\Omega)$, because problem (2.1) has *Dirichlet* boundary conditions, so that v_h in (2.22) can be replaced by (u_I-u_h). Using the well-known orthogonality property, for (2.2) and (2.5), that

$$a(u-u_h,v_h) = 0 \quad \forall \quad v_h \in S_0^h(\Omega) , \tag{2.24}$$

we have that

$$
\begin{aligned}
|u_I - u_h|^2_{1,\Omega} &= a(u_I - u_h, u_I - u_h) \\
&= a(u_I - u, u_I - u_h) \\
&\leq Ch^2 |u|_{3,\Omega} |u_I - u_h|_{1,\Omega}
\end{aligned} \tag{2.25}
$$

from which (2.20) follows immediately.

2.6 Mixed Boundary Conditions

We now change problem (2.1) to have mixed boundary conditions so that

$$
\begin{aligned}
-\Delta[u(\mathbf{x})] &= f(\mathbf{x}) , \quad \mathbf{x} \in \Omega , \\
u(\mathbf{x}) &= g^D(\mathbf{x}) , \quad \mathbf{x} \in \partial\Omega^D , \\
\frac{\partial u}{\partial n} &= g^N(\mathbf{x}) , \quad \mathbf{x} \in \partial\Omega^N ,
\end{aligned} \tag{2.26}
$$

where the boundary $\partial\Omega$ again consists of sides parallel to the x- and y-axes; hence Ω is again rectangular and is such that

$$
\partial\Omega \equiv \partial\Omega^D \cup \partial\Omega^N . \tag{2.27}
$$

It is again assumed that f, g^D and g^N are such that $u \in H^3(\Omega)$.

The definitions of H^1_E and S^h_E are generalised so that

$$
H^1_E \equiv \left\{ v : v \in H^1(\Omega) , v = g^D \text{ on } \partial\Omega^D \right\}
$$

$$
S^h_E \equiv \left\{ v_h : v_h \in S^h(\Omega) , v_h = g^D_I \text{ on } \partial\Omega^D \right\} .
$$

We also define

$$
H \equiv \left\{ v : v \in H^1(\Omega) , v = 0 \text{ on } \partial\Omega^D \right\}
$$

$$
H^h \equiv \left\{ v_h : v_h \in S^h(\Omega) , v_h = 0 \text{ on } \partial\Omega^D \right\} .
$$

In the weak formulation of (2.26) we seek

$u \in H^1_E \cap H^3(\Omega)$ such that

$$
a(u,v) = (f,v) + (g^N, v)_{\partial\Omega^N} , \quad \forall \; v \in H , \tag{2.28}
$$

where

$$
(g^N, v)_{\partial\Omega^N} \equiv \int_{\partial\Omega^N} g^N \, v \, ds \tag{2.29}
$$

and for the finite element approximation we seek $u_h \in S_E^h$ such that

$$a(u_h,v_h) = (f,v_h) + (g^N,v_h)_{\partial\Omega^N} , \quad \forall \ v_h \in H^h . \tag{2.30}$$

The orthogonality property follows from (2.28) and (2.30), as $H^h \subset H$, so that

$$a(u- u_h,v_h) = 0 , \quad \forall \ v_h \in H^h , \tag{2.31}$$

The introduction of mixed boundary conditions means that we can no longer, in general, obtain the superconvergent estimates (2.21). Although (2.19) can readily be shown still to hold, the difficulty occurs when we try to estimate the term $|u_I - u_h|_{1,\Omega}$. Clearly $(u_I - u_h) \notin S_0^h(\Omega)$, and so we can no longer use this term to replace v_h in (2.22), and therefore the proof of (2.20) breaks down. However, provided the Neumann boundary conditions are restricted to boundary segments which are parallel to a single axis (2.20) still holds, as does (2.21). The proofs, which are given in Goodsell [12], depend on the fact $a(u_I-u,v_h)$ can be shown to vanish for quadratic u for $v_h \notin S_0^h(\Omega)$ but taking non-zero values on boundary segments parallel to no more than one axis.

3. Poisson Problems: Superconvergence of Recovered Gradients on Subdomains of General Polygonal Regions

3.1 Problem, Subdomains, Finite Element Discretisation

We consider again problem (2.26) where now Ω is a "*general*" polygonal region, such that interior angles at points of $\partial\Omega$ where $\overline{\partial\Omega^D}$ and $\overline{\partial\Omega^N}$ intersect are convex. In this case the analysis of Section 2 is not applicable, so that we now work on a subdomain Ω_0 of Ω. We introduce further subdomains Ω_1 and Ω_2 such that

$$\Omega_0 < \Omega_1 < \Omega_2 \subseteq \Omega , \tag{3.1}$$

where the symbol '<' is defined as follows. Let

$$\operatorname{dist}_<(\Omega_i,\Omega_j) \equiv \inf_{\mathbf{x}\in\partial\Omega_i\backslash(\partial\Omega_i\cap\partial\Omega)} \operatorname{dist}\left\{\mathbf{x},\partial\Omega_j\backslash(\partial\Omega_j\cap\partial\Omega)\right\} ,$$

then $\Omega_i < \Omega_j$ signifies that $\Omega_i \subset \Omega_j$ and $\operatorname{dist}_<(\Omega_i,\Omega_j) > 0$.

The subdomains satisfy the following conditions, see Fig. 3.1.

Condition 1 Ω_0 and Ω_2 are rectangular and are partitioned uniformly into isosceles right-angled triangles as in Section 2.1. The remainder, $\Omega\backslash\Omega_2$, of Ω

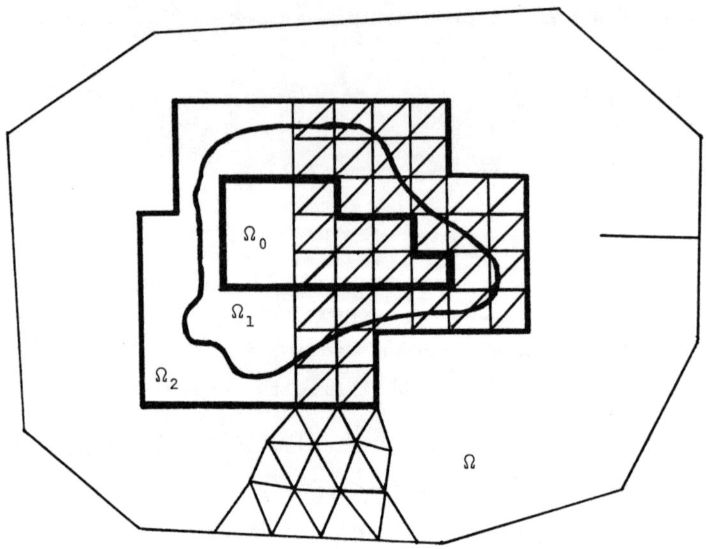

Fig. 3.1

is partitioned quasi-uniformly; i.e. there exist constants C_1, C_2, $0 < C_1 < C_2$, independent of h such that each element completely contains a circle of radius C_1h and is completely contained in a circle of radius C_2h.

Condition 2 Ω_1 is remote from all segments of $\partial\Omega$ on which Neumann boundary conditions are specified. Thus

$$\partial\Omega_1 \cap \partial\Omega^N = \emptyset \ . \tag{3.2}$$

Condition 3 Ω_2 is remote from points of $\partial\Omega$ at which u has a singularity so that $u \in H^3(\Omega_2)$.

We assume that f, g^D, and g^N are themselves sufficiently regular to guarantee the H^3-regularity of u away from $\partial\Omega$. The precise conditions for this will be given later in Section 3.3. Conditions 1-3 are not very restrictive and, for sufficiently small h, Ω_0 can represent most of Ω.

Estimates (2.19) can readily be shown to hold locally over Ω_0, so that the task of this section is to estimate the term $|u_I - u_h|_{1,\Omega_0}$. For this we employ a *cut-off* function.

The cut-off function $\omega \in W^2_\infty(\Omega)$ is defined such that

$$\omega(\mathbf{x}) \equiv \begin{cases} 1 & \text{in } \bar{\Omega}_0 \\ 0 & \text{in } (\bar{\Omega}\backslash\bar{\Omega}_1)\cup\left\{\partial\Omega_1\backslash(\partial\Omega_1\cap\partial\Omega)\right\} . \end{cases} \tag{3.3}$$

3.2 Lemmas Necessary for Superconvergence Estimates

Two lemmas are now proved, which show that the seminorm $|u_I - u_h|_{1,\Omega_0}$ is controlled by the norm $|u_I - u_h|_{0,\Omega_2}$. The proofs are much simplified versions of those of Wheeler and Whiteman [38].

Lemma 3.1 For $u_I \in S_E^h$, $u_h \in S_E^h$ the solution of (2.30)

$$|a(u_I-u_h,v_h)| \leq Ch^2|u|_{3,\Omega_2}|v_h|_{1,\Omega_2}\ , \quad \forall\ v_h \in S_0^h(\Omega_2) \tag{3.4}$$

Proof As $S_0^h(\Omega_2) \subset H^h$, we have from the orthogonality property (2.31) that

$$a(u_I-u_h,v_h) = a(u_I-u,v_h)\ , \quad \forall\ v_h \in S_0^h(\Omega_2)\ . \tag{3.5}$$

Then, as v_h vanishes outside Ω_2, the right-hand side of (3.5) is an integral over Ω_2 only. The proof then follows that of (2.22), which holds locally over Ω_2.

Lemma 3.2 For $u_I \in S_E^h$ and $u_h \in S_E^h$ the solution of (2.30)

$$|u_I - u_h|_{1,\Omega_0} \leq C(\omega)\left\{|u_I - u_h|_{0,\Omega_2} + h^2|u|_{3,\Omega_2}\right\}\ , \tag{3.6}$$

where $C(\omega)$ is a constant depending on the cut-off function ω.

Proof We let

$$\theta \equiv u_I - u_h \tag{3.7}$$

and $\psi \in S^h(\Omega)$ be the continuous piecewise linear interpolant to $\omega^2\theta$,

$$\psi \equiv (\omega^2\theta)_I\ . \tag{3.8}$$

Then

$$\psi \in S_0^h(\Omega_2)\ . \tag{3.9}$$

To see this we note that from (3.3), ω vanishes in $(\bar{\Omega}\backslash\bar{\Omega}_1)\cup\left\{\partial\Omega_1\backslash(\partial\Omega_1\cap\partial\Omega)\right\}$; and as Ω_1 is remote from $\partial\Omega^N$, recall (3.2), θ vanishes on $\partial\Omega_1 \cap \partial\Omega$. Hence $\omega^2\theta$ vanishes in $(\bar{\Omega}\backslash\bar{\Omega}_1) \cup \partial\Omega_1$, and hence (3.9) holds. Now

$$|\omega\theta|^2_{1,\Omega_2} \equiv \int_{\Omega_2} |\nabla(\omega\theta)|^2 d\mathbf{x}$$

$$= \int_{\Omega_2} |\theta|^2 |\nabla\omega|^2 d\mathbf{x} + \int_{\Omega_2} \nabla\theta . \nabla(\omega^2\theta - \psi) d\mathbf{x} + \int_{\Omega_2} \nabla\theta . \nabla\psi \, d\mathbf{x} = Q_1 + Q_2 + a(\theta,\psi). \quad (3.10)$$

We now estimate each of these three terms separately.

$$|Q_1| = \int_{\Omega_2} |\theta|^2 |\nabla\omega|^2 d\mathbf{x} \leq |\omega|^2_{1,\infty,\Omega_2} |\theta|^2_{0,\Omega_2} = C(\omega)|\theta|^2_{0,\Omega_2} \quad (3.11)$$

$$|Q_2| \leq \int_{\Omega_2} |\nabla\theta| \, |\nabla(\omega^2\theta - \psi)| d\mathbf{x} \leq |\theta|_{1,\Omega_2} |\omega^2\theta - \psi|_{1,\Omega_2} . \quad (3.12)$$

But from interpolation theory, since ψ interpolates to $\omega^2\theta$, see Ciarlet [5,p.133]

$$|\omega^2\theta - \psi|_{1,\Omega_2} \leq Ch \left\{ \sum_{T_k \in \Omega_2} |\omega^2\theta|^2_{2,T_k} \right\}^{\frac{1}{2}} \leq C(\omega)h\left\{ |\omega\theta|_{1,\Omega_2} + |\theta|_{0,\Omega_2} \right\} . \quad (3.13)$$

Now combining (3.12) and (3.13), and using the properties of functions in $S^h(\Omega)$, see Ciarlet [5,p.142], we have

$$|Q_2| \leq C(\omega)h|\theta|_{1,\Omega_2} \left\{ |\omega\theta|_{1,\Omega_2} + |\theta|_{0,\Omega_2} \right\} \leq C(\omega)|\theta|_{0,\Omega_2} \left\{ |\omega\theta|_{1,\Omega_2} + |\theta|_{0,\Omega_2} \right\} . \quad (3.14)$$

Finally from Lemma 3.1 and (3.9) we have

$$|a(\theta,\psi)| \leq Ch^2 |u|_{3,\Omega_2} |\psi|_{1,\Omega_2} . \quad (3.15)$$

But from (3.13)

$$|\psi|_{1,\Omega_2} = |\psi - \omega^2\theta + \omega^2\theta|_{1,\Omega_2} \leq |\omega^2\theta - \psi|_{1,\Omega_2} + |\omega^2\theta|_{1,\Omega_2}$$

$$\leq C(\omega)h\left\{ |\omega\theta|_{1,\Omega_2} + |\theta|_{0,\Omega_2} \right\} + |\omega^2\theta|_{1,\Omega_2} . \quad (3.16)$$

Now it readily follows that

$$|\omega^2\theta|_{1,\Omega_2} \leq C(\omega)\left\{ |\omega\theta|_{1,\Omega_2} + |\theta|_{0,\Omega_2} \right\} , \quad (3.17)$$

so that from (3.15) - (3.17)

$$|a(\theta,\psi)| \leq C(\omega)h^2 |u|_{3,\Omega_2} \left\{ |\omega\theta|_{1,\Omega_2} + |\theta|_{0,\Omega_2} \right\} . \quad (3.18)$$

We now combine (3.10), (3.11), (3.14) and (3.18) to obtain

$$|\omega\theta|^2_{1,\Omega_2} \leq |Q_1| + |Q_2| + |a(\theta,\psi)| \leq C(\omega)|\omega\theta|_{1,\Omega_2}\left\{|\theta|_{0,\Omega_2} + h^2|u|_{3,\Omega_2}\right\}$$

$$+ C(\omega)|\theta|_{0,\Omega_2}\left\{|\theta|_{0,\Omega_2}+h^2|u|_{3,\Omega_2}\right\} \leq C(\omega)\varepsilon|\omega\theta|^2_{1,\Omega_2} + \frac{C(\omega)}{\varepsilon}\left\{|\theta|^2_{0,\Omega_2}+h^4|u|^2_{3,\Omega_2}\right\}$$

$$+ C(\omega)\left\{|\theta|^2_{0,\Omega_2} + h^4|u|^2_{3,\Omega_2}\right\}, \tag{3.19}$$

where ε is an arbitrary small constant independent of h. From (3.19) we have immediately that

$$|\omega\theta|_{1,\Omega_2} \leq C(\omega)\left\{|\theta|_{0,\Omega_2} + h^2|u|_{3,\Omega_2}\right\}. \tag{3.20}$$

But

$$|\theta|_{1,\Omega_0} = |\omega\theta|_{1,\Omega_0} \leq |\omega\theta|_{1,\Omega_2}$$

and this together with (3.20) proves the result (3.6).

Remark 3.1

We recall that Condition 2 of Section 3.1 stipulated that (3.2) $\partial\Omega_1 \cap \partial\Omega^N = \emptyset$ and that the purpose of this was to ensure (3.9) in order that ψ could be substituted for v_h in Lemma 3.1. This condition can in fact be slightly weakened.

Following the arguments of Section 2.6, Lemma 3.1 can be shown to hold for $v_h \notin S^h_0(\Omega_2)$, but vanishing outside Ω_2, and taking non-zero values on sides of $\partial\Omega_2 \cap \partial\Omega$ parallel to one axis only. Then provided $\partial\Omega_1 \cap \partial\Omega^N$ is parallel to one axis only, $\psi \notin S^h_0(\Omega_2)$ will satisfy the above requirements on v_h, and Lemma 3.2 can readily be seen still to hold

3.3 Superconvergence Estimates for Recovered Approximations

3.3.1 Regularity Results

We first give the regularity requirements on f, g^D and g^N to guarantee the H^3-regularity of u, the solution of (2.26), away from $\partial\Omega$, in order that $u \in H^3(\Omega_2)$ as stipulated in Condition 3 of Section 3.1.

These requirements are, see Grisvard [14] and [15, pp.182-248],

$$f \in H^1(\Omega) \tag{3.21}$$

$$g^D \in H^{5/2}(\partial\Omega_i^D) \tag{3.22}$$

$$g^N \in H^{3/2}(\partial\Omega_i^N) \tag{3.23}$$

where the $\partial\Omega_i^D$ and $\partial\Omega_i^N$ are, respectively, the segments of $\partial\Omega^D$ and $\partial\Omega^N$.

In order to obtain estimates for the term $|u_I - u_h|_{0,\Omega_2}$ in (3.6) we must produce estimates over Ω because u_h is a *global* approximation to u. In this we make use of the Aubin-Nitsche Lemma, see Ciarlet [5,p.137], and Schatz [34], which we have modified slightly to cope with the non-homogeneous Dirichlet boundary conditions, $u(\mathbf{x}) = g^D(\mathbf{x})$, $\mathbf{x} \in \partial\Omega^D$, of (2.26). This lemma demands the use of a problem which is adjoint to (2.26),

$$\left.\begin{aligned} -\Delta[w(\mathbf{x})] &= f(\mathbf{x}) , && \mathbf{x} \in \Omega \\ w(\mathbf{x}) &= 0 , && \mathbf{x} \in \partial\Omega^D \\ \frac{\partial w}{\partial n} &= 0 , && \mathbf{x} \in \partial\Omega^N \end{aligned}\right\} \tag{3.24}$$

where

$$f \in L_2(\Omega) . \tag{3.25}$$

The *global* regularity of u and w is determined by the forms of the singularities of these functions at points of $\partial\Omega$. This regularity is now considered, see Grisvard [14] and [15,pp.182-248].

Let φ be the maximum of

(i) the largest interior angle at the corners of $\partial\Omega^D$,

(ii) the largest interior angle at the corners of $\partial\Omega^N$,

(iii) *twice* the largest interior angle at the points of intersection of $\overline{\partial\Omega^D}$ and $\overline{\partial\Omega^N}$ (including, if appropriate, the straight angle π).

There are two cases to consider.

Case 1: $0 < \varphi \leq \pi$

For this Ω is a convex polygon with interior angles at the points of intersection of $\overline{\partial\Omega^D}$ and $\overline{\partial\Omega^N}$ either acute or right-angles. We have

$$u,w \in H^2(\Omega) \tag{3.26}$$

and further

$$\|w\|_{2,\Omega} \leq C|f|_{0,\Omega} \, . \tag{3.27}$$

Case 2: $\pi < \varphi \leq 2\pi$

For this Ω is a general polygon and the interior angles at the points of intersection of $\overline{\partial\Omega^D}$ and $\overline{\partial\Omega^N}$ are convex, as was required in Section 3.1. We have for this case

$$u,w \in W^2_{p-\varepsilon}(\Omega) \tag{3.28}$$

for arbitrary small ε, and where

$$\frac{4}{3} \leq p = \frac{2\varphi}{(2\varphi-\pi)} < 2 \tag{3.29}$$

and further

$$\|w\|_{2,p-\varepsilon,\Omega} \leq C|f|_{0,p,\Omega} \, . \tag{3.30}$$

3.3.2 Aubin-Nitsche Lemma and Superconvergence Estimates

We first state and prove a modified version of the Aubin-Nitsche Lemma, see Ciarlet [5,p.137], and Schatz [34] with cases corresponding to those of Section 3.4.1.

Lemma 3.3 (Aubin-Nitsche)

Case 1 If $u \in H^2(\Omega)$ is the solution of (2.28) and $u_h \in S^h_E$ is the solution of (2.30), then

$$|u_I - u_h|_{0,\Omega_2} \leq Ch^2\{|u|_{2,\Omega} + |g^D|'_{2,\partial\Omega^D}\} \, . \tag{3.31}$$

Case 2 If $u \in W^2_{p-\varepsilon}(\Omega)$: $\frac{4}{3} \leq p < 2$ then

$$|u_I - u_h|_{0,\Omega_2} \leq Ch^{\gamma-\varepsilon}\left\{|u|_{2,p-\varepsilon,\Omega} + |g^D|'_{2,\partial\Omega^D}\right\} \tag{3.32}$$

where

$$1 \leqq \gamma = 4 - \frac{4}{p} < 2 \ . \tag{3.33}$$

Proof Céa's Lemma, see Ciarlet [5,pp.104-105] gives

$$|u - u_h|_{1,\Omega} \leqq |u - u_I|_{1,\Omega} \ . \tag{3.34}$$

Using some well-known results from interpolation theory and the Sobolev Embedding Theorem for traces of functions on $\partial\Omega$, see Adams [1,pp.113-115], we have that

$$|u_I - u_h|_{0,\Omega_2} \leq |u_I - u_h|_{0,\Omega} \leqq |u - u_h|_{0,\Omega} + |u - u_I|_{0,\Omega} \ . \tag{3.35}$$

Let w be the solution of the adjoint problem (3.24) with

$$f \equiv u - u_h \in L_2(\Omega) \ . \tag{3.36}$$

Now from (3.24), (3.36) and the orthogonality property (2.31), noting that $w_I \in H^h$, we have

$$\begin{aligned}
|u - u_h|^2_{0,\Omega} &\equiv \int_\Omega |u - u_h|^2 d\mathbf{x} \\
&= - \int_\Omega (u - u_h)\Delta w \ d\mathbf{x} \\
&= \int_\Omega \nabla(u - u_h).\nabla w \ d\mathbf{x} - \oint_{\partial\Omega} (u - u_h) \ \frac{\partial w}{\partial n} \ ds \\
&= a(u - u_h,w) - \int_{\partial\Omega^D} (g^D - g^D_I) \ \frac{\partial w}{\partial n} \ ds \\
&= a(u - u_h,w - w_I) - \int_{\partial\Omega^D} (g^D - g^D_I) \ \frac{\partial w}{\partial n} \ ds \\
&\leq |a(u - u_h,w - w_I)| + \left|\int_{\partial\Omega^D} (g^D - g^D_I) \ \frac{\partial w}{\partial n} \ ds\right| \ .
\end{aligned} \tag{3.37}$$

We now consider separately the two cases of Section 3.3.1.

Case 1 $(u \in H^2(\Omega))$

We have from (3.27), (3.34), (3.36) and (3.37)

$$|u - u_h|^2_{0,\Omega} \leq |u - u_h|_{1,\Omega}|w - w_I|_{1,\Omega} + |g^D - g^D_I|_{0,\partial\Omega^D}|\nabla w|_{0,\partial\Omega^D}$$

$$\leq |u - u_I|_{1,\Omega}|w - w_I|_{1,\Omega} + |g^D - g^D_I|_{0,\partial\Omega^D}|\nabla w|_{0,\partial\Omega^D}$$

$$\leq Ch^2\left\{|u|_{2,\Omega}|w|_{2,\Omega} + |g^D|'_{2,\partial\Omega^D}\|w\|_{2,\Omega}\right\}$$

$$\leq Ch^2\left\{|u|_{2,\Omega} + |g^D|'_{2,\partial\Omega^D}\right\}\|w\|_{2,\Omega}$$

$$\leq Ch^2\left\{|u|_{2,\Omega} + |g^D|'_{2,\partial\Omega^D}\right\}|u - u_h|_{0,\Omega}$$

thus

$$|u - u_h|_{0,\Omega} \leq Ch^2\left\{|u|_{2,\Omega} + |g^D|'_{2,\partial\Omega^D}\right\} . \tag{3.38}$$

Also, directly from interpolation theory, we have that

$$|u - u_I|_{0,\Omega} \leq Ch^2|u|_{2,\Omega} . \tag{3.39}$$

Result (3.31) then follows immediately from (3.35), (3.38) and (3.39). Otherwise

Case 2 $(u \in W^2_{p-\varepsilon}(\Omega),\ \frac{4}{3} \leq p < 2)$

We have from (3.30), (3.34), (3.36) and (3.37)

$$|u - u_h|^2_{0,\Omega} \leq |u - u_h|_{1,\Omega}|w - w_I|_{1,\Omega} + |g^D - g^D_I|_{0,\partial\Omega^D}|\nabla w|_{0,\partial\Omega^D}$$

$$\leq |u - u_I|_{1,\Omega}|w - w_I|_{1,\Omega} + |g^D - g^D_I|_{0,\partial\Omega^D}|\nabla w|_{0,\partial\Omega^D}$$

$$\leq C\left\{h^{\gamma-\varepsilon}|u|_{2,p-\varepsilon,\Omega}|w|_{2,p-\varepsilon,\Omega} + h^2|g^D|'_{2,\partial\Omega^D}\|w\|_{2,p-\varepsilon,\Omega}\right\}$$

$$\leq Ch^{\gamma-\varepsilon}\left\{|u|_{2,p-\varepsilon,\Omega} + |g^D|'_{2,\partial\Omega^D}\right\}\|w\|_{2,p-\varepsilon,\Omega}$$

$$\leq Ch^{\gamma-\varepsilon}\left\{|u|_{2,p-\varepsilon,\Omega} + |g^D|'_{2,\partial\Omega^D}\right\}|u-u_h|_{0,p,\Omega}$$

$$\leq Ch^{\gamma-\varepsilon}\left\{|u|_{2,p-\varepsilon,\Omega} + |g^D|'_{2,\partial\Omega^D}\right\}|u - u_h|_{0,\Omega} \quad ,$$

where $1 \leq \gamma = 4 - \frac{4}{p} < 2$ as in (3.33). Thus

$$|u - u_h|_{0,\Omega} \leq Ch^{\gamma-\varepsilon}\left\{|u|_{2,p-\varepsilon,\Omega} + |g^D|'_{2,\partial\Omega^D}\right\} . \tag{3.40}$$

In the limiting case, $p = \frac{4}{3}$, the above analysis is not strictly correct as $W^1_{4/3-\varepsilon}(\Omega) \not\longrightarrow L_2(\partial\Omega)$. Instead we now have

$$|u - u_h|^2_{0,\Omega} \leq |u - u_h|_{1,\Omega}|w - w_I|_{1,\Omega} + |g^D - g^D_I|_{0,2+\varepsilon,\partial\Omega^D}|\nabla w|_{0,2-\varepsilon,\partial\Omega^D}$$

which then leads to the same estimate (3.40) as before. Also, directly from interpolation theory, we have that

$$|u - u_I|_{0,\Omega} \leq Ch^{1+\gamma/2-\varepsilon}|u|_{2,p-\varepsilon,\Omega} . \tag{3.41}$$

Result (3.32) then follows immediately from (3.35), (3.40) and (3.41), noting that $\gamma < 1 + \frac{\gamma}{2}$.

Finally in this section we collect together our results in the following theorem.

Theorem 3.1

For u the solution of (2.28) and $u_h \in S^h_E$ the solution of (2.30) and the subdomain $\Omega_0 \subset \Omega$.

Case 1 If $u \in H^2(\Omega) \cap H^3(\Omega_2)$ then

$$|u - u_h|^{\circledast}_{1,\Omega_0} \leq C(\omega)h^2\left\{|u|_{2,\Omega}+|u|_{3,\Omega_2}+|g^D|'_{2,\partial\Omega^D}\right\} : \circledast = RM,\ RV,\ *,\ + , \quad (3.42)$$

where $C(\omega)$ is a constant depending on the cut-off function ω, and $|\cdot|^{\circledast}_{1,\Omega}$ is as in (2.19)

Case 2 If $u \in W^2_{p-\varepsilon}(\Omega) \cap H^3(\Omega_2) : \frac{4}{3} \leq p < 2$ then

$$|u - u_h|^{\circledast}_{1,\Omega_0} \leq C(\omega)h^{\gamma-\varepsilon}\left\{|u|_{2,p-\varepsilon,\Omega}+|u|_{3,\Omega_2}+|g^D|'_{2,\partial\Omega^D}\right\} : \circledast = RM,\ RV,\ *,\ + \quad (3.43)$$

where $1 \leq \gamma = 4 - \frac{4}{p} < 2$ as in (3.33).

Proof Estimates (3.42) and (3.43) follow from (3.6), (3.31) and (3.32).

3.4 Comparison with Standard Gradient Error Estimates

We compare the error estimates for the recovered gradients of the finite element approximation, u_h, obtained in Theorem 3.1, with the standard L_2-norm of the error of the gradient on a subdomain Ω_0 of a general polygonal region Ω. We use the result on subdomains due to Nitsche and Schatz [28],

$$|u - u_h|_{1,\Omega_0} \leq C\left\{|u - u_h|_{0,\Omega_2} + h|u|_{2,\Omega_2}\right\} . \quad (3.44)$$

We again consider separately the two cases of Section 3.4.1.

Case 1 If $u \in H^2(\Omega) \cap H^3(\Omega_2)$, we have from (3.42) $O(h^2)$ estimates for

$$|u - u_h|^{\circledast}_{1,\Omega_0} : \circledast = RM,\ RV,\ *,\ + .$$

However, we only have an $O(h)$ estimate for $|u - u_h|_{1,\Omega_0}$. Thus, in this case, we still have a full $O(h)$ faster convergence using recovered gradients.

Case 2 If $u \in W^2_{p-\varepsilon}(\Omega) \cap H^3(\Omega_2) : \frac{4}{3} \leq p < 2$, we have from (3.43) $O(h^{\gamma-\varepsilon})$ estimates for $|u - u_h|^{\circledast}_{1,\Omega_0} : \circledast = RM, RV, *, +$; where $1 \leq \gamma = 4 - \frac{4}{p} < 2$ as in

(3.33). However, we only have an O(h) estimate for $|u - u_h|_{1,\Omega_0}$ ($O(h^{1-\varepsilon})$ in the limiting case of $p = \frac{4}{3}$).

Thus, in Case 2, we still have a faster rate of convergence using recovered gradients, but with the improvement decreasing with the regularity of the solution u; except in the limiting case of $p = 4/3$, which corresponds to a region with a crack. In this case, for both recovered and standard gradients, we have $O(h^{1-\varepsilon})$ estimates.

3.5 The Effects of Numerical Integration

We have until now assumed the finite element approximation u_h, the solution of (2.30), to be calculated using analytic integration to evaluate the terms

(f,v_h) and $(g^N,v_h)_{\partial\Omega^N}$, $v_h \in H^h$.

We now examine the effect on our error estimates, (3.42) and (3.43), of using numerical integration to approximate these terms; in particular the centroid cubature rule on each element, and the mid-point quadrature rule on each element edge on $\partial\Omega^N$ respectively.

Denoting these approximations by $(f,v_h)^Q$ and $(g^N,v_h)^Q_{\partial\Omega^N}$, we obtain a perturbed finite element approximation, $u_h^Q \in S_E^h$ the solution of

$$a(u_h^Q,v_h) = (f,v_h)^Q + (g^N,v_h)^Q_{\partial\Omega^N}, \quad \forall\ v_h \in H^h \tag{3.45}$$

and it can be seen that the additional error in $|u - u_h^Q|^{\circledast}_{1,\Omega_0}$: $\circledast$ = RM, RV, $*$, $+$, is bounded by (a constant times) the seminorm $|u_h - u_h^Q|_{1,\Omega}$.

It can be shown that, see Goodsell [12], provided that f and g^N have the slightly greater regularity than that specified in (3.21) and (3.23),

$$f \in H^2(\Omega) \tag{3.46}$$

$$g^N \in H^2(\partial\Omega_i^N) \tag{3.47}$$

then

$$|u_h - u_h^Q|_{1,\Omega} \leq Ch^{2-\varepsilon}\left\{\|f\|_{2,\Omega} + \|g^N\|'_{2,\partial\Omega^N}\right\} \tag{3.48}$$

and hence our asymptotic rates of convergence are (almost) unaffected.

In fact the $O(h^{2-\varepsilon})$ term comes from the error in $(g^N,v_h)^Q_{\partial\Omega^N}$. If a higher order quadrature rule, e.g. Simpson's, is used instead, (3.48) becomes an $O(h^2)$ estimate. It is also very much easier to prove!

4. Recovered Gradient Superconvergence for Planar Linear Elasticity

4.1 Lamé Equations, Plane Stress and Plane Strain

We consider the problem of planar linear elasticity in which the displacement $\mathbf{u} \equiv \mathbf{u}(\mathbf{x}) \equiv (u_1(\mathbf{x}),u_2(\mathbf{x}))^T$ at the point $\mathbf{x} \equiv (x,y)^T \equiv (x_1,x_2)^T$ of the convex polygon $\Omega \subset \mathbb{R}^2$, with boundary $\partial\Omega \equiv \partial\Omega^C \cup \partial\Omega^T$, satisfies the Lamé equations

$$- \mu\Delta\mathbf{u}(\mathbf{x}) - (\lambda+\mu)\text{grad div } \mathbf{u}(\mathbf{x}) = \mathbf{f}(\mathbf{x}) , \quad \mathbf{x} \in \Omega , \tag{4.1}$$

where λ, μ; $\lambda,\mu > 0$ are the Lamé coefficients of the material, together with the boundary conditions, $\partial\Omega^C \neq \emptyset$, (later we also need $\partial\Omega^T \neq \emptyset$)

$$\mathbf{u}(\mathbf{x}) = \mathbf{0} , \quad \mathbf{x} \in \partial\Omega^C , \tag{4.2}$$

$$\sum_{j=1}^{2} \sigma_{ij}(\mathbf{u}(\mathbf{x}))n_j = g_i(\mathbf{x}) , \quad \mathbf{x} \in \partial\Omega^T , \quad i = 1,2 , \tag{4.3}$$

where $\mathbf{n} \equiv (n_1,n_2)^T$ is the unit outward normal vector to $\partial\Omega$. In (4.1) $\mathbf{f} \equiv (f_1,f_2)^T \in [L_2(\Omega)]^2$ is the vector of volumic forces. In (4.3) $\mathbf{g} \equiv (g_1,g_2)^T \in [L_2(\partial\Omega^T)]^2$ is the vector of boundary tractions and $\sigma_{ij}(\mathbf{u})$ is the stress tensor which satisfies Hooke's law

$$\sigma_{ij}(\mathbf{u}) = \sigma_{ji}(\mathbf{u}) = \lambda \sum_{k=1}^{2} \varepsilon_{kk}(\mathbf{u})\delta_{ij} + 2\mu\varepsilon_{ij}(\mathbf{u}) , \tag{4.4}$$

where the strain tensor $\varepsilon_{ij}(\mathbf{u})$ is defined as

$$\varepsilon_{ij}(\mathbf{u}) = \varepsilon_{ji}(\mathbf{u}) \equiv \frac{1}{2}\left(\partial_j u_i + \partial_i u_j\right) ; \tag{4.5}$$

(∂_j denotes differentiation with respect to x_j).

We note here that the stresses (and strains) are linear combinations of the components of the gradient of $\mathbf{u}$,

$$\nabla\mathbf{u} \equiv [(\nabla u_1)^T,(\nabla u_2)^T]^T \tag{4.6}$$

and hence are the quantities for which we seek superconvergent approximations.

The Lamé coefficients λ, μ can be written in terms of the Young's modulus E and Poisson's ratio ν of the material, see Wang [37], as

$$\lambda = \frac{\nu E}{1-\nu^2} \; ; \qquad \text{plane stress} \; , \tag{4.7}$$

$$\lambda = \frac{\nu E}{(1+\nu)(1-2\nu)} \; ; \qquad \text{plane strain} \; , \tag{4.8}$$

$$\mu = \frac{E}{2(1+\nu)} \; . \tag{4.9}$$

4.2 Weak Formulation and Finite Element Approximation

4.2.1 Weak Formulation

Let V be the space of admissible displacement vectors $\mathbf{v} \equiv (v_1, v_2)^T$, such that

$$V \equiv \{\mathbf{v} : \mathbf{v} \in [H^1(\Omega)]^2 \; , \; \mathbf{v} = \mathbf{0} \text{ on } \partial\Omega^c\} \; ,$$

and having the norm $\|\mathbf{v}\|_{1,\Omega}$. The weak form of problem (4.1)-(4.3) is that in which we seek $\mathbf{u} \in V$ such that

$$a(\mathbf{u},\mathbf{v}) = F(\mathbf{v}) \quad \forall \; \mathbf{v} \in V \; , \tag{4.10}$$

where the symmetric bilinear form a(.,.) is defined by

$$a(\mathbf{u},\mathbf{v}) \equiv \int_\Omega \Big\{\lambda \text{ div } \mathbf{u} \text{ div } \mathbf{v} + 2\mu \sum_{i,j=1}^{2} \varepsilon_{ij}(\mathbf{u})\varepsilon_{ij}(\mathbf{v})\Big\} d\mathbf{x} \; , \tag{4.11}$$

and the linear form F(.) by

$$F(\mathbf{v}) \equiv \int_\Omega \mathbf{f}.\mathbf{v} \, d\mathbf{x} + \int_{\partial\Omega^T} \mathbf{g}.\mathbf{v} \, ds \; . \tag{4.12}$$

In [40] Whiteman and Goodsell expressed the bilinear form a(.,.) as the sum of individual integrals, so that for $\mathbf{v},\mathbf{w} \in V$

$$a(\mathbf{v},\mathbf{w}) \equiv \sum_{i=1}^{2} \sum_{j=1}^{2} \sum_{k=1}^{2} \sum_{\ell=1}^{2} C_{ijk\ell} I_{ijk\ell}(\mathbf{v},\mathbf{w}) \; , \tag{4.13}$$

where

$$I_{ijk\ell}(\mathbf{v},\mathbf{w}) \equiv \int_\Omega \partial_i v_j \partial_k w_\ell \, d\mathbf{x} \, , \tag{4.14}$$

and the $C_{ijk\ell} \equiv C_{ijk\ell}(\lambda,\mu) \geq 0$ are constants such that $C_{ijk\ell} = C_{k\ell ij}$, $i,j,k,\ell = 1,2$. Specifically

$$\begin{aligned}
&C_{1111} = \lambda + 2\mu \, , \quad C_{1112} = 0 \, , \quad C_{1121} = 0 \, , \quad C_{1122} = \lambda \, , \\
&C_{1211} = 0 \, , \quad C_{1212} = \mu \, , \quad C_{1221} = \mu \, , \quad C_{1222} = 0 \, , \\
&C_{2111} = 0 \, , \quad C_{2112} = \mu \, , \quad C_{2121} = \mu \, , \quad C_{2122} = 0 \, , \\
&C_{2211} = \lambda \, , \quad C_{2212} = 0 \, , \quad C_{2221} = 0 \, , \quad C_{2222} = \lambda + 2\mu \, .
\end{aligned} \tag{4.15}$$

By the Lax-Milgram Lemma, see Ciarlet [5,Theorem 1.1.3], if a(.,.) is continuous on V × V and V-elliptic and if F(.) is continuous on V, then problem (4.10)-(4.12) has a unique solution, which coincides with the solution of the (classical) problem (4.1)-(4.3) when the latter exists.

As a result of (4.13)-(4.15) the proof of the continuity of a(.,.) is straightforward. The proof of the V-ellipticity of a(.,.) is much more complicated, see Ciarlet [5,p.24]. The linear functional F(.) of (4.12) can also be shown to be continuous, using

$$|F(\mathbf{v})| \leq |\mathbf{f}|_{0,\Omega}|\mathbf{v}|_{0,\Omega} + |\mathbf{g}|_{0,\partial\Omega^T}|\mathbf{v}|_{0,\partial\Omega^T} \quad \forall \ \mathbf{v} \in V \, , \tag{4.16}$$

and the Sobolev embedding $H^1(\Omega) \longrightarrow L_2(\partial\Omega)$, see Adams [1,pp.113-115].

4.2.2 Finite Element Approximation

It is again assumed that there exist subdomains Ω_0, Ω_1, Ω_2 such that $\Omega_0 < \Omega_1 < \Omega_2 \subseteq \Omega$ satisfying conditions analogous to Conditions 1, 2, 3 of Section 3.1 for the mesh and the subdomains, with $\partial\Omega^N$ in (3.2) replaced by $\partial\Omega^T$ and $\mathbf{u} \in [H^3(\Omega_2)]^2$. It is again assumed that **f** and **g** are sufficiently regular to guarantee this regularity of **u**, the solution of (4.10); thus $\mathbf{f} \in [H^1(\Omega)]^2$, $\mathbf{g} \in [H^{3/2}(\partial\Omega_i^T)]^2$, where $\partial\Omega^T \equiv \bigcup_i \partial\Omega_i^T$.

The finite dimensional space $V^h \subset V$ is the space of piecewise linear functions over the partition of triangles such that

$$V^h \equiv \{\mathbf{v}_h : \mathbf{v}_h \in [S^h(\Omega)]^2 \, , \, \mathbf{v}_h = \mathbf{0} \text{ on } \partial\Omega^C\} \, , \tag{4.17}$$

where $S^h(\Omega)$ is as in Section 2.1. The problem approximating (4.10) is that of finding $\mathbf{u}_h \in V^h$ such that

$$a(\mathbf{u}_h,\mathbf{v}_h) = F(\mathbf{v}_h) \quad \forall \quad \mathbf{v}_h \in V^h \ . \tag{4.18}$$

The orthogonality property

$$a(\mathbf{u} - \mathbf{u}_h,\mathbf{v}_h) = 0 \ , \quad \forall \quad \mathbf{v}_h \in V^h \tag{4.19}$$

analogous to (2.24) follows from (4.10) and (4.18).

4.3 Superconvergent Estimates for Recovered Gradients on a Subdomain

4.3.1 Extension of Results for Poisson Problems

Our task here is to obtain superconvergent estimates for the recovered gradients of the finite element approximation $\mathbf{u}_h$ of (4.18) on a subdomain Ω_0 of Ω; this extends the analysis of Section 3 to the case of planar linear elasticity.

With $\mathbf{Vu}$ defined as in (4.6) and with the definitions of the recovered gradients as in Section 2, the estimate (2.19) can readily be shown to hold locally over Ω_0 for the planar linear elasticity problem, i.e.

$$\left|\mathbf{u} - \mathbf{u}_h\right|^{\circledast}_{1,\Omega_0} \leq C\left\{\left|\mathbf{u}_I - \mathbf{u}_h\right|_{1,\Omega_0} + h^2\left|\mathbf{u}\right|_{3,\Omega_0}\right\} , \tag{4.20}$$

where $\circledast$ = RM, RV, *, + are as before. Once again we have to estimate $\left|\mathbf{u}_I - \mathbf{u}_h\right|_{1,\Omega_0}$. A simplified version of the analysis of Whiteman and Goodsell [40] is now given.

Lemma 4.1 If $\mathbf{u}_h \in V^h$ is the solution of problem (4.18) and $\mathbf{u}_I \in V^h$ is the piecewise linear interpolant to $\mathbf{u} \in V$, the solution of (4.10), then

$$\left|\mathbf{u}_I - \mathbf{u}_h\right|_{1,\Omega_0} \leq C(\omega)\left\{\left|\mathbf{u}_I - \mathbf{u}_h\right|_{0,\Omega_2} + h^2\left|\mathbf{u}\right|_{3,\Omega_2}\right\} \tag{4.21}$$

where $C(\omega)$ is again a constant depending on the cut-off function $\omega \in W^2_\infty(\Omega)$, as defined in (3.3).

Proof Let

$$\boldsymbol{\theta} \equiv \mathbf{u}_I - \mathbf{u}_h \ , \tag{4.22}$$

and

$$\mathbf{e} \equiv \mathbf{u}_I - \mathbf{u} \; . \tag{4.23}$$

We first obtain the estimate

$$\left|a(\boldsymbol{\theta},\mathbf{v}_h)\right| \leq Ch^2 \left|\mathbf{u}\right|_{3,\Omega_2} \left|\mathbf{v}_h\right|_{1,\Omega_2} \quad \forall \; \mathbf{v}_h \in [S_0^h(\Omega_2)]^2 \; , \tag{4.24}$$

which corresponds to (3.4), but with the much more complicated definition of the bilinear form $a(.,.)$. We note that, as $[S_0^h(\Omega_2)]^2 \subset V^h$, we have from (4.19) that

$$a(\boldsymbol{\theta},\mathbf{v}_h) = a(\mathbf{e},\mathbf{v}_h) \quad \forall \; \mathbf{v}_h \in [S_0^h(\Omega_2)]^2 \; , \tag{4.25}$$

and estimate (4.24) follows, see [40, Lemma 3.1].

Now let $\boldsymbol{\psi} \in [S^h(\Omega)]^2$ be the continuous piecewise linear interpolant to $\omega^2\boldsymbol{\theta}$,

$$\boldsymbol{\psi} \equiv [\omega^2\boldsymbol{\theta}]_I \tag{4.26}$$

Then, arguing as in Lemma 3.2,

$$\boldsymbol{\psi} \in [S_0^h(\Omega_2)]^2 \; . \tag{4.27}$$

Now, using the V-ellipticity of $a(.,.)$ and (4.13)-(4.15), we have

$$\begin{aligned}
C\left|\omega\boldsymbol{\theta}\right|^2_{1,\Omega_2} &\leq a(\omega\boldsymbol{\theta},\omega\boldsymbol{\theta}) \\
&= \sum_{i,j,k,\ell=1}^{2} C_{ijk\ell} \int_{\Omega_2} \partial_i(\omega\theta_j)\partial_k(\omega\theta_\ell)\mathbf{dx} \\
&= \sum_{i,j,k,\ell=1}^{2} C_{ijk\ell} \int_{\Omega_2} \theta_j\theta_\ell\partial_i\omega\partial_k\omega \; \mathbf{dx} \\
&+ \frac{1}{2} \sum_{i,j,k,\ell=1}^{2} C_{ijk\ell} \int_{\Omega_2} \left\{\partial_i\theta_j\partial_k(\omega^2\theta_\ell-\psi_\ell) + \partial_i(\omega^2\theta_j-\psi_j)\partial_k\theta_\ell\right\}\mathbf{dx} \\
&+ \frac{1}{2} \sum_{i,j,k,\ell=1}^{2} C_{ijk\ell} \int_{\Omega_2} \left\{\partial_i\theta_j\partial_k\psi_\ell+\partial_i\psi_j\partial_k\theta_\ell\right\}\mathbf{dx} \\
&\equiv Q_1 + Q_2 + \frac{1}{2}\left\{a(\boldsymbol{\theta},\boldsymbol{\psi})+a(\boldsymbol{\psi},\boldsymbol{\theta})\right\}
\end{aligned}$$

$$= Q_1 + Q_2 + a(\boldsymbol{\theta},\boldsymbol{\psi}) \; . \tag{4.28}$$

The proof then follows that of Lemma 3.2, noting that

$$|Q_1| \leq C \int_{\Omega_2} |\boldsymbol{\theta}|^2 |\nabla\omega|^2 d\mathbf{x} \tag{4.29}$$

$$|Q_2| \leq C \int_{\Omega_2} |\nabla\boldsymbol{\theta}| \, |\nabla(\omega^2\boldsymbol{\theta}-\boldsymbol{\psi})| d\mathbf{x} \tag{4.30}$$

and substituting $\boldsymbol{\psi} \in \left[S_0^h(\Omega_2)\right]^2$ for $\mathbf{v}_h$ in (4.24) to estimate $a(\boldsymbol{\theta},\boldsymbol{\psi})$.

Remark 4.1

Exactly as in Remark 3.1 the condition that $\partial\Omega_1 \cap \partial\Omega^T = \emptyset$ can be weakened to allow $\partial\Omega_1 \cap \partial\Omega^T$ to be parallel to a single axis.

4.3.2 Regularity Results

In order to obtain estimates for the term $|\mathbf{u}_I-\mathbf{u}_h|_{0,\Omega_2}$ in (4.21), we follow the procedure used for the Poisson problems and again produce global estimates over Ω, using the Aubin-Nitsche Lemma. This demands the use of a problem adjoint to (4.1)-(4.3), where $\mathbf{w}(\mathbf{x})$ satisfies

$$-\mu\Delta\mathbf{w}(\mathbf{x}) - (\lambda+\mu)\text{grad div } \mathbf{w}(\mathbf{x}) = \boldsymbol{f}(\mathbf{x}) \; , \quad \mathbf{x} \in \Omega \; , \tag{4.31}$$

$$\mathbf{w}(\mathbf{x}) = \mathbf{0} \; , \quad \mathbf{x} \in \partial\Omega^C \tag{4.32}$$

$$\sum_{j=1}^{2} \sigma_{ij}(\mathbf{w}(\mathbf{x}))n_j = 0 \; , \quad \mathbf{x} \in \partial\Omega^T \; , \quad i = 1,2 \; , \tag{4.33}$$

where

$$\boldsymbol{f} \in \left[L_2(\Omega)\right]^2 \; . \tag{4.34}$$

The global regularity of **u** and **w** is again controlled by the singularities of these functions at the corners of $\partial\Omega$, and now also by the Lamé coefficients λ and μ, see Grisvard [16, Section 4] and Williams [42].

Let φ be the maximum interior angle at the points of intersection of $\overline{\partial\Omega^C}$ and $\overline{\partial\Omega^T}$ (including, if appropriate, the "straight" angle π) and define

$$\beta(\varphi) = \mathrm{Re}\{\alpha(\varphi)\} \tag{4.35}$$

where α is that root, having minimum positive real part, of the transcendental equation

$$\sin^2(\alpha\varphi) = \frac{(\lambda+2\mu)^2 - (\lambda+\mu)^2\alpha^2\sin^2\varphi}{(\lambda+\mu)(\lambda+3\mu)} . \tag{4.36}$$

Further let φ_L, $0 < \varphi_L < \frac{\pi}{2}$ be such that

$$\sin^2\varphi_L = \frac{\lambda+2\mu}{2(\lambda+\mu)} , \tag{4.37}$$

so that φ_L is that value of φ for which $\beta(\varphi) = 1$.

There are again two cases to consider.

Case 1: $0 < \varphi \leqq \varphi_L$

For this case $\beta \geqq 1$ and as a result

$$\mathbf{u},\mathbf{w} \in [H^2(\Omega)]^2 \tag{4.38}$$

and further

$$\|\mathbf{w}\|_{2,\Omega} \leqq C|f|_{0,\Omega} . \tag{4.39}$$

Case 2: $\varphi_L < \varphi \leqq \pi$

For this case $\frac{1}{2} \leqq \beta < 1$, and as a result

$$\mathbf{u},\mathbf{w} \in \left[W^2_{p-\varepsilon}(\Omega)\right]^2 , \tag{4.40}$$

for arbitrary small ε, where

$$\frac{4}{3} \leqq p = \frac{2}{(2-\beta)} < 2 , \tag{4.41}$$

and further

$$\|\mathbf{w}\|_{2,p-\varepsilon,\Omega} \leqq C|f|_{0,p,\Omega} . \tag{4.42}$$

4.3.3 Superconvergence Estimates

The results for the problem of planar linear elasticity can be combined into a single theorem corresponding to the two cases of Section 4.3.2.

Theorem 4.1

If $\mathbf{u} \in V$ is the solution of (4.10), $\mathbf{u}_h$ is the solution of (4.18), and Ω_0 is a subdomain of a convex polygonal domain Ω, then:

Case 1

If $\mathbf{u} \in [H^2(\Omega)]^2 \cap [H^3(\Omega_2)]^2$, then

$$|\mathbf{u}-\mathbf{u}_h|^{\circledast}_{1,\Omega_0} \leqq C(\omega)h^2\left\{|\mathbf{u}|_{2,\Omega} + |\mathbf{u}|_{3,\Omega_2}\right\} ,$$

$$: \circledast = RM,\ RV,\ *,\ + , \tag{4.43}$$

where $C(\omega)$ is a constant depending on the cut-off function ω.

Case 2

If $\mathbf{u} \in \left[W^2_{p-\varepsilon}(\Omega)\right]^2 \cap \left[H^3(\Omega_2)\right]^2 : \frac{4}{3} \leqq p < 2$, then

$$|\mathbf{u}-\mathbf{u}_h|^{\circledast}_{1,\Omega_0} \leqq C(\omega)h^{\gamma-\varepsilon}\left\{|\mathbf{u}|_{2,p-\varepsilon,\Omega} + |\mathbf{u}|_{3,\Omega_2}\right\}$$

$$: \circledast = RM,\ RV,\ *,\ + \tag{4.44}$$

where $1 \leqq \gamma = 4 - \frac{4}{p} < 2$.

Proof From (4.20) and (4.21) we have the estimate

$$|\mathbf{u}-\mathbf{u}_h|^{\circledast}_{1,\Omega_0} \leqq C(\omega)\left\{|\mathbf{u}_I-\mathbf{u}_h|_{0,\Omega_2} + h^2|\mathbf{u}|_{3,\Omega_2}\right\} ,$$

$$: \circledast = RM,\ RV,\ *,\ + . \tag{4.45}$$

The term $|\mathbf{u}_I-\mathbf{u}_h|_{0,\Omega_2}$ is controlled by the term $|\mathbf{u}-\mathbf{u}_h|_{0,\Omega}$, which must now be estimated.

Let $\mathbf{w}$ be the solution of the adjoint problem (4.31)-(4.33) with

$$\boldsymbol{f} \equiv (\mathbf{u}-\mathbf{u}_h) \in [L_2(\Omega)]^2 . \tag{4.46}$$

In the weak form of the adjoint problem we seek $\mathbf{w} \in V$ such that

$$a(\mathbf{w},\mathbf{v}) = \int_\Omega (\mathbf{u}-\mathbf{u}_h).\mathbf{v}\,dx\ , \qquad \forall\ \ \mathbf{v} \in V\ . \tag{4.47}$$

Noting that $(\mathbf{u}-\mathbf{u}_h) \in V$, we have that

$$\left|\mathbf{u}-\mathbf{u}_h\right|^2_{0,\Omega} \equiv \int_\Omega \left|\mathbf{u}-\mathbf{u}_h\right|^2 dx = a(\mathbf{u}-\mathbf{u}_h,\mathbf{w}) = a(\mathbf{u}-\mathbf{u}_h,\mathbf{w}-\mathbf{w}_I)$$

$$\leq C\left\|\mathbf{u}-\mathbf{u}_h\right\|_{1,\Omega}\left\|\mathbf{w}-\mathbf{w}_I\right\|_{1,\Omega}\ , \tag{4.48}$$

where the continuity of $a(.,.)$, (4.19) and the fact that $\mathbf{w}_I \in V^h$, $\mathbf{w}_I$ the interpolant to $\mathbf{w}$, have been used.

The proof then follows that of Lemma 3.3 using again the Céa Lemma, see Ciarlet [5,pp.104-105].

4.4 Comparison with Standard Gradient Error Estimates

As in Section 3.5 for Poisson problems, we now compare the error estimates for the recovered gradients of the finite element approximation, $\mathbf{u}_h$, obtained in Theorem 4.1, with the standard L_2-norm of the error of the gradient on a subdomain Ω_0 of a convex polygonal region Ω. We again use the result on subdomains due to Nitsche and Schatz [28], which remains valid for the planar elasticity problem,

$$\left|\mathbf{u}-\mathbf{u}_h\right|_{1,\Omega_0} \leq C\left\{\left|\mathbf{u}-\mathbf{u}_h\right|_{0,\Omega_2} + h\left|\mathbf{u}\right|_{2,\Omega_2}\right\}\ . \tag{4.49}$$

Considering again the two cases of Section 4.3.2.

Case 1 If $\mathbf{u} \in [H^2(\Omega)]^2 \cap [H^3(\Omega_2)]^2$, we have $O(h^2)$ estimates for $\left|\mathbf{u}-\mathbf{u}_h\right|^{\circledast}_{1,\Omega_0}$, thus giving a full $O(h)$ faster convergence over standard techniques.

Case 2 If $\mathbf{u} \in \left[W^2_{p-\varepsilon}(\Omega)\right]^2 \cap \left[H^3(\Omega_2)\right]^2$, $\frac{4}{3} \leq p < 2$, we have $O(h^{\gamma-\varepsilon})$ estimates for $\left|\mathbf{u}-\mathbf{u}_h\right|^{\circledast}_{1,\Omega_0}$, where $1 \leq \gamma = 4 - \frac{4}{p} < 2$ as above. As for Case 2 for the Poisson problem there is an improvement in the rate of convergence over that for standard techniques, but the improvement decreases with the regularity of $\mathbf{u}$; except in the limiting case $p = 4/3$, which arises for example in a Mode I linear elastic fracture problem.

4.5 The Effects of Numerical Integration

Estimates of the type (3.48) for $|\mathbf{u}_h-\mathbf{u}_h^Q|_{1,\Omega}$, with $\mathbf{u}_h^Q \in V^h$ the numerical approximation calculated using quadrature and cubature formulae, hold for the linear elasticity case.

5. Problem of Linear Elastic Fracture

5.1 Introduction and Stress Intensity Factors

We now consider a two-dimensional Mode I linear elastic fracture problem, as a particular case of the planar linear elasticity problem (4.1)-(4.3) of Section 4. Under the assumption of linear elasticity theory, the presence of a crack in a solid induces singular stress fields local to the crack tip. In reality (local yielding) plasticity occurs near the tip. The fundamental assumption of linear elastic fracture mechanics is that the region of local yielding is small by comparison with the dimension of the crack. The discussion here is restricted to a pure Mode I linear elastic fracture problem, see Rice [33].

For this problem the region Ω is a rectangle with a crack along the negative x_1-axis as in Fig. 5.1. The region is symmetric about the crack and is loaded symmetrically as shown. In terms of polar coordinates local to the crack tip the

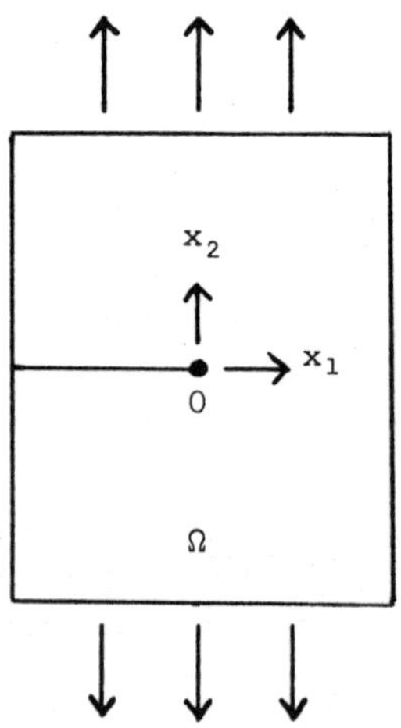

Fig. 5.1

near-tip stress field can be expressed as

$$\begin{bmatrix} \sigma_{11} \\ \\ \sigma_{22} \\ \\ \sigma_{12} \end{bmatrix} = \frac{K_I}{(2\pi r)^{\frac{1}{2}}} \cos\frac{\theta}{2} \begin{bmatrix} 1 - \sin\frac{\theta}{2}\sin\frac{3\theta}{2} \\ \\ 1 + \sin\frac{\theta}{2}\sin\frac{3\theta}{2} \\ \\ \sin\frac{\theta}{2}\cos\frac{3\theta}{2} \end{bmatrix}, \tag{5.1}$$

where K_I, the mode I stress intensity factor, is a measure of the near-tip stresses. We note that the crack faces are stress free. The corresponding near-tip displacement field is given by

$$\begin{bmatrix} u_1 \\ \\ u_2 \end{bmatrix} = \frac{K_I\, r^{\frac{1}{2}}}{2\mu(2\pi)^{\frac{1}{2}}} \begin{bmatrix} \cos\frac{\theta}{2}\,(\kappa-1+2\sin^2\frac{\theta}{2}) \\ \\ \sin\frac{\theta}{2}\,(\kappa+1-2\cos^2\frac{\theta}{2}) \end{bmatrix}, \tag{5.2}$$

where, for plane stress,

$$\kappa = (3-\nu)/(1+\nu) \tag{5.3}$$

and, for plane strain

$$\kappa = 3-4\nu. \tag{5.4}$$

It is the stress intensity factor K_I, a function of crack geometry and loading, which determines whether a crack will propagate; propagation occurs if K_I attains a critical value, K_c say, which is a function of the material properties only, and is independent of geometry and loading.

In this Section we describe two distinct methods for calculating K_I; the first, as proposed by Destuynder, Djaoua and Lescure [7], [8], involves an area integral over a subdomain of Ω remote from the crack tip, whilst the second involves a contour integral (the well-known J-integral of Rice [32]) on a contour remote from the crack tip. The approximation of each integral using the recovered (and standard) gradients of the finite element approximation $\mathbf{u}_h$ to the displacement $\mathbf{u}$ is then considered.

5.2 Area Integral Formulation for the Stress Intensity Factor

With the previous notation for subdomains, for the region of Fig. 5.1 we define the subdomain Ω_0 to be the subdomain bounded by two non-intersecting contours running from the lower to the upper crack faces and the relevant parts of the crack faces, see Fig. 5.2. Thus Ω_0 is remote from the crack tip. The remainder of Ω, $\Omega\backslash\Omega_0$, consists of two subdomains, Ω_{INT} and Ω_{EXT}, as shown so that $\Omega \equiv \Omega_{INT} \cup \Omega_0 \cup \Omega_{EXT}$.

For these subdomains we define the cut-off function $\zeta \in W^1_\infty(\Omega)$ to be such that

$$\zeta(\mathbf{x}) \equiv \begin{Bmatrix} 1 & \text{in} & \bar{\Omega}_{INT} \\ & & \\ 0 & \text{in} & \bar{\Omega}_{EXT} \end{Bmatrix} . \tag{5.5}$$

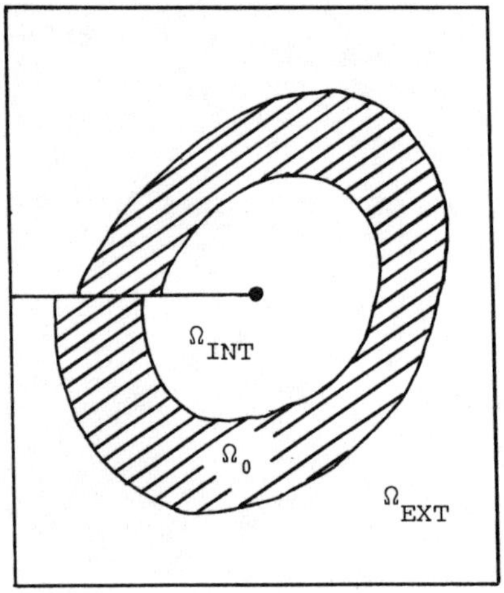

Fig. 5.2

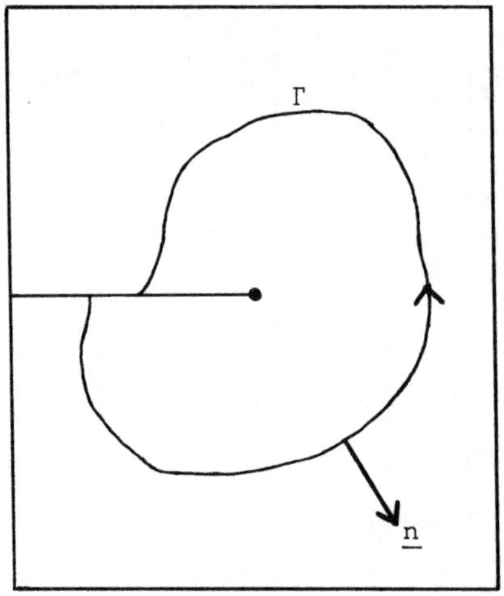

Fig. 5.3

Denoting by P the strain energy release rate, that is the rate of decrease in potential energy with respect to crack length, there are the following relationships between P and K_I for the Mode I problem:

$$P = \frac{1}{E} K_I^2 \ , \quad \text{for plane stress} \ , \tag{5.6}$$

$$P = \frac{1-\nu^2}{E} K_I^2 \ , \quad \text{for plane strain} \ . \tag{5.7}$$

Our approach to K_I is via approximations of P.

It has been shown by Destuynder and Djaoua [7] that P can be written as an integral over Ω involving derivatives of ζ. However, from the definition of ζ in (5.5) the integral is in fact over the subdomain Ω_0, so that

$$P = \int_{\Omega_0} \Big[\frac{1}{2}\frac{\partial \zeta}{\partial x_1}\left\{\sigma_{11}\frac{\partial u_1}{\partial x_1} + \sigma_{12}\left(\frac{\partial u_2}{\partial x_1} - \frac{\partial u_1}{\partial x_2}\right) - \sigma_{22}\frac{\partial u_2}{\partial x_2}\right\}$$

$$+ \frac{\partial \zeta}{\partial x_2}\left\{\sigma_{12}\frac{\partial u_1}{\partial x_1} + \sigma_{22}\frac{\partial u_2}{\partial x_1}\right\}\Big] d\mathbf{x} \,. \tag{5.8}$$

We emphasise that in (5.8) Ω_0 can be any subdomain as in Fig. 5.2 and $\zeta \in W^1_\infty(\Omega)$ any cut-off function satisfying (5.5).

5.3 The J-Integral

Let Γ be a contour running anticlockwise from the lower to the upper faces of the crack, remote from the crack tip as in Fig. 5.3. Rice [32] defined the J-integral to be

$$J \equiv \int_\Gamma \sum_{i,j=1}^{2} \left\{\frac{1}{2}\sigma_{ij}\varepsilon_{ij} dx_2 - \sigma_{ij}\frac{\partial u_i}{\partial x_1} n_j ds\right\} , \tag{5.9}$$

where $\mathbf{n}$ is the unit outward normal vector to Γ. For linear elastic fracture J is path independent, so that Γ can be any contour defined as above. Further, using the Lamé equations (4.1) and the fact that the crack faces are stress free, it follows immediately that J = P. This can readily be shown from (5.8) by integrating by parts and regarding Γ as the inner contour of $\partial\Omega_0$; i.e. $\Gamma = \partial\Omega_0 \cap \partial\Omega_{INT}$. Thus from (5.6) and (5.7) we have that

$$J = \frac{1}{E} K_I^2 \,, \quad \text{for plane stress} \,, \tag{5.10}$$

$$J = \frac{1-\nu^2}{E} K_I^2 \,, \quad \text{for plane strain} \,. \tag{5.11}$$

5.4 Finite Element Approximation of Fracture Problem

By exploiting the symmetry of the problem, see Fig. 5.1, we need only consider the problem in the upper half-plane, Fig. 5.4, which we now term Ω, and we introduce $\partial\Omega^F$ which is the upper crack face. The displacement $\mathbf{u}$ then satisfies the Lamé equations (4.1) with zero volumic forces in Ω, together with the boundary conditions

$$\mathbf{u}(\mathbf{0}) = \mathbf{0} \tag{5.12}$$

$$u_2 = \frac{\partial u_1}{\partial x_2} = 0 \quad \text{on} \quad \partial\Omega^C \tag{5.13}$$

$$\sum_{j=1}^{2} \sigma_{ij}(\mathbf{u}) n_j = 0 \text{ on } \partial\Omega^F \,, \quad i = 1,2 \tag{5.14}$$

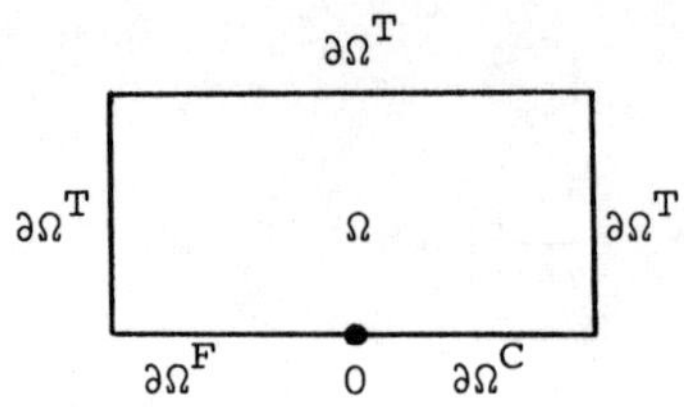

Fig. 5.4

$$\sum_{j=1}^{2} \sigma_{ij}(\mathbf{u})n_j = g_i \quad \text{on} \quad \partial\Omega^T , \quad i = 1,2 . \tag{5.15}$$

We note from the near-tip displacement field (5.2) that $\mathbf{u}$ has global regularity

$$\mathbf{u} \in \left[W^2{}_{4/3-\varepsilon}(\Omega)\right]^2 , \tag{5.16}$$

which we would expect from Section 4.3.2, with an interior angle π at the crack tip, the point of intersection of $\overline{\partial\Omega^F}$ and $\overline{\partial\Omega^c}$. The weak formulation of (4.1) and (5.12)-(5.15) is given by (4.10), but now in the space $\tilde{V}$, where

$$\tilde{V} \equiv \left\{\mathbf{v} : \mathbf{v} \in [H^1(\Omega)]^2 , \mathbf{v}(\mathbf{0}) = \mathbf{0} , v_2 = 0 \text{ on } \partial\Omega^c\right\}$$

with norm $\|\mathbf{v}\|_{1,\Omega}$ as on V. In the absence of volumic forces the linear form F is now given by

$$F(\mathbf{v}) \equiv \int_{\partial\Omega^T} \mathbf{g}.\mathbf{v} \, ds . \tag{5.17}$$

The finite element approximation $\mathbf{u}_h \in \tilde{V}^h$ to $\mathbf{u} \in \tilde{V}$ is given by (4.18), but now in the space of piecewise linear functions $\tilde{V}^h \subset \tilde{V}$, where

$$\tilde{V}^h \equiv \left\{\mathbf{v}_h : \mathbf{v}_h \in \left[S^h(\Omega)\right]^2 , \mathbf{v}_h(\mathbf{0}) = \mathbf{0} , (\mathbf{v}_h)_2 = 0 \text{ on } \partial\Omega^c\right\} \tag{5.18}$$

5.5 Approximation of the Stress Intensity Factor

5.5.1 Area Integral Formulation

We construct subdomains $\Omega_0 < \Omega_1 < \Omega_2 \subset \Omega$ of the region Ω, as shown (together with part of the finite element mesh) in Fig. 5.5; which satisfy the three conditions of Section 3.1, but with Condition 2 slightly weakened as in Remark 4.1.

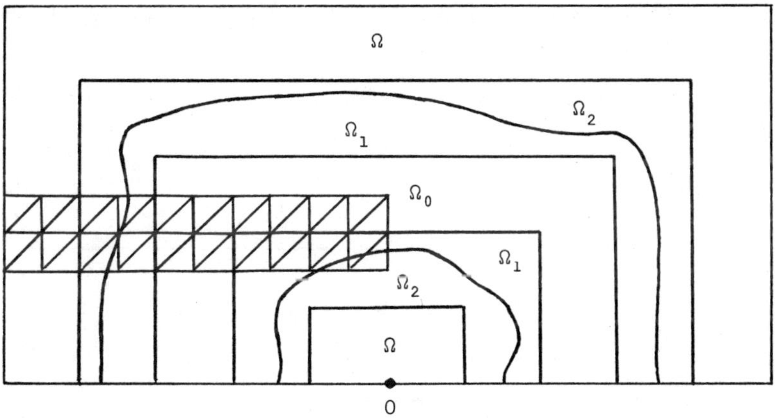

Fig. 5.5

For this case we have from (5.8) that

$$P = \int_{\Omega_0} \left[\frac{\partial\zeta}{\partial x_1} \left\{\sigma_{11} \frac{\partial u_1}{\partial x_1} + \sigma_{12} \left(\frac{\partial u_2}{\partial x_1} - \frac{\partial u_1}{\partial x_2}\right) - \sigma_{22} \frac{\partial u_2}{\partial x_2}\right\} \right.$$

$$\left. + 2 \frac{\partial\zeta}{\partial x_2} \left\{\sigma_{12} \frac{\partial u_1}{\partial x_1} + \sigma_{22} \frac{\partial u_2}{\partial x_1}\right\}\right] d\mathbf{x} , \qquad (5.19)$$

where $\zeta \in W^1_\infty(\Omega)$ is a cut-off function defined as in (5.5). This follows by symmetry: regarding the cut-off function ζ as symmetric about the x_1-axis, the integrand is then an *even* function of x_2.

We have similar expressions for the approximations, P^h, P^*, and P^+ to P, in which the components of $\nabla\mathbf{u}$ in (5.19) are replaced respectively by those of $\nabla\mathbf{u}_h$, $\nabla^*\mathbf{u}_h$, and $\nabla^+\mathbf{u}_h$.

In order to minimise the amount of work needed to evaluate the above area integrals, it is essential to construct the cut-off function ζ as simply as possible in Ω_0. We consider Ω_0 to be the union of the three subdomains, $\Omega_0^{(1)}$, $\Omega_0^{(2)}$, $\Omega_0^{(3)}$, as in Fig. 5.6, and let ζ be linear (and hence $\nabla\zeta$ constant) in each $\Omega_0^{(i)}$, i = 1,2,3. From (5.6) and (5.7) we have, respectively for plane stress and plane strain, that

$$K_I = (EP)^{\frac{1}{2}} \text{ or } K_I = \left(\frac{E}{1-\nu^2} P\right)^{\frac{1}{2}} \tag{5.20}$$

with similar expressions for the approximations K_I^h, K_I^* and K_I^+ to K_I.

We now consider the error estimates for $\left|K_I - K_I^{\circledast}\right|$: $\circledast$ = *,+ obtained using the recovered gradients, $\nabla^* \mathbf{u}_h$ and $\nabla^+ \mathbf{u}_h$, of the finite element approximation $\mathbf{u}_h$, and compare these with the error estimate for $\left|K_I - K_I^h\right|$ obtained using the standard gradient $\nabla\mathbf{u}_h$. We first give a Theorem.

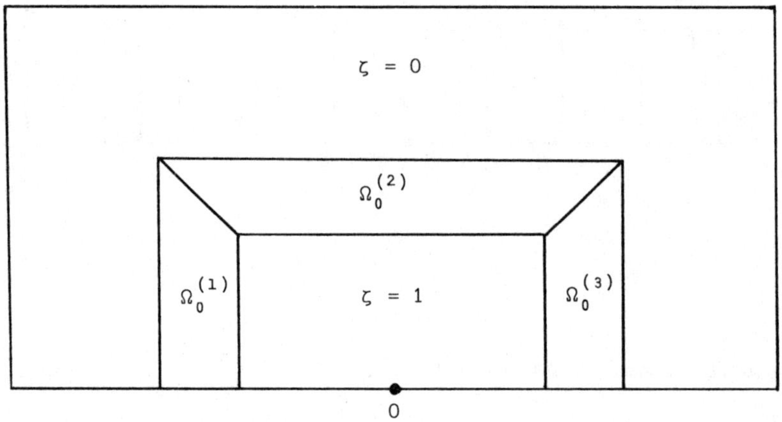

Fig. 5.6

Theorem 5.1 If $\mathbf{u} \in \tilde{V}$ is the weak solution of the problem (4.1) and (5.12) - (5.15) defined in the region of Fig. 5.4, and ω and ζ are cut-off functions defined respectively in (3.3) and (5.5), then

$$\left|K_I - K_I^{\circledast}\right| \le \frac{C(\omega,\zeta)}{K_I} h^{1-\varepsilon}\left\{\left|\mathbf{u}\right|^2_{2,4/3-\varepsilon,\Omega} + \left\|\mathbf{u}\right\|^2_{3,\Omega_2}\right\}, \quad \circledast = *,+ , \tag{5.21}$$

Proof We begin the proof by showing that $\left|K_I - K_I^{\circledast}\right|$ has the same asymptotic rate of convergence as $|P-P^{\circledast}|$. Thus from (5.20) we have

$$|K_I - K_I^{\circledast}| = C|P^{\frac{1}{2}} - (P^{\circledast})^{\frac{1}{2}}| = C|P - P^{\circledast}| \Big/ \{P^{\frac{1}{2}} + (P^{\circledast})^{\frac{1}{2}}\}$$

$$\leq C|P - P^{\circledast}| \Big/ P^{\frac{1}{2}} = \frac{C}{K_I}|P - P^{\circledast}| \ . \tag{5.22}$$

Now we observe from (5.19) that $P-P^{\circledast}$ is a linear combination of the terms

$$\int_{\Omega_0} \{(\nabla\mathbf{u})_i(\nabla\mathbf{u})_j - (\nabla^{\circledast}\mathbf{u}_h)_i(\nabla^{\circledast}\mathbf{u}_h)_j\}\partial_k\zeta \, d\mathbf{x} : i,j = 1,2,3,4 : k = 1,2 \tag{5.23}$$

and thus to bound $P-P^{\circledast}$ we must bound the above terms.

Now

$$\left|\int_{\Omega_0} \{(\nabla\mathbf{u})_i(\nabla\mathbf{u})_j - (\nabla^{\circledast}\mathbf{u}_h)_i(\nabla^{\circledast}\mathbf{u}_h)_j\}\partial_k\zeta \, d\mathbf{x}\right|$$

$$\leq \left|\int_{\Omega_0} (\nabla\mathbf{u})_i \{(\nabla\mathbf{u})_j - (\nabla^{\circledast}\mathbf{u}_h)_j\}\partial_k\zeta \, d\mathbf{x}\right|$$

$$+ \left|\int_{\Omega_0} (\nabla^{\circledast}\mathbf{u}_h)_j \{(\nabla\mathbf{u})_i - (\nabla^{\circledast}\mathbf{u}_h)_i\}\partial_k\zeta \, d\mathbf{x}\right|$$

$$\leq \int_{\Omega_0} \{|\nabla\mathbf{u}| + |\nabla^{\circledast}\mathbf{u}_h|\} |\nabla\mathbf{u} - \nabla^{\circledast}\mathbf{u}_h||\nabla\zeta| \, d\mathbf{x}$$

$$\leq \int_{\Omega_0} \{2|\nabla\mathbf{u}| + |\nabla\mathbf{u} - \nabla^{\circledast}\mathbf{u}_h|\} |\nabla\mathbf{u} - \nabla^{\circledast}\mathbf{u}_h||\nabla\zeta| \, d\mathbf{x}$$

$$\leq |\zeta|_{1,\infty,\Omega_0} \int_{\Omega_0} \{2|\nabla\mathbf{u}| + |\nabla\mathbf{u} - \nabla^{\circledast}\mathbf{u}_h|\} |\nabla\mathbf{u} - \nabla^{\circledast}\mathbf{u}_h| \, d\mathbf{x}$$

$$\leq C(\zeta)\{|\nabla\mathbf{u}|_{0,\Omega_0}|\nabla\mathbf{u} - \nabla^{\circledast}\mathbf{u}_h|_{0,\Omega_0} + |\nabla\mathbf{u} - \nabla^{\circledast}\mathbf{u}_h|^2_{0,\Omega_0}\} \ . \tag{5.24}$$

Thus from (5.22)-(5.24), and using the result of Theorem 4.1 (Case 2 with $p = 4/3$, see (5.16)) we have

$$|K_I - K_I^{\circledast}| \leq \frac{C(\omega,\zeta)}{K_I} h^{1-\varepsilon}\{|\mathbf{u}|_{1,\Omega_0}[|\mathbf{u}|_{2,4/3-\varepsilon,\Omega} + |\mathbf{u}|_{3,\Omega_2}]$$

$$+ |\mathbf{u}|^2_{2,4/3-\varepsilon,\Omega} + |\mathbf{u}|^2_{3,\Omega_2}\}$$

$$\leq \frac{C(\omega,\zeta)}{K_I} h^{1-\varepsilon}\{|\mathbf{u}|^2_{2,4/3-\varepsilon,\Omega} + \|\mathbf{u}\|^2_{3,\Omega_2}\} \qquad \square$$

Using the result of Nitsche and Schatz (4.49), we can obtain, using the arguments of the above proof, the error estimate corresponding to (5.21) for standard gradients

$$\left|K_I - K_I^h\right| \leq \frac{C(\zeta)}{K_I} h^{1-\varepsilon}\left\{\left|\mathbf{u}\right|^2_{2,4/3-\varepsilon,\Omega} + \left\|\mathbf{u}\right\|^2_{2,\Omega_2}\right\} . \tag{5.25}$$

Comparison of estimates (5.21) and (5.25) shows that there is unfortunately no improvement in the asymptotic rate of convergence using recovered gradients, the rates being $O(h^{1-\varepsilon})$ for both recovered and standard gradients. This is caused by the low regularity of the displacement **u** (5.16), see Section 4.4

5.5.2 Contour Integral Formulation

We now consider a method of approximating the stress intensity factor, K_I, using the J integral of Section 5.3, see Whiteman and Goodsell [39].

In the region Ω we construct a contour Γ from the union of three straight lines parallel to the x_1- and x_2-axes, where we suppose $\Gamma \subset \overline{\Omega_0}$, where $\Omega_0 < \Omega_1 < \Omega_2 \subset \Omega$ are subdomains of the region Ω as in Section 5.5.1. The situation is shown in Figs. 5.7a and 5.7b, which show also the triangles of the finite element mesh surrounding the contour Γ; the union of which we denote by $R(\Gamma)$.

The mesh size h is chosen in Fig. 5.7a so that Γ passes between elements at element edge mid-points, and in Fig. 5.7b so that Γ travels along element edges.

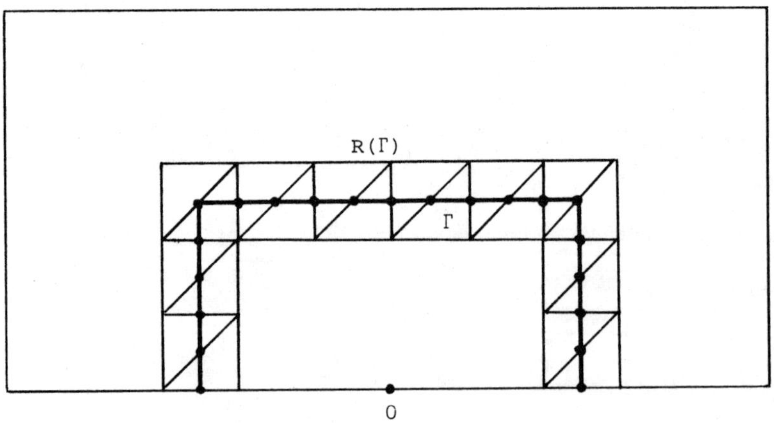

Fig. 5.7a

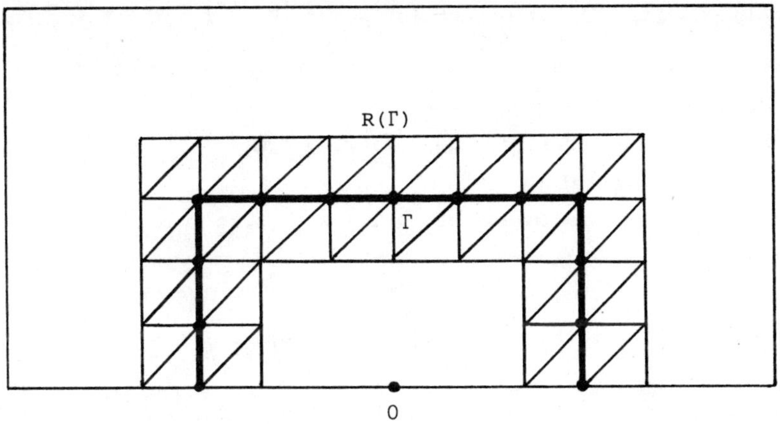

Fig. 5.7b

Then we have from (5.9) the following expression for the J integral

$$J = \int_{\Gamma} \sum_{i,j=1}^{2} \left\{ \sigma_{ij}\varepsilon_{ij}\,dx_2 \quad 2\sigma_{ij}\frac{\partial u_i}{\partial x_1} n_j\,ds \right\} . \tag{5.26}$$

Once again we see that this follows by symmetry, the integrand being an *even* function of x_2.

We have similar expressions for the approximations, J^h, J^* and J^+ to J, in which the components of $\nabla \mathbf{u}$ in (5.26) are replaced respectively by those of $\nabla \mathbf{u}_h$, $\nabla^* \mathbf{u}_h$ and $\nabla^+ \mathbf{u}_h$.

We note here that we choose the mesh size h so that the situation is as in Fig. 5.7a when calculating J^h and J^*, and so that it is as in Fig. 5.7b when calculating J^+.

Now from (5.10)-(5.11) we have respectively for plane stress and plane strain, that

$$K_I = (EJ)^{\frac{1}{2}} \text{ or } K_I = \left(\frac{E}{1-\nu^2} J\right)^{\frac{1}{2}} \tag{5.27}$$

and we have similar expressions for the approximations K_Γ^h, K_Γ^* and K_Γ^+ to K_I.

The notation K_Γ^h, K_Γ^* and K_Γ^+ is used to distinguish these quantities from K_I^h, K_I^* and K_I^+: for although from above we have J=P; however

$$J^h \neq P^h , \; J^* \neq P^* , \; J^+ \neq P^+ . \tag{5.28}$$

We again consider the error estimates for $\left|K_I - K_\Gamma^{\circledast}\right|$: $\circledast$ = *,+ obtained using the recovered gradients, and compare these with the error estimate for $\left|K_I - K_\Gamma^h\right|$ obtained using the standard gradient.

We shall need the result of the following lemma.

Lemma 5.1 For $\underset{\sim}{\mathbf{u}}$ the solution of (4.10) with zero volumic forces and in the space $\underset{\sim}{V}$ and $\mathbf{u}_h \in \underset{\sim}{V}^h$,

$$\left|\nabla\mathbf{u} - \nabla^{\circledast}\mathbf{u}_h\right|_{0,\Gamma} \leq C(\omega)h^{\frac{1}{2}-\varepsilon}\left\{\left|\mathbf{u}\right|_{2,4/3-\varepsilon,\Omega} + \left|\mathbf{u}\right|_{3,\Omega_2}\right\} \; : \; \circledast = *,+ \tag{5.29}$$

where $C(\omega)$ is a constant depending on the cut-off function ω, defined in (3.3).

Proof We firstly use the triangle inequality to give

$$\left|\nabla\mathbf{u} - \nabla^{\circledast}\mathbf{u}_h\right|_{0,\Gamma} \leq \left|\nabla^{\circledast}(\mathbf{u}_I - \mathbf{u}_h)\right|_{0,\Gamma} + \left|\nabla\mathbf{u} - \nabla^{\circledast}\mathbf{u}_I\right|_{0,\Gamma} \, . \tag{5.30}$$

Then, with $R(\Gamma)$ as in Fig. 5.7a for $\circledast$ = *, or Fig. 5.7b for $\circledast$ = +, we have, using the techniques of Section 2,

$$\left|\nabla^{\circledast}(\mathbf{u}_I - \mathbf{u}_h)\right|_{0,\Gamma} \leq Ch^{-\frac{1}{2}}\left|\mathbf{u}_I - \mathbf{u}_h\right|_{1,R(\Gamma)} \tag{5.31}$$

and

$$\left|\nabla\mathbf{u} - \nabla^{\circledast}\mathbf{u}_I\right|_{0,\Gamma} \leq Ch^{3/2}\left|\mathbf{u}\right|_{3,R(\Gamma)} \, . \tag{5.32}$$

Thus from (5.30)-(5.32) we have

$$\begin{aligned}\left|\nabla\mathbf{u} - \nabla^{\circledast}\mathbf{u}_h\right|_{0,\Gamma} &\leq C\left\{h^{-\frac{1}{2}}\left|\mathbf{u}_I - \mathbf{u}_h\right|_{1,R(\Gamma)} + h^{3/2}\left|\mathbf{u}\right|_{3,R(\Gamma)}\right\} \\ &\leq C\left\{h^{-\frac{1}{2}}\left|\mathbf{u}_I - \mathbf{u}_h\right|_{1,\Omega_0} + h^{3/2}\left|\mathbf{u}\right|_{3,\Omega_0}\right\} \, . \end{aligned} \tag{5.33}$$

The result (5.29) then follows using Case 2 of Theorem 4.1 with $p = \frac{4}{3}$, see (5.16)

Using the result of Nitsche and Schatz (4.49), we can obtain, using the arguments of the above proof, the estimate corresponding to (5.29) for standard gradients.

$$\left|\nabla(\mathbf{u} - \mathbf{u}_h)\right|_{0,\Gamma} \leq Ch^{\frac{1}{2}-\varepsilon}\left\{\left|\mathbf{u}\right|_{2,4/3-\varepsilon,\Omega} + \left|\mathbf{u}\right|_{2,\Omega_2}\right\} \, . \tag{5.34}$$

We are now in a position to prove the following theorem.

Theorem 5.2

$$\left|K_I - K_\Gamma^{\circledast}\right| \leq \frac{C(\omega)}{K_I} h^{\frac{1}{2}-\varepsilon}\left\{\left|\mathbf{u}\right|^2_{2,4/3-\varepsilon,\Omega} + \left\|\mathbf{u}\right\|^2_{3,\Omega_2}\right\} \; : \; \circledast = *,+ \tag{5.35}$$

where $C(\omega)$ again depends on the cut-off function ω.

Proof Arguing as in the proof of Theorem 5.1, we have from (5.27)

$$\left|K_I - K_\Gamma^{\circledast}\right| \leq \frac{C}{K_I}\left|J - J^{\circledast}\right| \tag{5.36}$$

and observing from (5.26) that $J - J^{\circledast}$ is a linear combination of the terms

$$\int_\Gamma \left\{(\nabla\mathbf{u})_i(\nabla\mathbf{u})_j - (\nabla^{\circledast}\mathbf{u}_h)_i(\nabla^{\circledast}\mathbf{u}_h)_j\right\}ds \; : \; i,j = 1,2,3,4 \tag{5.37}$$

we must bound these terms in order to bound $J-J^{\circledast}$, and hence $K_I-K_\Gamma^{\circledast}$.

Now, again arguing as in the proof of Theorem 5.1 we have

$$\left|\int_\Gamma \left\{(\nabla\mathbf{u})_i(\nabla\mathbf{u})_j - (\nabla^{\circledast}\mathbf{u}_h)_i(\nabla^{\circledast}\mathbf{u}_h)_j\right\}ds\right|$$

$$\leq C\left\{\left|\nabla\mathbf{u}\right|_{0,\Gamma}\left|\nabla\mathbf{u}-\nabla^{\circledast}\mathbf{u}_h\right|_{0,\Gamma} + \left|\nabla\mathbf{u}-\nabla^{\circledast}\mathbf{u}_h\right|^2_{0,\Gamma}\right\}$$

$$\leq C\left\{\left\|\mathbf{u}\right\|_{2,\Omega_0}\left|\nabla\mathbf{u}-\nabla^{\circledast}\mathbf{u}_h\right|_{0,\Gamma} + \left|\nabla\mathbf{u}-\nabla^{\circledast}\mathbf{u}_h\right|^2_{0,\Gamma}\right\} \tag{5.38}$$

and the result (5.35) then follows immediately from (5.36)-(5.38), together with (5.29).

Using (5.34) we can obtain, using the arguments of the above proof, the error estimate corresponding to (5.35) for standard gradients.

$$\left|K_I-K_\Gamma^h\right| \leq \frac{C}{K_I} h^{\frac{1}{2}-\varepsilon}\left\{\left|\mathbf{u}\right|^2_{2,4/3-\varepsilon,\Omega} + \left\|\mathbf{u}\right\|^2_{2,\Omega_2}\right\} . \tag{5.39}$$

In the next section, we shall give evidence, both numerical and theoretical, which suggests that estimates (5.35) and (5.39) are not, however, optimal; but that they can both be improved to $O(h^{1-\varepsilon})$.

5.5.3 Numerical Results for Model Fracture Problem

The piecewise linear finite element approximations for problems of planar linear elasticity were computed using the program PEARS, see Colyer and

Trowbridge [6].

We consider the (plane stress) problem in which $\mathbf{u}(\mathbf{x})$ satisfies (4.1) with zero volumic forces and the boundary conditions (5.12)-(5.15) where in Fig. 5.4 Ω is the region $[-1,1] \times [0,1]$, see Figs. 5.8a,b. The material constants are $E = 10^4$, $\nu = \frac{1}{4}$ and the displacement vector $\mathbf{u} \in \left[W^2_{4/3-\varepsilon}(\Omega)\right]^2$ is given by

$$\mathbf{u} = \frac{r^{\frac{1}{2}}}{2000} \begin{bmatrix} \cos\frac{\theta}{2}\,(3+5\sin^2\frac{\theta}{2}) \\ \\ \sin\frac{\theta}{2}\,(8-5\cos^2\frac{\theta}{2}) \end{bmatrix} \text{ in } \overline{\Omega} \tag{5.40}$$

in plane polar coordinates local to the origin; which thus corresponds to the near-tip displacement field (5.2) for a general theoretical problem.

In the restriction of $F(\mathbf{v})$ in (5.17) to $\tilde{V}_h$, the terms $\int_{\partial\Omega^T} \mathbf{g}.\mathbf{v}_h ds$ are approximated using the mid-point rule.

Now with the contour Γ as shown in Fig. 5.8a we define

$$K = \left| \int_\Gamma \left\{(\nabla\mathbf{u})_1 + (\nabla\mathbf{u})_4\right\}\left\{(\nabla\mathbf{u})_2 - (\nabla\mathbf{u})_3\right\} ds \right|^{\frac{1}{2}} = \frac{1}{1000}\left\{3\ell n2 + \frac{3\pi}{2}\right\}^{\frac{1}{2}} \tag{5.41}$$

with the approximations K^h, K^* and K^+ to K defined similarly. We note that K is *not* the stress intensity factor K_I, but it is the square root of a 'J type'

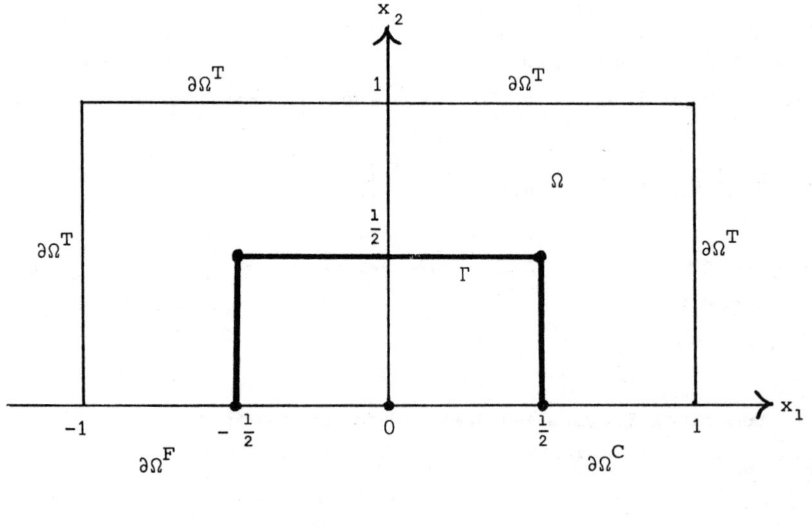

Fig. 5.8a

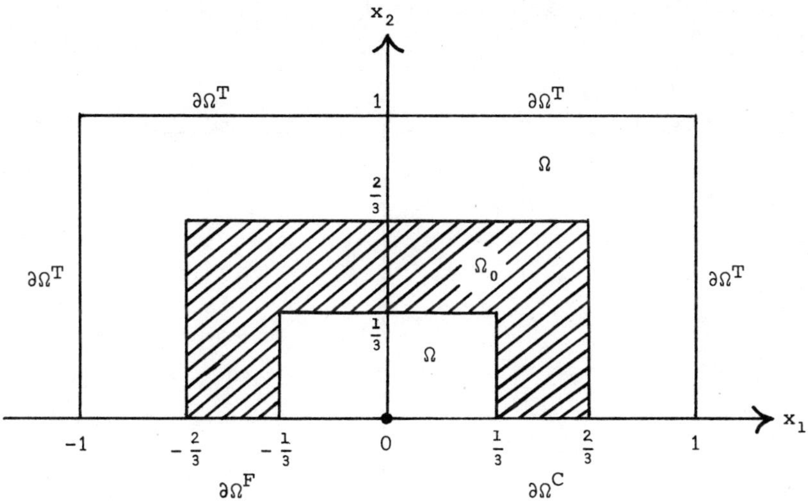

Fig. 5.8b

integral, see (5.37), and so the error analysis of Section 5.5.2 will apply.

We investigate the rates of convergence of $|K-K^h|$, $|K-K^*|$ and $|K-K^+|$ for decreasing values of mesh size h.

In our numerical results we proceed as follows. We define EK_i as the error in $|K-K^h|$, etc. (or S_i for a seminorm) using mesh size h_i, where

$$\lim_{i\to\infty} h_i = 0 . \tag{5.42}$$

We then plot the values of

$$L_i = \ell n\{EK_{i+1}/EK_i\}/\ell n\{h_{i+1}/h_i\} \tag{5.43}$$

against the values of h_i.

When the asymptotic rate of convergence is $O(h^\gamma)$, we therefore expect

$$\lim_{i\to\infty} L_i = \gamma . \tag{5.44}$$

Plots follow for each of these quantities, which suggest higher asymptotic rates of convergence than our theoretical $O(h^{\frac{1}{2}-\varepsilon})$; and we now seek to justify this.

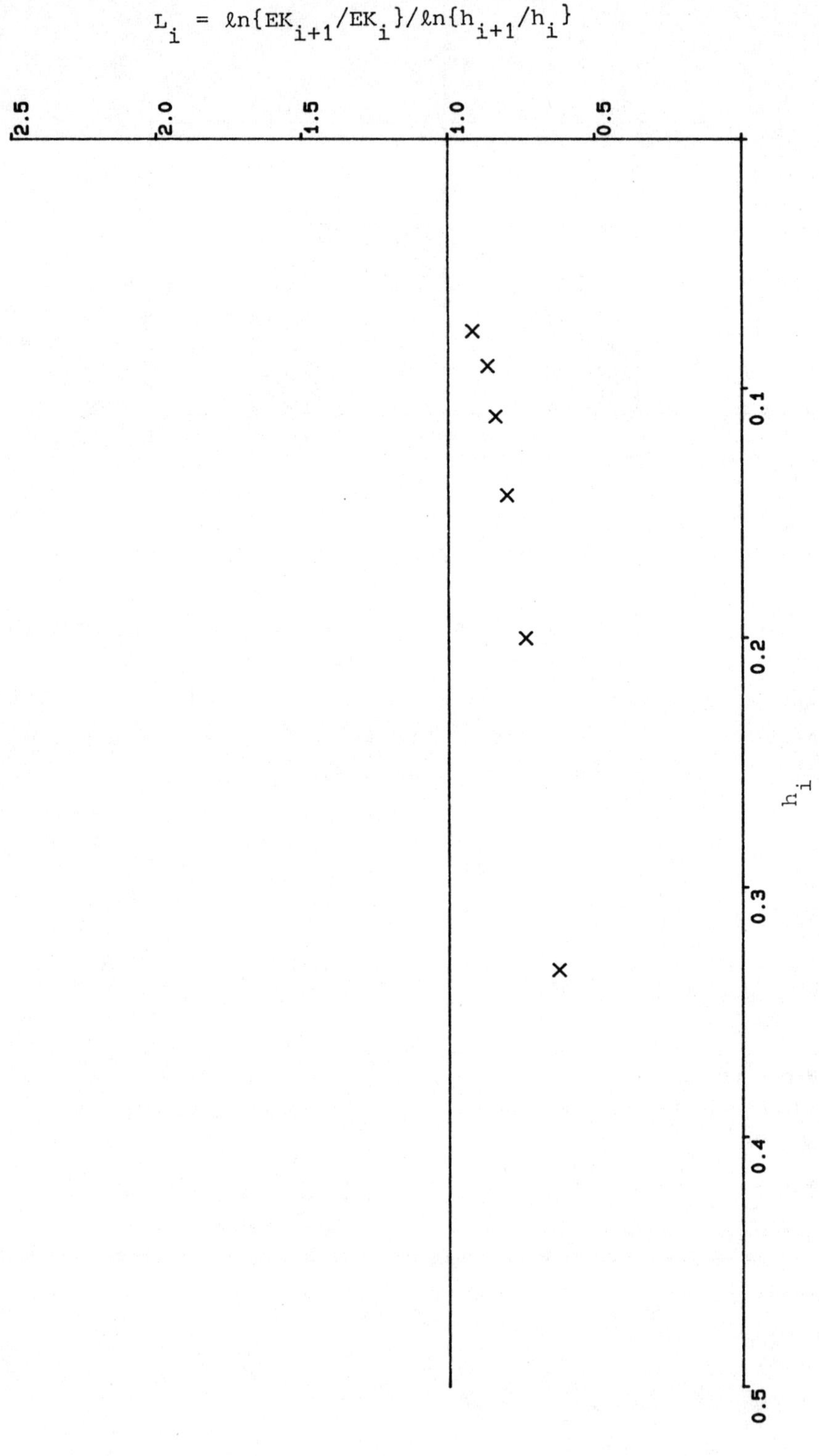

MODEL FRACTURE PROBLEM
$|K - K(h)|$

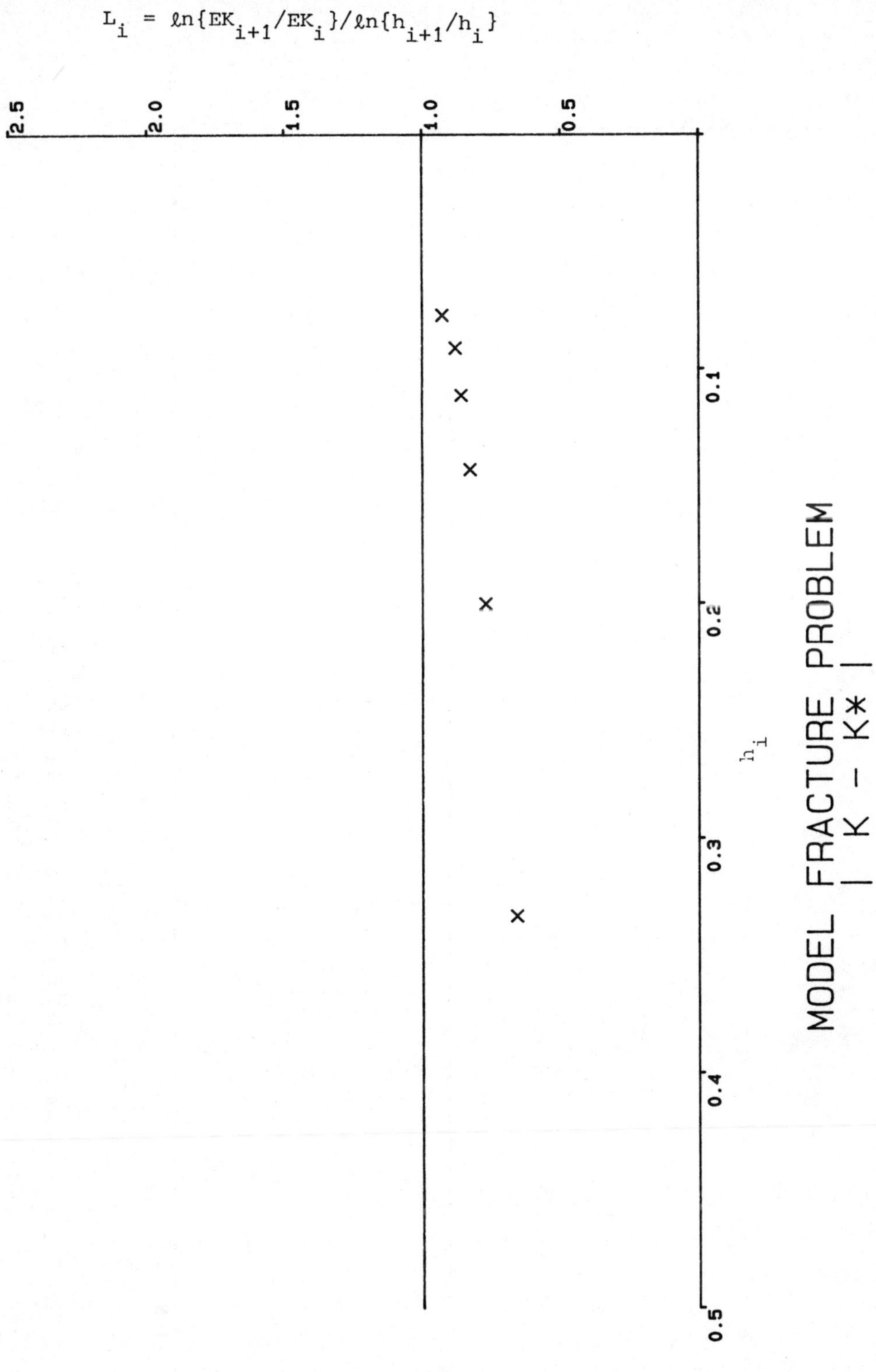
L_i = ℓn{EK_i+1/EK_i}/ℓn{h_i+1/h_i}
2.5
2.0
1.5
1.0
0.5
0.1
0.2
0.3
0.4
0.5
h_i
MODEL FRACTURE PROBLEM
| K - K* |

$L_i = \ell n\{EK_{i+1}/EK_i\}/\ell n\{h_{i+1}/h_i\}$

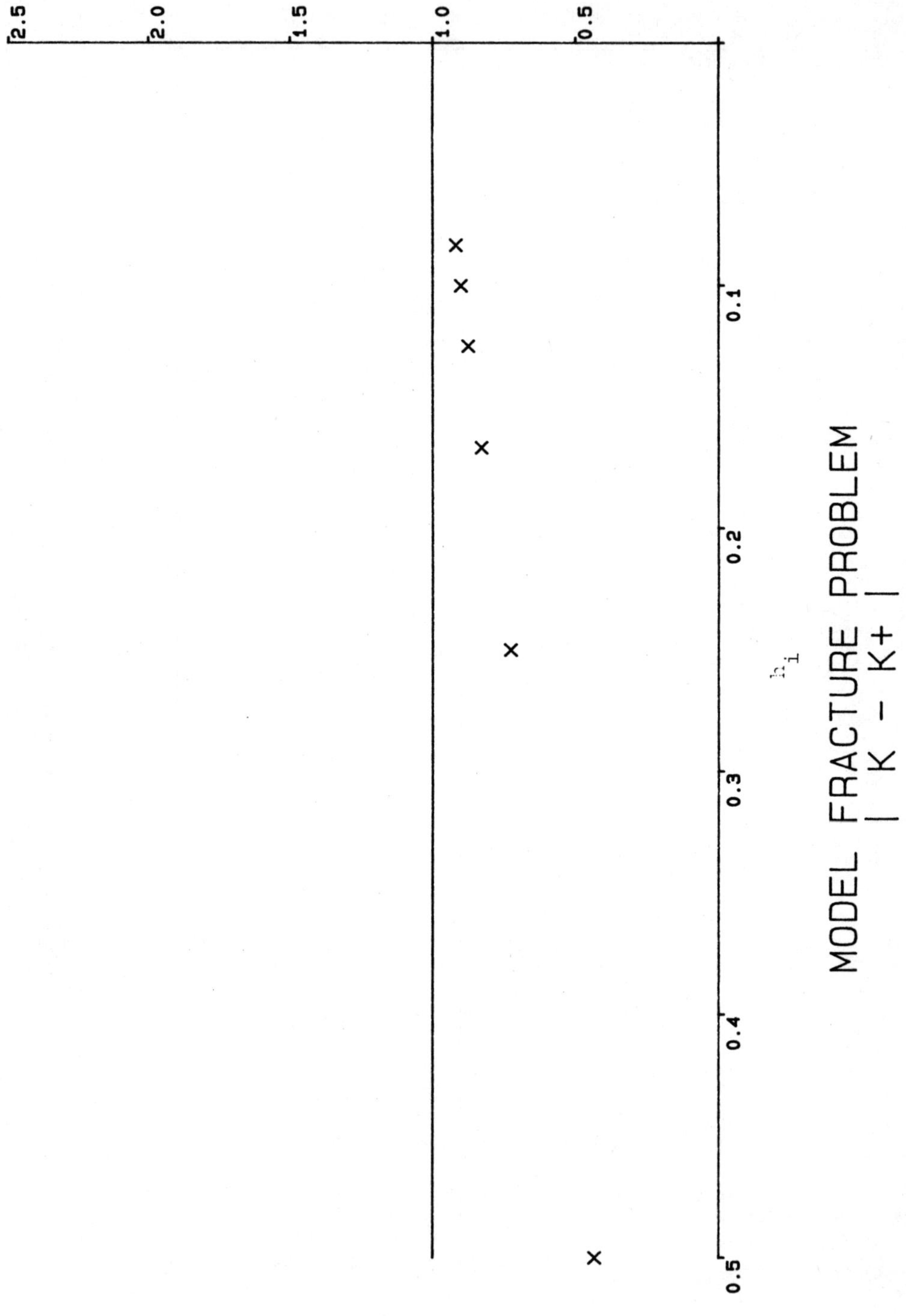

We can rewrite (5.33) of Lemma 5.1, exploiting the fact that $\text{meas}(R(\Gamma)) = O(h)$, as

$$\left|\nabla \mathbf{u}-\nabla^{\circledast}\mathbf{u}_h\right|_{0,\Gamma} \leq C\left\{\left|\mathbf{u}_I-\mathbf{u}_h\right|_{1,\infty,\Omega_0} + h^2\left|\mathbf{u}\right|_{3,\infty,\Omega_0}\right\} \;:\; \circledast = *,+ \tag{5.45}$$

and the corresponding result for standard gradients as

$$\left|\nabla(\mathbf{u}-\mathbf{u}_h)\right|_{0,\Gamma} \leq C\left\{\left|\mathbf{u}_I-\mathbf{u}_h\right|_{1,\infty,\Omega_0} + h\left|\mathbf{u}\right|_{2,\infty,\Omega_0}\right\} . \tag{5.46}$$

If we assume that, on a subdomain Ω_0, the asymptotic rate of convergence of $\mathbf{u}_I-\mathbf{u}_h$ is the same in the W^1_∞ norm as in the H^1 norm, i.e. $O(h^{1-\varepsilon})$ in this case, the $O(h^{1-\varepsilon})$ estimates for $\left|K_I-K^h_\Gamma\right|$, $\left|K_I-K^*_\Gamma\right|$ and $\left|K_I-K^+_\Gamma\right|$, suggested by our numerical results, follow immediately.

We shall investigate this assumption in Section 6; and will show that it does indeed hold for Poisson problems. We encounter difficulties when trying to extend the result to problems of planar linear elasticity, which we have not yet overcome; however, numerical evidence strongly supports the assumption.

This is seen from the following plots for the quantities $\left|\mathbf{u}_I-\mathbf{u}_h\right|_{1,\Omega_0}$ and $\left|\mathbf{u}_I-\mathbf{u}_h\right|_{1,\infty,\Omega_0}$, with the subdomain Ω_0 as shaded in Fig. 5.8b. These plots respectively confirm and support asymptotic rates of convergence of $O(h^{1-\varepsilon})$.

5.5.4 Conclusions

We conclude this section on the approximation of the stress intensity factor, K_I, on a somewhat mixed note. We have found, disappointingly, that, for both area and contour integral approximations, there is no improvement in the asymptotic rate of convergence using recovered gradients. This is due to the low global regularity of the displacement, **u**, caused by the stress singularity at the crack tip. Unless, therefore, there is a significant reduction in the *actual* error of the approximations using recovered gradients, the extra work involved in their computation cannot be justified.

As an example, we consider the numerical results for our Model Fracture Problem, and compare the actual errors in $|K-K^h|$ and $|K-K^*|$ for mesh sizes varying from $h = 1/3$ to $h = 1/15$, see Table 5.1. We see that the errors are fairly similar, with those for $|K-K^h|$ slightly smaller!

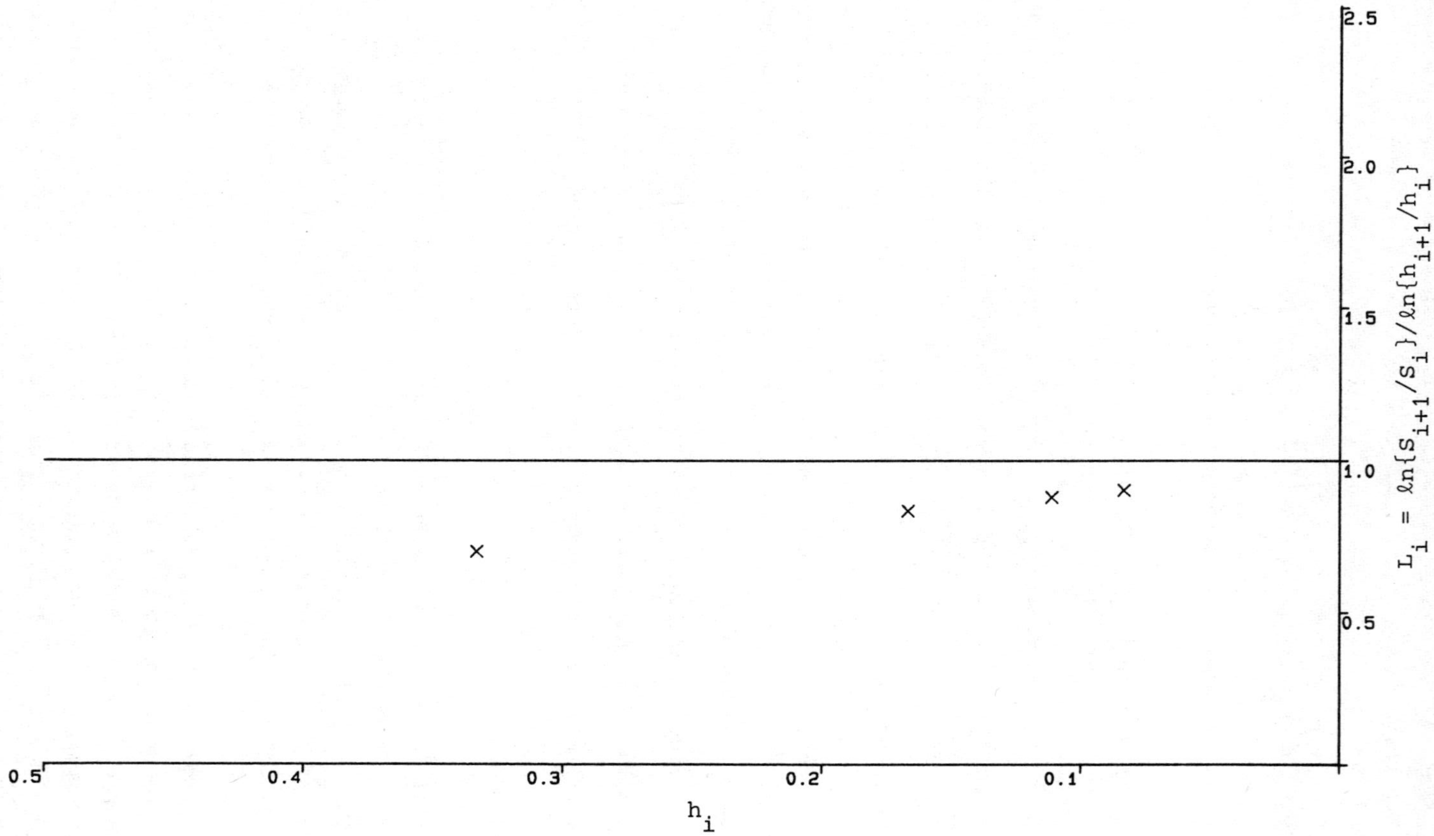

MODEL FRACTURE PROBLEM
H1 seminorm : u(I) - u(h) over Subdomain

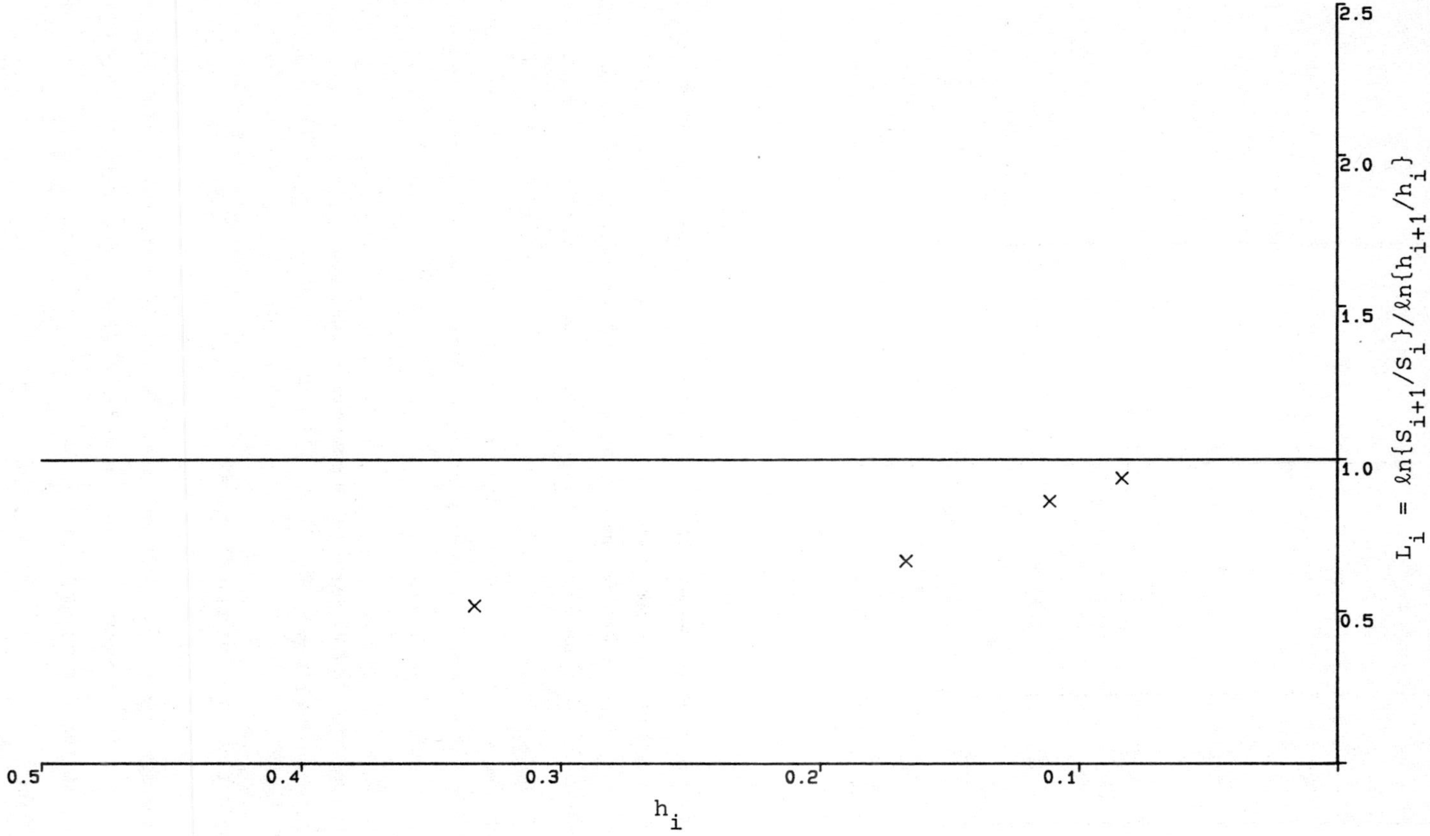

MODEL FRACTURE PROBLEM
W1, infinity seminorm : u(I) - u(h) over Subdomain

h	$10^4\lvert K-K^h\rvert$	$10^4\lvert K-K^*\rvert$
$\frac{1}{3}$	7.01	7.39
$\frac{1}{5}$	5.09	5.27
$\frac{1}{7}$	3.97	4.07
$\frac{1}{9}$	3.25	3.31
$\frac{1}{11}$	2.75	2.78
$\frac{1}{13}$	2.38	2.40
$\frac{1}{15}$	2.08	2.10

Table 5.1

However, let us be more optimistic. The approximations of K_I using contour integrals involve significantly less work to compute than those using area integrals; in fact $O(h)$ times the amount of work. Thus ideally we would wish to use them. However, from our initial theoretical results, these integrals appear to have a slower rate of convergence. In spite of this we have produced numerical evidence suggesting that this rate is not optimal, and the contour integral approximations have the same asymptotic rate of convergence, $O(h^{1-\varepsilon})$, as the area integral approximations.

6. Pointwise Superconvergence of Recovered Gradients for Poisson Problems

6.1 Introduction and Poisson Problem

We now consider again the Poisson problem (2.1), with Dirichlet boundary conditions, and first restrict Ω to be a rectangle with sides parallel to the x- and y-axes. It is further assumed that f and g are such that for this problem the solution $u \in W^3_\infty(\Omega)$.

In Sections 2.2 and 2.4 respectively the (discontinuous and continuous) piecewise linear recovered gradient functions $\nabla^* u_h$ and $\nabla^+ u_h$ were defined for the piecewise linear Galerkin approximation u_h to u, the solution of (2.2). It was shown that both these functions have faster asymptotic rates of convergence to ∇u in the L_2-norm than does ∇u_h; over the whole of Ω when $u \in H^3(\Omega)$, and over a subdomain Ω_0, $\Omega_0 < \Omega_1 < \Omega_2 \subseteq \Omega$, when $u \in H^3(\Omega_2)$. The purpose of Section 6 is to show that the recovered gradient functions are superconvergent approximations to ∇u pointwise in the L_∞-norm as well as in the L_2-norm.

6.2 Superconvergent $W^1_\infty(\Omega)$ Error Estimates

Following the definitions (2.10) and (2.15) we define the $W^1_\infty(\Omega)^*$ and $W^1_\infty(\Omega)^+$ error seminorms

$$\left|u - u_h\right|^{\circledast}_{1,\infty,\Omega} \equiv \left|\nabla u - \nabla^{\circledast} u_h\right|_{0,\infty,\Omega} : \circledast = *,+ . \tag{6.1}$$

In order to derive superconvergent estimates for (6.1) for the case of the weak problem (2.2) with Ω, f, g as described in Section 6.1, we need the following lemma.

Lemma 6.1

Let $u \in W^3_\infty(\Omega)$ be the solution of problem (2.2) with the restrictions on Ω, f, g as above, and let $u_h \in S^h_E$ be the solution of (2.5), then

$$\left|u - u_h\right|^{\circledast}_{1,\infty,\Omega} \leq C\left\{\left|u_I - u_h\right|_{1,\infty,\Omega} + h^2\left|u\right|_{3,\infty,\Omega}\right\} , \tag{6.2}$$

where $u_I \in S^h_E$ is the piecewise linear interpolant to u.

Proof Using the triangle inequality, we have for $\circledast = *,+$,

$$\left|u - u_h\right|^{\circledast}_{1,\infty,\Omega} \equiv \left|\nabla u - \nabla^{\circledast} u_h\right|_{0,\infty,\Omega} \leq \left|\nabla^{\circledast}(u_I - u_h)\right|_{0,\infty,\Omega} + \left|\nabla u - \nabla^{\circledast} u_I\right|_{0,\infty,\Omega} . \tag{6.3}$$

Using the definitions of $\nabla^{\circledast} v_h$ for $v_h \in S^h(\Omega)$, the first term on the right-hand side of (6.3) can be bounded as follows,

$$\left|\nabla^{\circledast}(u_I - u_h)\right|_{0,\infty,\Omega} \leq C\left|u_I - u_h\right|_{1,\infty,\Omega} . \tag{6.4}$$

The second term is bounded using the Sobolev and Bramble-Hilbert Lemmas, see Sections 1.2.3 and 1.2.4, as follows; using the fact that $\nabla^{\circledast} u_I \equiv \nabla u$ for u *quadratic* (and thus obtaining an extra power of h than we would have if this were only the case for linear u)

$$|\nabla u - \nabla^{\circledast} u_I|_{0,\infty,\Omega} \leq Ch^2 |u|_{3,\infty,\Omega} \, . \tag{6.5}$$

The required result (6.2) follows immediately from (6.3)-(6.5). □

The task of estimating $|u - u_h|^{\circledast}_{1,\infty,\Omega}$ is again that of estimating the difference between the gradients of the interpolant u_I and the Galerkin approximation u_h, but now in the L_∞-norm. Levine [24, Section 5.1] proved, for the Poisson problem (2.2) with the restrictions as stated and $u \in W^3_\infty(\Omega)$, that

$$|u_I - u_h|_{1,\infty,\Omega} \leq Ch^{2-\varepsilon} |u|_{3,\infty,\Omega} \, ; \tag{6.6}$$

see also Zhu [43] and Lin, Lü and Shen [25, Section 3].

From (6.2) and (6.6) we have the following theorem.

Theorem 6.1

Let $u \in W^3_\infty(\Omega)$ be the solution of problem (2.2) with the restrictions on Ω, f, g as above, $u_h \in S^h_E$ be the solution of (2.5) and $\nabla^* u_h$, $\nabla^+ u_h$ be the recovered gradient functions. Then

$$|u - u_h|^{\circledast}_{1,\infty,\Omega} \leq Ch^{2-\varepsilon} |u|_{3,\infty,\Omega} \, : \quad \circledast = *,+ \, . \tag{6.7}$$

Remark 6.1

For the case where $u \in W^{3+\varepsilon}_\infty(\Omega)$, Levine [24, Section 5.2] showed that

$$|u_I - u_h|_{1,\infty,\Omega} \leq Ch^2 \|u\|_{3+\varepsilon,\infty,\Omega} \tag{6.8}$$

so that our estimate (6.7) is improved to become

$$|u - u_h|^{\circledast}_{1,\infty,\Omega} \leq Ch^2 \|u\|_{3+\varepsilon,\infty,\Omega} \, : \quad \circledast = *,+ \, . \tag{6.9}$$

Using the theory of Rannacher and Scott [31] and Levine [24], we can obtain

$$|u - u_h|_{1,\infty,\Omega} \leq Ch |u|_{2,\infty,\Omega} \, , \tag{6.10}$$

for the present Poisson problem and Galerkin approximation. Comparison of (6.9) with (6.10) shows that with recovered gradients we have a full extra O(h) estimate, provided u has sufficient regularity.

6.3 Pointwise Superconvergence on Subdomains of General Polygonal Regions

6.3.1 Generalisation of Problem

In this Section we shall derive superconvergent pointwise estimates for Poisson problems in a more general context than that of Sections 6.1 and 6.2. We therefore now consider the mixed boundary value problem (2.26) with Ω a general polygonal region such that the interior angles at the points of intersection of $\overline{\partial\Omega}^D$ and $\overline{\partial\Omega}^N$ are convex. As in the L_2 case, we similarly cannot for this problem obtain the pointwise superconvergent estimates of Section 6.2 over the whole of Ω and thus, as in Section 3.1, we work with subdomains of Ω.

For the present analysis we need the subdomains Ω_0, Ω_1, Ω_2, Ω_3 of Ω such that

$$\Omega_0 \subset\subset \Omega_1 \subset\subset \Omega_2 \subset\subset \Omega_3 \subset \Omega , \tag{6.11}$$

where $\Omega_i \subset\subset \Omega_j$ signifies that $\Omega_i \subset \Omega_j$ and $\text{dist}(\Omega_i,\Omega_j) > 0$, which satisfy the following conditions, see Fig. 6.1,

Condition 1 Ω_0, Ω_1 and Ω_2 are rectangular, and partitioned by a uniform isosceles right-triangular mesh, as described in Section 2.1.

Condition 2 Ω_3 is a *rectangle*, and partitioned uniformly as above.

Condition 3 Ω_3 is remote from sufficient sections of the boundary $\partial\Omega$ on which Neumann boundary conditions are specified, so that $\partial\Omega_3 \cap \partial\Omega^N$ is parallel to no more than one axis.

Condition 4 Ω_3 is remote from singularities of the solution u at the corners of the boundary $\partial\Omega$, in order that $u \in W_\infty^3(\Omega_3)$. We assume, of course that f, g^D and g^N are sufficiently regular to guarantee this; thus, see Grisvard [14] and [15,pp.182-248],

$$f \in W_\infty^1(\Omega) \tag{6.12}$$

$$g^D \in W_\infty^3(\partial\Omega_i^D) \tag{6.13}$$

$$g^N \in W_\infty^2(\partial\Omega_i^N) \tag{6.14}$$

where we recall the $\partial\Omega_i^D$ and $\partial\Omega_i^N$ are the sides of $\partial\Omega^D$ and $\partial\Omega^N$.

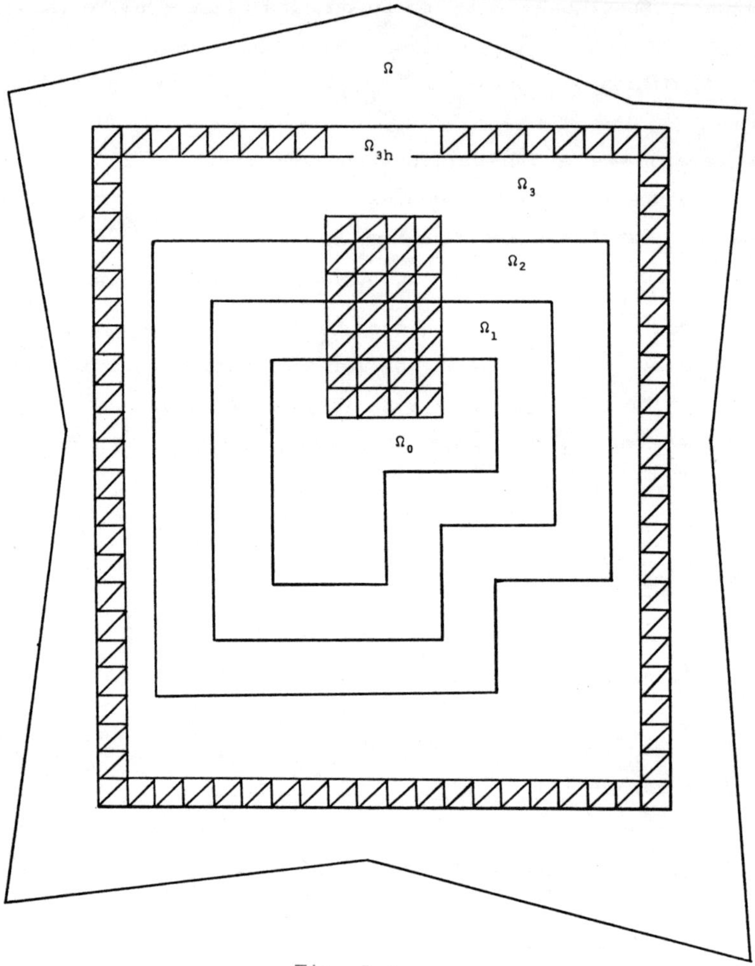

Fig. 6.1

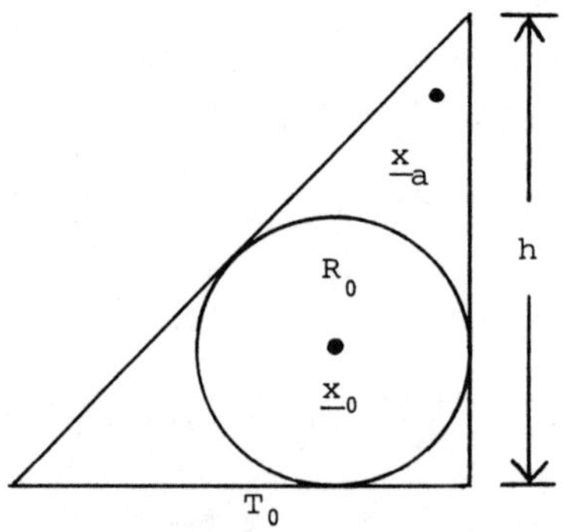

Fig. 6.2

Our analysis will also require the definition of the region $\Omega_{3h} \subset \Omega_3 \setminus \Omega_2$, see Fig. 6.1, which is defined as the outermost layer of triangles of the finite element mesh in Ω_3.

6.3.2 Smooth Delta and Derivative Green's Functions

The estimates (6.2) can readily be shown to hold locally over the subdomain Ω_0, so that

$$\left|u - u_h\right|^{\circledast}_{1,\infty,\Omega_0} \leq C\left\{\left|u_I-u_h\right|_{1,\infty,\Omega_0}+h^2\left|u\right|_{3,\infty,\Omega_0}\right\} \quad : \quad \circledast = *,+ \ . \tag{6.15}$$

The task, similar to that previously, is thus to estimate the term $\left|u_I - u_h\right|_{1,\infty,\Omega_0}$.

We define $\mathbf{x}_a \in \Omega_0$ to be an *arbitrary* point. We shall estimate $\left|\frac{\partial}{\partial x}(u_I-u_h)(\mathbf{x}_a)\right|$ and, since $\mathbf{x}_a$ is arbitrary, this will estimate $\left|\frac{\partial}{\partial x}(u_I-u_h)\right|_{0,\infty,\Omega_0}$. A similar analysis for $\left|\frac{\partial}{\partial y}(u_I-u_h)\right|_{0,\infty,\Omega_0}$ leads to an estimate for $\left|u_I-u_h\right|_{1,\infty,\Omega_0}$.

Let $T_0 \subset \Omega_0$ be a triangle of the regular partition having short side length h, see Fig. 6.1. Let $\mathbf{x}_a \in T_0$ and define R_0 to be the inscribed circle of T_0 with centre $\mathbf{x}_0$; see Fig. 6.2. Using plane polar coordinates (r,θ) local to $\mathbf{x}_0$, we then define the function, $\delta \in C_0^\infty(\bar{R}_0)$, to vanish outside R_0, and to integrate to unity over R_0. More formally

$$\left.\begin{aligned} &\delta \equiv \delta(r) \ , \ 0 \leq \delta \leq Ch^{-2} \ , \\ &\int_{R_0} \delta \ d\mathbf{x} = 1 \ . \end{aligned}\right\} \tag{6.16}$$

This function, which is like a smoothed Dirac-delta function, is essentially the same as that considered by Rannacher and Scott [31], and Levine [24, Section 5.1].

We now define the function G as the solution of the following problem

$$\left.\begin{aligned} -\Delta[G(\mathbf{x})] &= -\frac{\partial\delta}{\partial x} \ , \ \mathbf{x} \in \Omega_3 \ , \\ G(\mathbf{x}) &= 0 \ , \quad \mathbf{x} \in \partial\Omega_3 \ , \end{aligned}\right\} \tag{6.17}$$

and let G vanish outside Ω_3. The corresponding weak formulation is given by $G \in \overset{\circ}{H}{}^1(\Omega_3)$ such that

$$\int_{\Omega_3} \nabla G.\nabla v \ d\mathbf{x} = -\int_{\Omega_3} v\,\frac{\partial\delta}{\partial x}\,d\mathbf{x} \ , \quad \forall \quad v \in \overset{\circ}{H}{}^1(\Omega_3) \tag{6.18}$$

and the finite element approximation by

$G_h \in S_0^h(\Omega_3)$ such that

$$\int_{\Omega_3} \nabla G_h . \nabla v_h \, d\mathbf{x} = - \int_{\Omega_3} v_h \frac{\partial \delta}{\partial x} \, d\mathbf{x} \ , \quad \forall \quad v_h \in S_0^h(\Omega_3) \ . \tag{6.19}$$

We shall also need the orthogonality property, which follows from (6.18)-(6.19), as $S_0^h(\Omega_3) \subset \mathring{H}^1(\Omega_3)$,

$$\int_{\Omega_3} \nabla(G-G_h) . \nabla v_h \, d\mathbf{x} = 0 \ , \quad \forall \quad v_h \in S_0^h(\Omega_3) \ . \tag{6.20}$$

Like δ, the function G is essentially the same as that considered by Rannacher and Scott [31] and Levine [24], but now defined only over the subdomain Ω_3. As Ω_3 satisfies the geometry and mesh requirements of Condition 2 of Section 6.3.1 we have the following, see [31],

$$|G - G_h|_{1,1,\Omega_3} \leq C \tag{6.21}$$

and, see [24],

$$|G|_{1,1,\Omega_3} \leq Ch^{-\varepsilon} \tag{6.22}$$

so that, using the triangle inequality,

$$|G_h|_{1,1,\Omega_3} \leq Ch^{-\varepsilon} \ . \tag{6.23}$$

We can split the function G and write

$$G \equiv G^\delta + G^H \ , \tag{6.24}$$

where, in polar coordinates local to $\mathbf{x}_0$,

$$G^\delta = \frac{\cos \theta}{r} \int_0^r \rho \ \delta(\rho) \ d\rho \ . \tag{6.25}$$

The integral, $\int_0^r \rho \ \delta(\rho) \ d\rho$, is a smoothing term which, from (6.16) takes the value $\frac{1}{2\pi}$ outside the circle R_0. Thus

$$G^\delta = \frac{\cos \theta}{2\pi r} \text{ in } (\bar{\Omega}_3 \backslash \bar{R}_0) \cup \partial R_0 . \tag{6.26}$$

We observe that G^δ is just a smoothed x-derivative of the free-space Green's function for the Laplace operator

$$\frac{\partial}{\partial x}\left\{\frac{\ln r}{2\pi}\right\} = \frac{\cos\theta}{2\pi r} . \tag{6.27}$$

Thus *outside* Ω_1 G^δ is bounded in all norms independently of h. Now G^δ satisfies the differential equation (6.17) but *not* the boundary conditions. Thus G^H of (6.24) is a *harmonic* function in Ω_3, which satisfies the boundary conditions

$$G^H = \frac{-\cos\theta}{2\pi r} \quad \text{on} \quad \partial\Omega_3 \tag{6.28}$$

and is thus bounded independently of h in all its norms.

From the above we have that *outside* Ω_1 the function G is bounded independently of the mesh size h in all its norms. This property is needed later.

6.3.3 Lemmas Necessary for Superconvergence Estimates

Two lemmas are now proved which show that (u_I-u_h) has the same asymptotic rate of convergence in the $W^1_\infty(\Omega_0)$ seminorm as it has in the $H^1(\Omega_3)$ norm. This result should be compared with the assumption of Section 5.5.3 for the linear elastic fracture problem.

Lemma 6.2 If $u \in H^1_E \cap W^3_\infty(\Omega_3)$ is the solution of problem (2.28) in the polygonal region of Section 6.3.1, $u_h \in S^h_E$ is the Galerkin approximation and $u_I \in S^h_E$ is the piecewise linear interpolant to u, then

$$|u_I - u_h|_{1,\infty,\Omega_0} \leq C\left\{\|u_I - u_h\|_{1,\Omega_3} + h^{2-\varepsilon}|u|_{3,\infty,\Omega_3}\right\} . \tag{6.29}$$

Proof Following the technique outlined in Section 6.3.2 it is sufficient to show that (6.29) holds for $\left|\frac{\partial}{\partial x}(u_I-u_h)(\mathbf{x}_a)\right|$. As $(u_I-u_h) \in S^h(\Omega)$, $\frac{\partial}{\partial x}(u_I-u_h)$ is a constant in each triangle of the partition. Thus as $R_0 \subset T_0$, we have from (6.16) -(6.17) integrating by parts twice

$$\frac{\partial}{\partial x}(u_I-u_h)(\mathbf{x}_a) = \left[\frac{\partial}{\partial x}(u_I-u_h)\right]_{R_0} = \left[\frac{\partial}{\partial x}(u_I-u_h)\right]_{R_0}\int_{R_0}\delta\, d\mathbf{x}$$

$$= \int_{R_0}\frac{\partial}{\partial x}(u_I-u_h)\delta\, d\mathbf{x} = \int_{\Omega_3}\frac{\partial}{\partial x}(u_I-u_h)\delta\, d\mathbf{x}$$

$$= -\int_{\Omega_3} (u_I - u_h) \frac{\partial \delta}{\partial x}\, d\mathbf{x} = -\int_{\Omega_3} (u_I - u_h) \Delta G\, d\mathbf{x}$$

$$= \int_{\Omega_3} \nabla(u_I - u_h) . \nabla G\, d\mathbf{x} - \oint_{\partial\Omega_3} (u_I - u_h) \frac{\partial G}{\partial n}\, ds$$

$$= \int_{\Omega_3} \nabla(u_I - u_h) . \nabla G_h\, d\mathbf{x} + \int_{\Omega_3} \nabla(u_I - u_h) . \nabla(G - G_h) d\mathbf{x} - \oint_{\partial\Omega_3} (u_I - u_h) \frac{\partial G}{\partial n}\, ds$$

$$\equiv Q_1 + Q_2 + Q_3 \ . \tag{6.30}$$

Each of the terms Q_i, $i = 1,2,3$ will now be estimated.

As $G_h \in S_0^h(\Omega_3) \subset H^h$, we have from the orthogonality property (2.31) that

$$Q_1 = \int_{\Omega_3} \nabla(u_I - u_h) . \nabla G_h\, d\mathbf{x}$$

$$= \int_{\Omega_3} \nabla(u_I - u) . \nabla G_h\, d\mathbf{x} \ . \tag{6.31}$$

We then use (2.23), which holds locally over Ω_3, together with (6.23), to give

$$|Q_1| = \left|\int_{\Omega_3} \nabla(u_I - u) . \nabla G_h d\mathbf{x}\right|$$

$$\leq Ch^2 |u|_{3,\infty,\Omega_3} |G_h|_{1,1,\Omega_3}$$

$$\leq Ch^{2-\varepsilon} |u|_{3,\infty,\Omega_3} \ . \tag{6.32}$$

$$Q_2 = \int_{\Omega_3} \nabla(u_I - u_h) . \nabla(G - G_h) d\mathbf{x}$$

$$= \int_{\Omega_3} \nabla(u_I - u_{Ih}) . \nabla(G - G_h) d\mathbf{x} + \int_{\Omega_3} \nabla(u_{Ih} - u_h) . \nabla(G - G_h) d\mathbf{x} \tag{6.33}$$

where we define the function, $u_{Ih} \in S_E^h$, by

$$u_{Ih} \equiv \left\{ \begin{array}{ll} u_I & \text{in} \ \ (\bar{\Omega}_3 \backslash \bar{\Omega}_{3h}) \cup (\partial\Omega_{3h} \backslash \partial\Omega_3) \\ & \\ u_h & \text{in} \ \ (\bar{\Omega} \backslash \bar{\Omega}_3) \cup \partial\Omega_3 \end{array} \right\} . \tag{6.34}$$

Obviously u_{Ih} is uniquely defined in Ω_{3h} by interpolation to nodal values.

The motivation for splitting Q_2 as in (6.33) is as follows. It would be ideal if Q_2 were to vanish by orthogonality property (6.20): unfortunately this is not the case as $(u_I - u_h) \notin S_0^h(\Omega_3)$. However, by replacing $(u_I - u_h)$ by $(u_{Ih} - u_h) \in S_0^h(\Omega_3)$, a function identical to $(u_I - u_h)$ in $\Omega_3\backslash\Omega_{3h}$, we are left with an integral restricted to Ω_{3h}, the outermost layer of triangles of the finite element mesh in Ω_3. Thus from (6.33)-(6.34), orthogonality property (6.20), and using the fact that $(u_I - u_{Ih})$ vanishes in $\Omega_3 \setminus \Omega_{3h}$, we have

$$\begin{aligned}|Q_2| &= \left|\int_{\Omega_3} \nabla(u_I - u_{Ih}).\nabla(G - G_h)d\mathbf{x}\right| \\ &= \left|\int_{\Omega_{3h}} \nabla(u_I - u_{Ih}).\nabla(G - G_h)d\mathbf{x}\right| \\ &\leq |u_I - u_{Ih}|_{1,\Omega_{3h}} |G - G_h|_{1,\Omega_{3h}} . \end{aligned} \tag{6.35}$$

Now using the properties of functions in $S^h(\Omega)$, see [5], and of u_{Ih} in particular; and also exploiting the fact that $\text{meas}(\Omega_{3h}) = O(h)$, we have

$$\begin{aligned}|u_I - u_{Ih}|_{1,\Omega_{3h}} &\leq Ch^{\frac{1}{2}}|u_I - u_{Ih}|_{1,\infty,\Omega_{3h}} \\ &\leq Ch^{-\frac{1}{2}}|u_I - u_{Ih}|_{0,\infty,\Omega_{3h}} = Ch^{-\frac{1}{2}} \max_{\text{vertices of } \Omega_{3h}} |u_I - u_{Ih}| \\ &\leq Ch^{-\frac{1}{2}} \max_{\text{vertices of } \Omega_{3h}} |u_I - u_h| = Ch^{-\frac{1}{2}}|u_I - u_h|_{0,\infty,\Omega_{3h}} \\ &\leq Ch^{-\frac{1}{2}}|u_I - u_h|_{0,\infty,\Omega_3} . \end{aligned} \tag{6.36}$$

Now for arbitrary small $\varepsilon > 0$, $\exists q \equiv q(\varepsilon) = \dfrac{2}{\varepsilon} < \infty$ such that

$$|u_I - u_h|_{0,\infty,\Omega_3} \leq Ch^{-\varepsilon}|u_I - u_h|_{0,q,\Omega_3} \tag{6.37}$$

and so by the Sobolev Embedding Theorem, see Adams [1,pp.95-99],

$$|u_I - u_h|_{0,\infty,\Omega_3} \leq Ch^{-\varepsilon}\|u_I - u_h\|_{1,\Omega_3} . \tag{6.38}$$

Thus from (6.36) and (6.38) we have

$$\left|u_I - u_{Ih}\right|_{1,\Omega_{3h}} \leq Ch^{-\frac{1}{2}-\varepsilon}\left\|u_I - u_h\right\|_{1,\Omega_3} . \tag{6.39}$$

To estimate the second term, $\left|G - G_h\right|_{1,\Omega_{3h}}$, on the right-hand side of (6.35), we use the result of Nitsche and Schatz [28], which holds for the subdomains $\Omega_3 \setminus \Omega_2 \subset \Omega_3 \setminus \Omega_1$, of Ω_3, to give

$$\begin{aligned}\left|G - G_h\right|_{1,\Omega_{3h}} &\leq \left|G - G_h\right|_{1,\Omega_3\setminus\Omega_2} \\ &\leq C\left\{\left\|G - G_h\right\|_{-1,\Omega_3\setminus\Omega_1} + h\left|G\right|_{2,\Omega_3\setminus\Omega_1}\right\} \\ &\leq C\left\{\left\|G - G_h\right\|_{-1,\Omega_3} + h\left|G\right|_{2,\Omega_3\setminus\Omega_1}\right\} .\end{aligned} \tag{6.40}$$

The space $H^{-1}(\Omega_3)$ is the ***dual space*** of $\overset{\circ}{H}{}^1(\Omega_3)$. In Lemma 6.3 we shall prove the estimate

$$\left\|G - G_h\right\|_{-1,\Omega_3} \equiv \sup_{v\in\overset{\circ}{H}{}^1(\Omega_3)} \left\|v\right\|^{-1}_{1,\Omega_3}\left|\int_{\Omega_3} (G - G_h)v\, d\mathbf{x}\right| \leq Ch^{1-\varepsilon} . \tag{6.41}$$

Let us, however, assume it in the meantime. Now from the discussion of Section 6.3.2, we see that the term $\left|G\right|_{2,\Omega_3\setminus\Omega_1}$ is bounded independently of the mesh size h, i.e.

$$\left|G\right|_{2,\Omega_3\setminus\Omega_1} \leq C \tag{6.42}$$

and thus from (6.40)-(6.42) we have

$$\left|G - G_h\right|_{1,\Omega_{3h}} \leq Ch^{1-\varepsilon} . \tag{6.43}$$

Combining the results (6.35), (6.39) and (6.43), we thus have

$$\left|Q_2\right| \leq Ch^{\frac{1}{2}-\varepsilon}\left\|u_I - u_h\right\|_{1,\Omega_3} . \tag{6.44}$$

Finally, to estimate Q_3, we use the Sobolev embedding, see Section 1.2.2, $H^1(\Omega_3) \longrightarrow L_2(\partial\Omega_3)$, to give

$$\begin{aligned}\left|Q_3\right| &\leq \oint_{\partial\Omega_3} \left|u_I - u_h\right|\left|\nabla G\right| ds \\ &\leq \left|u_I - u_h\right|_{0,\partial\Omega_3}\left|\nabla G\right|_{0,\partial\Omega_3} \\ &\leq C\left\|u_I - u_h\right\|_{1,\Omega_3}\left|\nabla G\right|_{0,\partial\Omega_3} .\end{aligned} \tag{6.45}$$

Again, from the discussion of Section 6.3.2, we see that $|\nabla G|_{0,\partial\Omega_3}$ is bounded independently of the mesh size h, i.e.

$$|\nabla G|_{0,\partial\Omega_3} \leqq C \tag{6.46}$$

and thus from (6.45)-(6.46), we have

$$|Q_3| \leqq C\|u_I - u_h\|_{1,\Omega_3} . \tag{6.47}$$

We now combine (6.30), (6.32), (6.44) and (6.47) to obtain

$$\left|\frac{\partial}{\partial x}(u_I - u_h)(\mathbf{x}_a)\right| \leqq |Q_1| + |Q_2| + |Q_3|$$

$$\leqq C\left\{\|u_I - u_h\|_{1,\Omega_3} + h^{2-\varepsilon}|u|_{3,\infty,\Omega_3}\right\} \tag{6.48}$$

from which the result (6.29) follows immediately. Our proof, of course, relies upon the assumption of (6.41), which we now prove.

Lemma 6.3 If $G \in \overset{\circ}{H}{}^1(\Omega_3)$ is the solution of (6.18) and $G_h \in S_0^h(\Omega_3)$ is the solution of (6.19), then

$$\|G - G_h\|_{-1,\Omega_3} \leqq Ch^{1-\varepsilon} .$$

Proof We recall the definition

$$\|G - G_h\|_{-1,\Omega_3} \equiv \sup_{v\in \overset{\circ}{H}{}^1(\Omega_3)} \|v\|_{1,\Omega_3}^{-1} \left|\int_{\Omega_3} (G - G_h)v \, d\mathbf{x}\right| . \tag{6.49}$$

Now for *every* $v \in \overset{\circ}{H}{}^1(\Omega_3)$, there exists a function $w \equiv w(v)$, the solution of the *adjoint problem*

$$\left.\begin{aligned} -\Delta[w(\mathbf{x})] &= v(\mathbf{x}) , \quad \mathbf{x} \in \Omega_3 \\ w(\mathbf{x}) &= 0 , \quad \mathbf{x} \in \partial\Omega_3 \end{aligned}\right\} \tag{6.50}$$

Let us now give some regularity results for the functions v and w. From the Sobolev Embedding Theorem, see Section 1.2.2,

$$H^1(\Omega_3) \longrightarrow L_p(\Omega_3) , \quad \forall \; p < \infty , \tag{6.51}$$

so we have

$$v \in L_p(\Omega_3)\ ,\quad \forall\ p < \infty\ , \tag{6.52}$$

and further that

$$|v|_{0,p,\Omega_3} \leq C\|v\|_{1,\Omega_3}\ ,\quad \forall\ p < \infty\ , \tag{6.53}$$

We recall from Condition 2 of Section 6.3.1, that Ω_3 is a *rectangle*, and hence from (6.52), see Grisvard, [14] and [15,pp.182-248], the function w has global regularity

$$w \in W_p^2(\Omega_3)\ ,\quad \forall\ p < \infty \tag{6.54}$$

and further

$$\|w\|_{2,p,\Omega_3} \leq C|v|_{0,p,\Omega_3}\ ,\quad \forall\ p < \infty\ . \tag{6.55}$$

Thus immediately from (6.53) and (6.55)

$$\|w\|_{2,p,\Omega_3} \leq C\|v\|_{1,\Omega_3}\ ,\quad \forall\ p < \infty\ . \tag{6.56}$$

Using the shorthand notation, $\sup_v$, to represent $\sup_{v\in \overset{\circ}{H}{}^1(\Omega_3)}$, we have from (6.49)-(6.50), integrating by parts, and remembering that $(G - G_h) \in \overset{\circ}{H}{}^1(\Omega_3)$,

$$\begin{aligned}
\|G - G_h\|_{-1,\Omega_3} &\equiv \sup_v \|v\|_{1,\Omega_3}^{-1} \left|\int_{\Omega_3} (G - G_h)v\ d\mathbf{x}\right| \\
&= \sup_v \|v\|_{1,\Omega_3}^{-1} \left|\int_{\Omega_3} (G - G_h)\Delta w\ d\mathbf{x}\right| \\
&= \sup_v \|v\|_{1,\Omega_3}^{-1} \left|\int_{\Omega_3} \nabla(G - G_h).\nabla w\ d\mathbf{x}\right| \\
&= \sup_v \|v\|_{1,\Omega_3}^{-1} \left|\int_{\Omega_3} \nabla(G - G_h).\nabla(w - w_I) d\mathbf{x}\right|\ ,
\end{aligned} \tag{6.57}$$

which follows from orthogonality property (6.20), as $w_I \in S_0^h(\Omega_3)$. Thus from (6.21)

$$\begin{aligned}
\|G - G_h\|_{-1,\Omega_3} &\leq |G - G_h|_{1,1,\Omega_3} \sup_v \|v\|_{1,\Omega_3}^{-1} |w - w_I|_{1,\infty,\Omega_3} \\
&\leq C \sup_v \|v\|_{1,\Omega_3}^{-1} |w - w_I|_{1,\infty,\Omega_3}\ .
\end{aligned} \tag{6.58}$$

Now from interpolation theory, for arbitrary small $\varepsilon > 0$, $\exists q \equiv q(\varepsilon) = \frac{2}{\varepsilon} < \infty$ such that

$$|w - w_I|_{1,\infty,\Omega_3} \leq Ch^{1-\varepsilon}|w|_{2,q,\Omega_3}\ . \tag{6.59}$$

Thus finally from (6.56)-(6.59)

$$\|G - G_h\|_{-1,\Omega_3} \leq Ch^{1-\varepsilon} \sup_v \|v\|^{-1}_{1,\Omega_3} |w|_{2,q,\Omega_3}$$

$$\leq Ch^{1-\varepsilon} \sup_v \|v\|^{-1}_{1,\Omega_3} \|v\|_{1,\Omega_3}$$

$$= Ch^{1-\varepsilon} .$$

which concludes the proof of the Lemma.

6.3.4 Generalisation of Subdomains

We now describe two ways in which the subdomains Ω_0, Ω_1, Ω_2, Ω_3 of Ω, introduced in Section 6.3.1, can be generalised.

Firstly requirement (6.11) can be relaxed to permit

$$\Omega_0 < \Omega_1 < \Omega_2 < \Omega_3 \subset \Omega , \qquad (6.60)$$

where the symbol "<" is as in (3.1), provided Ω_2 is remote from all sections of $\partial\Omega^N$; i.e., see Fig. 6.3,

$$\partial\Omega_2 \cap \partial\Omega^N = \emptyset . \qquad (6.61)$$

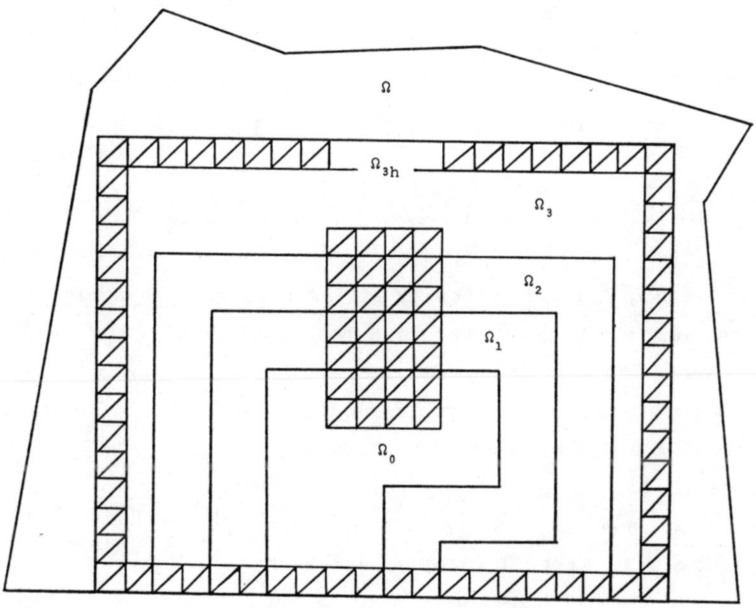

Fig. 6.3

For $\mathbf{x}_a \in \Omega_0$ an arbitrary point the estimates (6.21)-(6.23) still hold and, provided $\mathbf{x}_a$ is bounded away from $\partial\Omega$, we have as in Section 6.3.2 that *outside* Ω_1 the function G is bounded in all its norms independently of the mesh size h. If, however $\mathbf{x}_a$ is not bounded away from $\partial\Omega$, our argument no longer holds as, recall (6.28), the function G^H is no longer bounded independently of h on $\partial\Omega_3$.

However, by using the "image function" technique of Levine [24,Section 5.1] we are able to retain the original result. For this we suppose, for simplicity, that $\mathbf{x}_a$ is not bounded away from the single segment Γ_1 of $\partial\Omega$. (A slightly more involved argument is required if $\mathbf{x}_a$ is not bounded away from a corner, and hence two sides, of $\partial\Omega$.)

Let the point $\mathbf{x}_1$ be the reflection of $\mathbf{x}_0$ in Γ_1 and rewrite (6.24) as

$$G \equiv G^\delta + G_1^\delta + \tilde{G}^H \tag{6.62}$$

where the function G_1^δ is given by

$$G_1^\delta(r,\theta) \equiv G^\delta(r_1,\theta_1) \tag{6.63}$$

with (r_1,θ_1) being plane polar coordinates local to $\mathbf{x}_1$, with the radius $\theta_1 = 0$ chosen so that

$$G^\delta + G_1^\delta = 0 \quad \text{on} \quad \Gamma_1 \ . \tag{6.64}$$

Fairly obviously outside Ω_1, G^δ and G_1^δ are bounded in all norms independently of the mesh size h.

Now as $\mathbf{x}_1 \notin \Omega$, so G_1^δ is a function *harmonic* in Ω, and therefore $\tilde{G}^H$ is a function *harmonic* in Ω_3, and satisfying the boundary conditions

$$\tilde{G}^H = \left\{\begin{array}{ll} 0 & \text{on } \partial\Omega_3 \cap \Gamma_1 \ . \\ -\dfrac{1}{2\pi}\left(\dfrac{\cos\theta}{r} + \dfrac{\cos\theta_1}{r_1}\right) & \text{on } \partial\Omega_3 \backslash (\partial\Omega_3 \cap \Gamma_1) \ . \end{array}\right\} \tag{6.65}$$

Thus $\tilde{G}^H$ is bounded independently of the mesh size h in all its norms.

From the above, we have therefore that outside Ω_1, the function G is again bounded independently of the mesh size h in all its norms. The proof of Lemma 6.2 then follows as before to yield estimate (6.29). In order to see this we consider the terms Q_1, Q_2 and Q_3 of (6.30).

Estimate (6.32) for Q_1 readily follows as before.

Although we now have $\Omega_{3h} \not\subset \Omega_3 \setminus \Omega_2$, see Fig. 6.3, the function $u_I - u_{Ih}$ vanishes in $\Omega_{3h} \cap \Omega_2$ (recall (6.61)), and thus

$$Q_2 = \int_{\Omega_{3h}\backslash(\Omega_{3h}\cap\Omega_2)} \nabla(u_I - u_{Ih}).\nabla(G - G_h)d\mathbf{x} \tag{6.66}$$

and is bounded as before in (6.44)

Finally (recall again (6.61))

$$Q_3 = -\int_{\partial\Omega_3\backslash(\partial\Omega_2\cap\partial\Omega)} (u_I - u_h)\,\frac{\partial G}{\partial n}\,ds \tag{6.67}$$

and, following the discussion above, the term $|\nabla G|_{0,\partial\Omega_3\backslash(\partial\Omega_2\cap\partial\Omega)}$ is bounded independently of the mesh size h; thus Q_3 is bounded as before in (6.47)

The second generalisation of the subdomains is more easily justified. The requirement (Condition 2 of Section 6.3.1) that Ω_3 be a *rectangle*, and not merely rectangular, can be very restrictive on the choice of Ω_0; consider, for example, the case when Ω is an L-shaped region. However, by considering Ω_0 as the union of a *fixed number* of smaller subdomains, and obtaining estimate (6.29) in each of these, we see that this restriction can readily be disregarded.

With these two generalisations, we note that, for sufficiently small h, the subdomain Ω_0 can in effect cover *most* of Ω.

6.3.5 Pointwise Superconvergence Estimates

Having shown in Lemma 6.2 that the asymptotic rate of convergence of $|u_I - u_h|_{1,\infty,\Omega_0}$ is controlled by that of $\|u_I - u_h\|_{1,\Omega_3}$, we must now estimate this latter term. In order to do this we apply the results of Section 3 directly.

Two further subdomains Ω_4 and Ω_5 such that

$$\Omega_3 < \Omega_4 < \Omega_5 \subseteq \Omega \tag{6.68}$$

are required, which must satisfy the following conditions.

Condition 1 Ω_5 is rectangular and is partitioned by a uniform isosceles right-triangular mesh; the remainder of Ω, $\Omega\backslash\Omega_5$, need only be partitioned quasi-uniformly.

Condition 2 Ω_4 is remote from sufficient segments of $\partial\Omega$ on which Neumann boundary conditions are specified, so that $\partial\Omega_4 \cap \partial\Omega^N$ is parallel to no more than one axis. (This is the reason for requiring Condition 3 of Section 6.3.1).

Condition 3 Ω_5 is remote from singularities of the solution u at the corners of the boundary $\partial\Omega$, in order that $u \in W^3_\infty(\Omega_5)$.

We see that the subdomains Ω_3, Ω_4 and Ω_5 of this section satisfy the conditions on the subdomains Ω_0, Ω_1 and Ω_2 respectively of Section 3. Thus, defining a suitable cut-off function $\omega \in W^2_\infty(\Omega)$, as in Section 3.1, which takes the value unity in Ω_3 and vanishes outside Ω_4, we have the following theorem for which there are two cases.

Theorem 6.2

Case 1 If $u \in H^2(\Omega) \cap W^3_\infty(\Omega_5)$ is the solution of (2.28) in the polygonal region of Section 6.3.1 and $u_h \in S^h_E$ is the Galerkin approximation, then

$$|u - u_h|^{\circledast}_{1,\infty,\Omega_0} \leq C(\omega)h^{2-\varepsilon}\left\{|u|_{2,\Omega} + |u|_{3,\infty,\Omega_5} + |g^D|'_{2,\partial\Omega^D}\right\} : \circledast = *,+ \ . \quad (6.69)$$

where $C(\omega)$ is a constant depending on the cut-off function ω.

Case 2 If $u \in W^2_{p-\varepsilon}(\Omega) \cap W^3_\infty(\Omega_5)$, $\frac{4}{3} \leq p < 2$, is the solution of (2.28) and $u_h \in S^h_E$ is the Galerkin approximation, then

$$|u - u_h|^{\circledast}_{1,\infty,\Omega_0} \leq C(\omega)h^{\gamma-\varepsilon}\left\{|u|_{2,p-\varepsilon,\Omega} + |u|_{3,\infty,\Omega_5} + |g^D|'_{2,\partial\Omega^D}\right\} : \circledast = *,+ \quad (6.70)$$

where $1 \leq \gamma = 4 - \frac{4}{p} < 2$ as in (3.33).

Proof These estimates readily follow from (6.15), (6.29), (3.6), (3.31) and (3.32).

Remark 6.2 The estimate (6.32) for Q_1 in Lemma 6.2 can be improved, using Levine [24, Section 5.2], if $u \in W^{3+\varepsilon}_\infty(\Omega_3)$, so that (compare (6.8))

$$|Q_1| \leq Ch^2\|u\|_{3+\varepsilon,\infty,\Omega_3} \ . \quad (6.71)$$

As a result of (6.71) we have that, if $u \in H^2(\Omega) \cap W^{3+\varepsilon}_\infty(\Omega_5)$ is the solution of (2.28), then

$$|u - u_h|^{\circledast}_{1,\infty,\Omega_0} \leq C(\omega)h^2\left\{|u|_{2,\Omega} + \|u\|_{3+\varepsilon,\infty,\Omega_5} + |g^D|'_{2,\partial\Omega^D}\right\} : \circledast = *,+ \ . \quad (6.72)$$

6.3.6 Comparison with Standard Gradient Error Estimates

We now compare the recovered gradient pointwise error estimates of Theorem 6.2 with standard L_∞-norm estimates of the gradient error on a subdomain Ω_0 of a general polygonal region Ω. The estimate

$$|u - u_h|_{1,\infty,\Omega_0} \leq C\{|u - u_h|_{0,\Omega_5} + h|u|_{2,\infty,\Omega_5}\} . \tag{6.73}$$

follows the result of Nitsche and Schatz [28] for subdomains, together with the alternative non-optimal estimate to (6.29),

$$|u_I - u_h|_{1,\infty,\Omega_0} \leq C\{\|u_I - u_h\|_{1,\Omega_3} + h|u|_{2,\infty,\Omega_3}\} . \tag{6.74}$$

The estimate (6.74) in turn follows from the alternative non-optimal estimate for Q_1 of Lemma 6.2 whereby

$$|Q_1| \leq Ch|u|_{2,\infty,\Omega_3} . \tag{6.75}$$

This can be shown as follows. From (6.31)

$$Q_1 = \int_{\Omega_3} \nabla(u_I - u).\nabla G_h \, d\mathbf{x}$$

$$= \int_{\Omega_3} \nabla(u - u_I).\nabla(G - G_h) d\mathbf{x} + \int_{\Omega_3} \nabla(u_I - u).\nabla G \, d\mathbf{x} . \tag{6.76}$$

Integration by parts twice of the latter term in (6.76) gives

$$\int_{\Omega_3} \nabla(u_I - u).\nabla G \, d\mathbf{x} = \int_{\Omega_3} (u - u_I)\Delta G \, d\mathbf{x} + \oint_{\partial\Omega_3} (u_I - u) \frac{\partial G}{\partial n} ds$$

$$= \int_{\Omega_3} (u - u_I) \frac{\partial\delta}{\partial x} d\mathbf{x} + \oint_{\partial\Omega_3} (u_I - u) \frac{\partial G}{\partial n} ds$$

$$= \int_{\Omega_3} \frac{\partial}{\partial x} (u_I - u)\delta \, d\mathbf{x} + \oint_{\partial\Omega_3} (u_I - u) \frac{\partial G}{\partial n} ds . \tag{6.77}$$

Thus from (6.76) and (6.77),

$$|Q_1| \leq |u - u_I|_{1,\infty,\Omega_3}\{|G - G_h|_{1,1,\Omega_3} + |\delta|_{0,1,\Omega_3}\} + |u - u_I|_{0,\partial\Omega_3}|\nabla G|_{0,\partial\Omega_3}$$

$$\leq C\|u - u_I\|_{1,\infty,\Omega_3} \leq Ch|u|_{2,\infty,\Omega_3},$$

as required for (6.75).

We again consider the two cases of Section 3.4.

Case 1 If $u \in H^2(\Omega) \cap W_\infty^{3+\varepsilon}(\Omega_5)$ is the solution of (2.28) in the polygonal region of Section 6.3.1 and $u_h \in S_E^h$ is the Galerkin approximation, then

$$\left|u - u_h\right|^{\circledast}_{1,\infty,\Omega_0} = O(h^2) \; : \; \circledast = *,+ \; , \tag{6.78}$$

whilst

$$\left|u - u_h\right|_{1,\infty,\Omega_0} = O(h) \; . \tag{6.79}$$

Thus for this case use of recovered gradients produces a full O(h) faster convergence

Case 2 If $u \in W^2_{p-\varepsilon}(\Omega) \cap W^3_{\infty}(\Omega_5) \; : \; \frac{4}{3} \leqq p < 2$ is the solution of (2.28) as above and $u_h \in S^h_E$ is the Galerkin approximation, then

$$\left|u - u_h\right|^{\circledast}_{1,\infty,\Omega_0} = O(h^{\gamma-\varepsilon}) \; : \; \circledast = *,+ \; , \tag{6.80}$$

where $1 \leqq \gamma = 4 - \frac{4}{p} < 2$ as in (3.33), whilst

$$\left|u - u_h\right|_{1,\infty,\Omega_0} = \begin{cases} O(h) \; , & \frac{4}{3} < p < 2 \\ O(h^{1-\varepsilon}) \; , & p = \frac{4}{3} \end{cases} \; . \tag{6.81}$$

In this case the recovered gradients in general still produce a faster rate of convergence, but the improvement decreases with decrease in the regularity of the solution u. For the limiting case $p = 4/3$ both (6.80) and (6.81) give the same $O(h^{1-\varepsilon})$ estimates.

In both Cases 1 and 2, for a given *global* regularity of the solution u, the asymptotic rates of convergence of ∇u_h, $\nabla^* u_h$ and $\nabla^+ u_h$ to ∇u over a subdomain Ω_0 are the same in the L_∞-norm as in the L_2-norm.

6.3.7 The Effects of Numerical Integration

As in Section 3.5 we again examine the effect on the estimates (6.69), (6.70) and (6.72) of using numerical integration to evaluate inner products. Once more for $v_h \in H^h$ we use the centroid cubature rule on each element for approximating (f,v_h) and the mid-point quadrature rule on each element edge on $\partial\Omega^N$ for approximating $(g^N,v_h)_{\partial\Omega^N}$.

With $u_h^Q \in S^h_E$ the perturbed finite element approximation of (3.45), the additional error in $\left|u - u_h^Q\right|^{\circledast}_{1,\infty,\Omega_0} : \; \circledast = *,+$ is controlled by the additional error in $\left|u_I - u_h^Q\right|_{1,\infty,\Omega_0}$, see (6.15). In order to estimate this we consider each of the terms Q_1, Q_2, Q_3 of Lemma 6.2 As in Section 3.5 we require extra regularity of f over that of (6.12), so that now

$$f \in W^2_\infty(\Omega) \ . \tag{6.82}$$

From (6.30) the additional error in Q_1 is bounded by

$$\left|\int_{\Omega_3} \nabla(u_h - u_h^Q).\nabla G_h \, d\mathbf{x}\right|$$

$$\leq \left|(f,G_h) - (f,G_h)^Q\right| + \left|(g^N,G_h)_{\partial\Omega^N} - (g^N,G_h)^Q_{\partial\Omega^N}\right|$$

$$= \left|(f,G_h) - (f,G_h)^Q\right| \ , \tag{6.83}$$

as $G_h \in S^h_0(\Omega_3) \subset H^h$ and using (2.30) and (3.45). Now

$$\left|(f,v_h) - (f,v_h)^Q\right| \leq Ch^2\|f\|_{2,\infty,\Omega_3}|v_h|_{1,1,\Omega_3} \ , \quad \forall \ v_h \in S^h_0(\Omega_3) \ , \tag{6.84}$$

so that from (6.83) and (6.23)

$$\left|\int_{\Omega_3} \nabla(u_h - u_h^Q).\nabla G_h \, d\mathbf{x}\right| \leq Ch^2\|f\|_{2,\infty,\Omega_3}|G_h|_{1,1,\Omega_3}$$

$$\leq Ch^{2-\varepsilon}\|f\|_{2,\infty,\Omega_3} \ . \tag{6.85}$$

The additional errors in Q_2 and Q_3 are, from (6.44) and (6.47), controlled by the term $|u_h - u_h^Q|_{1,\Omega}$ which has been estimated in (3.48). Thus from (6.85) and (3.48) the additional error in $|u - u_h|^{\circledast}_{1,\infty,\Omega_0}$: $\circledast = *,+$ is bounded by

$$C\left\{h^{2-\varepsilon}\|f\|_{2,\infty,\Omega_3} + |u_h - u_h^Q|_{1,\Omega}\right\} \leq Ch^{2-\varepsilon}\left\{\|f\|_{2,\infty,\Omega} + \|g^N\|'_{2,\partial\Omega^N}\right\} \ . \tag{6.86}$$

The asymptotic rates of convergence of $\nabla^{\circledast}u_h$ to ∇u in the L_∞-norm over the subdomain Ω_0 are therefore (almost) unaffected by the use of numerical integration. For the $O(h^{2-\varepsilon})$ estimate (6.86) to be improved to $O(h^2)$ higher quadrature and cubature rules are needed.

6.4 The Problem of Planar Linear Elasticity

The extension of the superconvergence results of Sections 4 and 5 to the L_∞-context is clearly of interest. Whilst we have as yet been unable to achieve this, we make the following remarks.

With the subdomains as defined in Sections 6.3.1 and 6.3.4, we follow the approach of Section 6.3.2 but in the context of the linear elastic problem (4.1)-(4.3). We thus define **G** to be the solution of the problem

$$\boldsymbol{L}(\mathbf{G}) \equiv -\mu\Delta\mathbf{G} - (\lambda+\mu)\text{grad div } \mathbf{G} = -\begin{bmatrix} \frac{\partial\delta}{\partial x} \\ 0 \end{bmatrix} \text{ in } \Omega_3 , \tag{6.87}$$

together with boundary conditions

$$\mathbf{G} = \mathbf{0} \quad \text{on} \quad \partial\Omega_3 . \tag{6.88}$$

With $\mathbf{G}_h \in \left[S_0^h(\Omega_3)\right]^2$ the finite element approximation to $\mathbf{G}$, we need to establish the estimates corresponding to (6.21) and (6.22); respectively

$$\left|\mathbf{G} - \mathbf{G}_h\right|_{1,1,\Omega_3} \leqq C \tag{6.89}$$

and

$$\left|\mathbf{G}\right|_{1,1,\Omega_3} \leqq Ch^{-\varepsilon} \tag{6.90}$$

(from which the estimate corresponding to (6.23),

$$\left|\mathbf{G}_h\right|_{1,1,\Omega_3} \leqq Ch^{-\varepsilon} \tag{6.91}$$

would follow immediately.)

Now estimate (6.89) follows from Rannacher and Scott [31], as (recall Ω_3 is a *rectangle*, and see [16,Section 4]) the mapping

$$\boldsymbol{L} : \left[\overset{\circ}{W}{}_p^1(\Omega_3)\right]^2 \cap \left[W_p^2(\Omega_3)\right]^2 \longrightarrow \left[L_p(\Omega_3)\right]^2$$

is a homeomorphism for $p \in (1,2+\varepsilon]$.

However, estimate (6.90), and also the boundedness, ***outside*** Ω_1, of all norms of $\mathbf{G}$ independently of the mesh size h, require knowledge of the form of a function $\mathbf{G}^\delta$, satisfying the system of differential equations (6.87), i.e.

$$\boldsymbol{L}(\mathbf{G}^\delta) = -\begin{bmatrix} \frac{\partial\delta}{\partial x} \\ 0 \end{bmatrix} \quad \text{in} \quad \Omega_3 \tag{6.92}$$

(but **not** however the boundary conditions (6.88)).

From the discussion of Section 6.3.2 we might expect $\mathbf{G}^\delta$ to depend upon both the x-derivative of the free-space Green's function, see Hsiao, Stephan and Wendland [18],

$$\frac{\partial}{\partial x}\left\{\frac{\lambda+3\mu}{4\pi\mu(\lambda+2\mu)}\begin{bmatrix} \ell n\, r - \frac{(\lambda+\mu)}{(\lambda+3\mu)}\cos^2\theta \\ -\frac{(\lambda+\mu)}{2(\lambda+3\mu)}\sin 2\theta \end{bmatrix}\right\}$$

$$= \frac{\lambda+3\mu}{4\pi\mu(\lambda+2\mu)r}\begin{bmatrix} \cos\theta - \frac{(\lambda+\mu)}{(\lambda+3\mu)}\sin\theta\sin 2\theta \\ \frac{(\lambda+\mu)}{(\lambda+3\mu)}\sin\theta\cos 2\theta \end{bmatrix} \tag{6.93}$$

and also the smoothing term $\int_0^r \rho\,\delta(\rho)\,d\rho$.

Unfortunately we have been able to show that no solution to (6.92) of the form

$$\mathcal{G}_1(r,\theta) + \mathcal{G}_2(r,\theta)\int_0^r \rho\,\delta(\rho)\,d\rho$$

exists. We have as yet been unable to find a suitable function $\mathbf{G}^\delta$. However, encouraged by the results of numerical experiments, we are trying to estimate $\mathbf{G}$ without using $\mathbf{G}^\delta$, see Goodsell [12]

7. Brief Concluding Remarks

Apart from the suggestions made at the end of Section 6, the work on this type of recovered gradient superconvergence for *piecewise linear* finite element approximations to second-order elliptic problems with constant coefficients would seem to be more or less complete. There seems to be little to be gained from constructing more elaborate recovered gradient functions from the piecewise linear approximations in pursuit of yet higher accuracy. For example, if we constructed a C° piecewise quadratic recovered gradient function, $\nabla^{**}u_h$ say, which was such that the error in $(\nabla u - \nabla^{**}u_I)$ was $O(h^3)$ in any L_p-norm, we would be constrained by the fact that $\|\nabla u - \nabla^{**}u_h\|_{L_p}$ is controlled by the term $\|\nabla(u_I - u_h)\|_{L_p}$, which can never be better than $O(h^2)$.

Some recent work of Andreev and Lazarov [2] has investigated gradient superconvergence properties of *piecewise quadratic* finite element approximations in two dimensions. For such (six-node triangular) elements, they show that the second-order Gauss points on the element edges are points of exceptional accuracy for derivatives of the finite element approximation *tangential* to the edge. For a uniform triangular mesh they present a scheme for recovering to superconvergent $O(h^3)$ accuracy (in the discrete ℓ_2 sense) the full gradient at an element vertex by taking a (weighted) average of the tangential derivatives at the Gauss points on the edges meeting at the vertex. Clearly a topic for possible future study is the construction of recovered gradient functions, defined at every point, which have superconvergent $O(h^3)$ rates of convergence to ∇u in the L_2- or L_∞-norms.

We finally comment on the finite element mesh. It has here been assumed throughout that either Ω, or a subdomain Ω_0, is rectangular and that the partition consists of isosceles right angled triangles. Whilst in the former case this can be unnecessarily restrictive (see [24] for a discussion of more general meshes satisfying the six-element property), in the latter case it ceases to be a restriction as, see Sections 3.1 and 6.3.4, for sufficiently small mesh size h, we can construct a rectangular subdomain Ω_0 to include "most" of the general polygonal region Ω.

Acknowledgement

Some of the work by Whiteman was done whilst he visited the Department of Mathematics, Texas A and M University, in 1986 and 1988. Throughout this work Goodsell was supported by a Research Studentship from the Science and Engineering Research Council. All this support is gratefully acknowledged.

References

1. Adams, R.A. (1975) Sobolev Spaces. Academic Press, New York.
2. Andreev, A.B. and Lazarov, R.D. (1988) Superconvergence of the gradient for quadratic triangular finite elements. Numer. Meth. Partial Differential Equations, **4**, 15-32.
3. Barlow, J. (1976) Optimal stress locations in finite element models. Int. J. Numer. Meth. Eng., **10**, 243-251.
4. Bramble, J.H. and Hilbert, S.R. (1970) Estimation of linear functionals on Sobolev spaces with applications to Fourier transforms and spline interpolation. SIAM J. Numer. Anal., **7**, 112-124.
5. Ciarlet, P.G. (1978) The Finite Element Method for Elliptic Problems. North-Holland, Amsterdam.
6. Colyer, B. and Trowbridge, C.W. (1985) Finite element analysis using a single-user computer. Computer-Aided Design, **17**, 142-148.
7. Destuynder, P. and Djaoua, M. (1981) Sur une interprétation Mathématique de l'intégrale de Rice en théorie de la rupture fragile. Math. Meth. in the Appl. Sci., **3**, 70-87.
8. Destuynder, P., Djaoua, M. and Lescure, S. (1984) On a numerical method for fracture mechanics. pp.69-84 of P. Grisvard, W.L. Wendland, and J.R. Whiteman (eds.), Singularities and Constructive Methods for Their Treatment, Lecture Notes in Mathematics 1121. Springer-Verlag, Berlin.
9. Duvaut, G. and Lions, J.L. (1972) Les Inéquations en Mécanique et en Physique. Dunod, Paris.
10. Fichera, G. (1972) Existence theorems in elasticity - boundary value problems of elasticity with unilateral constraints. pp.347-424 of C. Truesdell, (ed.) Mechanics of Solids II, Vol.VIa/2 of S. Flügge, (Chief ed.). Encyclopedia of Physics. Springer-Verlag, Berlin.
11. Goodsell, G. (1985) Gradient superconvergence properties for finite element approximations to two-dimensional Poisson problems. M.Sc. Dissertation, Brunel University.
12. Goodsell, G. (1988) Gradient Superconvergence in the Finite Element Method with Applications to Planar Linear Elasticity. Ph.D. Thesis, Brunel University.
13. Goodsell, G. and Whiteman, J.R. (1987) A Unified Treatment of Superconvergent Recovered Gradient Functions for Piecewise Linear Finite Element Approximations. Technical Report BICOM 87/1, Institute of Computational Mathematics, Brunel University. (to appear in Acta Applicandae Mathematicae)
14. Grisvard, P. (1976) Behavior of the solutions of an elliptic boundary value problem in a polygonal or polyhedral domain. pp.207-274 of B. Hubbard (ed.), Numerical Solution of Partial Differential Equations III, (SYNSPADE 1975). Academic Press, New York.
15. Grisvard, P. (1985) Elliptic Problems in Nonsmooth Domains. Pitman, Boston.

16. Grisvard, P. (1986) Problèmes aux Limites dans les Polygones. Mode d'Emploi. E.D.F. Bulletin de la Direction des Etudes et Recherches. Serie C, Mathématiques. Informatique No.1, pp.21-59.
17. Hlaváček, I. and Křížek, M. (1987) On a superconvergent finite element scheme for elliptic systems, Parts I-III. Apl. Mat., **32**, 131-154, 200-213, 276-289.
18. Hsiao, G.C., Stephan, E.P. and Wendland, W.L. (1984) An integral equation formulation for a boundary value problem of elasticity in the domain exterior to an arc. pp.153-165 of P. Grisvard, W.L. Wendland and J.R. Whiteman (eds.), Singularities and Constructive Methods for Their Treatment. Lecture Notes in Mathematics 1121. Springer-Verlag, Berlin.
19. Křížek, M. and Neittaanmäki, P. (1984) Superconvergence phenomenon in the finite element method arising from averaging gradients. Numer. Math., **45**, 105-116.
20. Křížek, M. and Neittaanmäki, P. (1987) On a global superconvergence of the gradient of linear triangular elements. J. Comput. Appl. Math., **18**, 221-233.
21. Křížek, M. and Neittaanmäki, P. (1987) On superconvergence techniques. Acta Applicandae Mathematicae, **9**.
22. Levine, N.D. (1982) Stress sampling points for linear triangles in the finite element method. Numerical Analysis Report 10/82, University of Reading.
23. Levine, N.D. (1985) Superconvergent recovery of the gradient from piecewise linear finite element approximations. IMA J. Numer. Anal., **5**, 407-427.
24. Levine, N.D. (1985) Superconvergent estimation of the gradient from linear finite element approximations on triangular elements. Numerical Analysis Report 3/85. Ph.D. Thesis, University of Reading.
25. Lin, Q., Lü, T. and Shen, S. (1983) Maximum norm estimate, extrapolation and optimal points of stresses for the finite element methods on the strongly regular triangulation. J. Comput. Math., **1**, 376-383.
26. Nečas, J. and Hlaváček, I. (1981) Mathematical Theory of Elastic and Elasto-Plastic Bodies: an Introduction. Elsevier, Amsterdam.
27. Neittaanmäki, P. and Křížek, M. (1984) Superconvergence of the finite element schemes arising from the use of averaged gradients. pp.169-178 of Accuracy Estimates and Adaptive Refinements in Finite Element Computations, Proc. Int. Conf., Lisbon, 1984.
28. Nitsche, J.A. and Schatz, A.H. (1974) Interior estimates for Ritz-Galerkin methods. Math. Comp., **28**, 937-958.
29. Oden, J.T. and Reddy, J.N. (1976) An Introduction to the Mathematical Theory of Finite Elements. Wiley-Interscience, New York.
30. Oganesjan, L.A. and Ruchovec, L.A. (1969) An investigation of the rate of convergence of variational difference schemes for second-order elliptic equations in a two-dimensional region with a smooth boundary. Z. Vycisl. Mat. i Mat. Fiz., **9**, 1102-1120.
31. Rannacher, R. and Scott, R. (1982) Some optimal error estimates for piecewise linear finite element approximations. Math. Comp., **38**, 437-445.
32. Rice, J.R. (1968) A path independent integral and the approximate analysis of strain concentration by notches and cracks. J. Appl. Mech., **34**, 379-386.
33. Rice, J.R. (1968) Mathematical analysis in the mechanics of fracture. pp.191-311 of H. Liebowitz (ed.) Fracture, Vol.II. Academic Press, New York.
34. Schatz, A.H. (1985) An introduction to the analysis of the error in the finite element method for second-order elliptic boundary value problems. pp.94-139 of P.R. Turner (ed.) Proc. Numerical Analysis Summer School, Lancaster, 1984. Lecture Notes in Mathematics 1129. Springer-Verlag, Berlin.
35. Strang, G. and Fix, G.J. (1973) An Analysis of the Finite Element Method. Prentice-Hall, New Jersey.
36. Veryard, D.A. (1971) Problems Associated with the Convergence of Isoparametric and Mixoparametric Finite Elements. M.Sc. Thesis, University of Wales.
37. Wang, C.T. (1953) Applied Elasticity. McGraw-Hill, New York.
38. Wheeler, M.F. and Whiteman, J.R. (1987) Superconvergent recovery of gradients on subdomains from piecewise linear finite element approximations. Numer. Meth. Partial Differential Equations, **3**, 65-82.

39. Whiteman, J.R. and Goodsell, G. (1987) On some finite element error estimates for stress intensity factors in Mode I linear elastic fracture problems. Technical Report BICOM 87/8, Institute of Computational Mathematics, Brunel University.
40. Whiteman, J.R. and Goodsell, G. (1988) Superconvergent recovery for stresses from finite element approximations on subdomains for planar problems of linear elasticity. pp.29-53 of J.R. Whiteman (ed.) The Mathematics of Finite Elements and Applications VI, (MAFELAP 1987). Academic Press, London.
41. Whiteman, J.R. and Thompson, G.M. (1985) Finite element calculations of parameters for singularities in problems of fracture. pp.27-47 of J.R. Whiteman (ed.) The Mathematics of Finite Elements and Applications V, (MAFELAP 1984). Academic Press, London.
42. Williams, M.L. (1952) Stress singularities resulting from various boundary conditions in angular corners of plates in extension. J. Appl. Mech., **24**, 526-528.
43. Zhu, Q.D. (1983) Uniform superconvergence estimates of derivatives for the finite element method. Numer. Math. J. Chinese Univ., **5**, 311-318.
44. Zienkiewicz, O.C. and Cheung, Y.K. (1967) The Finite Element Method in Structural and Continuum Mechanics. McGraw-Hill, London.
45. Zlámal, M. (1977) Some superconvergence results in the finite element method. pp.353-362 of C. de Boor (ed.) Mathematical Aspects of Finite Element Methods, Proc. Conf., Rome, 1975. Lecture Notes in Mathematics 606. Springer-Verlag, Berlin.
46. Zlámal, M. (1978) Superconvergence and reduced integration in the finite element method. Math. Comp., **32**, 663-685.

Programme of the Meeting

Week 1 *Ten lecture course*:

LARGE SPARSE SYSTEMS J.A.George

Five lecture course:

PARALLEL COMPUTATION IN OPTIMIZATION ALGORITHMS L.C.W.Dixon

Three lecture course:

CONJUGATE GRADIENT TECHNIQUES FOR LINEAR SYSTEMS B.Parlett

Seminar program:

Pole assignment problems and the perturbation analysis of EVPs K-W.E.Chu

Low thrust trajectory optimization Z.A.Maany

The ABS class for linear and nonlinear equations and optimization E.Spedicato

Week 2 *Five lecture course*:

ALGORITHMIC ASPECTS OF PARALLEL COMPUTATION D.Parkinson

Four lecture course:

FLOATING-POINT ARITHMETIC A.Feldstein

Seven lecture course:

LEVEL-INDEX ARITHMETIC: AN INTRODUCTORY SURVEY

C.W.Clenshaw, F.W.J.Olver and P.R.Turner

Seminar program:

Parallelisation and vectorisation in the numerical solution of ODEs

I. Initial value problems and II. Boundary value problems I.Gladwell

Complex data-fitting in antenna array design J.C.Mason

Functional programming and numerical computation G.P.McKeown

Residual calculations using FFTs C.Phillips

An alternative to Chan's deflation algorithm J.D.Pryce

Expected conditioning R.Fletcher

SIMD least squares, or, simplicity wins the day M.J.Stanford

Stable modification of explicit LU factors for simplex updates R.Fletcher

Numerical solution of viscous flow on the Cyber 205 J.S.Bramley

Week 3 *Ten lecture course*:

SUPERCONVERGENCE AND RETRIEVAL IN THE NUMERICAL SOLUTION OF DIFFERENTIAL EQUATIONS T.Dupont

Three lecture course:

GRADIENT SUPERCONVERGENCE OF FINITE ELEMENT APPROXIMATIONS TO ELLIPTIC PROBLEMS J.R.Whiteman and G.Goodsell

Seminar programme:

A remarkable singular-value inequality J.D.Pryce

The finite element approximation of singular elliptic problems using dual singular functions J.W.Barrett

Recovery in characteristic Galerkin methods for hyperbolic problems K.W.Morton

Optimal recovery in the finite element method K.W.Morton

List of participants

Prof. J.Abathy	Budapest and Hatfield Polytechnic
Mr S.H.Abbas	University of Liverpool
Mr M.Ainsworth	Durham University
Dr P.W.Banks	British Gas
Dr J.W.Barrett	Imperial College, London
Dr J.S.Bramley	University of Strathclyde
Mr A.S.Broad	Meteorological Office, Bracknell
Dr B.M.Brown	University College, Cardiff
Mr D.Burton	University of Lancaster
Dr D.B.Clegg	Liverpool Polytechnic
Prof. C.W.Clenshaw	University of Lancaster
Dr A.W.Craig	Durham University
Dr K-W.E.Chu	University of Reading
Mr P.Davis	Bristol University
Dr M.S.Derakshan	University of Manchester
Prof. L.C.W.Dixon	Hatfield Polytechnic
Dr P.Dolan	Imperial College, London
Dr R.Dow	University of Lancaster
Prof. T.Dupont	University of Chicago
Mr S.Ellis	RMCS, Shrivenham
Prof. A.Feldstein	Arizona State University
Prof. R.Fletcher	University of Dundee
Prof. J.A.George	University of Tennessee and Oak Ridge National Laboratory
Mr J.Gilbert	University of Lancaster
Dr I.Gladwell	University of Manchester
Dr J.Gleghorn	University of Lancaster
Dr E.Godoy	University of Santiago de Compostella, Spain
Mr G.Goodsell	Brunel University
Dr W.Govaerts	Gent University, Belgium
Dr D.C.Handscomb	Oxford University
Mr P.J.Harris	Plymouth Polytechnic
Mr S.Hurley	University College, Cardiff
Mr P.Jimack	Bristol University
Mr Chen Ke	Plymouth Polytechnic
Dr D.Kershaw	University of Lancaster
Mr S.M.Kirkup	Brighton Polytechnic
Dr W.A.Light	University of Lancaster
Dr Z.A.Maany	Hatfield Polytechnic

Dr J.R.Mallinson	British Gas
Prof. J.C.Mason	RMCS, Shrivenham
Dr G.P.McKeown	University of East Anglia
Dr J.T.Modi	Cambridge University
Dr G.Moore	Imperial College, London
Prof. K.W.Morton	Oxford University
Prof. J.Oliver	University of Essex
Prof. F.W.J.Olver	University of Maryland, College Park
Prof. D.Parkinson	Queen Mary College, London
Prof. B.Parlett	University of California, Berkeley
Prof. L.Pasquini	University of Aquia, Italy
Dr C.Phillips	University of Newcastle-upon-Tyne
Dr D.C.Pountney	Liverpool Polytechnic
Dr J.D.Pryce	Bristol University
Mr I.Reid	University of Lancaster
Dr A.N.Richmond	Liverpool Polytechnic
Dr J.D.Santos	University of Kent and Pernambuco, Brazil
Dr N.Shahtamasebi	University of Lancaster
Prof. E.Spedicato	Hatfield Polytechnic and Milan, Italy
Dr M.J.Stanford	National Physical Laboratory
Dr P.R.Turner	University of Lancaster
Prof. J.R.Whiteman	Brunel University
Mr S.Wilde	RMCS, Shrivenham
Mr M.A.Wolfe	University of St. Andrews

Vol. 1232: P.C. Schuur, Asymptotic Analysis of Soliton Problems. VIII, 180 pages. 1986.

Vol. 1233: Stability Problems for Stochastic Models. Proceedings, 1985. Edited by V.V. Kalashnikov, B. Penkov and V.M. Zolotarev. VI, 223 pages. 1986.

Vol. 1234: Combinatoire énumérative. Proceedings, 1985. Edité par G. Labelle et P. Leroux. XIV, 387 pages. 1986.

Vol. 1235: Séminaire de Théorie du Potentiel, Paris, No. 8. Directeurs: M. Brelot, G. Choquet et J. Deny. Rédacteurs: F. Hirsch et G. Mokobodzki. III, 209 pages. 1987.

Vol. 1236: Stochastic Partial Differential Equations and Applications. Proceedings, 1985. Edited by G. Da Prato and L. Tubaro. V, 257 pages. 1987.

Vol. 1237: Rational Approximation and its Applications in Mathematics and Physics. Proceedings, 1985. Edited by J. Gilewicz, M. Pindor and W. Siemaszko. XII, 350 pages. 1987.

Vol. 1238: M. Holz, K.-P. Podewski and K. Steffens, Injective Choice Functions. VI, 183 pages. 1987.

Vol. 1239: P. Vojta, Diophantine Approximations and Value Distribution Theory. X, 132 pages. 1987.

Vol. 1240: Number Theory, New York 1984–85. Seminar. Edited by D.V. Chudnovsky, G.V. Chudnovsky, H. Cohn and M.B. Nathanson. V, 324 pages. 1987.

Vol. 1241: L. Gårding, Singularities in Linear Wave Propagation. III, 125 pages. 1987.

Vol. 1242: Functional Analysis II, with Contributions by J. Hoffmann-Jørgensen et al. Edited by S. Kurepa, H. Kraljević and D. Butković. VII, 432 pages. 1987.

Vol. 1243: Non Commutative Harmonic Analysis and Lie Groups. Proceedings, 1985. Edited by J. Carmona, P. Delorme and M. Vergne. V, 309 pages. 1987.

Vol. 1244: W. Müller, Manifolds with Cusps of Rank One. XI, 158 pages. 1987.

Vol. 1245: S. Rallis, L-Functions and the Oscillator Representation. XVI, 239 pages. 1987.

Vol. 1246: Hodge Theory. Proceedings, 1985. Edited by E. Cattani, F. Guillén, A. Kaplan and F. Puerta. VII, 175 pages. 1987.

Vol. 1247: Séminaire de Probabilités XXI. Proceedings. Edité par J. Azéma, P.A. Meyer et M. Yor. IV, 579 pages. 1987.

Vol. 1248: Nonlinear Semigroups, Partial Differential Equations and Attractors. Proceedings, 1985. Edited by T.L. Gill and W.W. Zachary. IX, 185 pages. 1987.

Vol. 1249: I. van den Berg, Nonstandard Asymptotic Analysis. IX, 187 pages. 1987.

Vol. 1250: Stochastic Processes – Mathematics and Physics II. Proceedings 1985. Edited by S. Albeverio, Ph. Blanchard and L. Streit. VI, 359 pages. 1987.

Vol. 1251: Differential Geometric Methods in Mathematical Physics. Proceedings, 1985. Edited by P.L. García and A. Pérez-Rendón. VII, 300 pages. 1987.

Vol. 1252: T. Kaise, Représentations de Weil et GL_2 Algèbres de division et GL_n. VII, 203 pages. 1987.

Vol. 1253: J. Fischer, An Approach to the Selberg Trace Formula via the Selberg Zeta-Function. III, 184 pages. 1987.

Vol. 1254: S. Gelbart, I. Piatetski-Shapiro, S. Rallis. Explicit Constructions of Automorphic L-Functions. VI, 152 pages. 1987.

Vol. 1255: Differential Geometry and Differential Equations. Proceedings, 1985. Edited by C. Gu, M. Berger and R.L. Bryant. XII, 243 pages. 1987.

Vol. 1256: Pseudo-Differential Operators. Proceedings, 1986. Edited by H.O. Cordes, B. Gramsch and H. Widom. X, 479 pages. 1987.

Vol. 1257: X. Wang, On the C*-Algebras of Foliations in the Plane. V, 165 pages. 1987.

Vol. 1258: J. Weidmann, Spectral Theory of Ordinary Differential Operators. VI, 303 pages. 1987.

Vol. 1259: F. Cano Torres, Desingularization Strategies for Three-Dimensional Vector Fields. IX, 189 pages. 1987.

Vol. 1260: N.H. Pavel, Nonlinear Evolution Operators and Semigroups. VI, 285 pages. 1987.

Vol. 1261: H. Abels, Finite Presentability of S-Arithmetic Groups. Compact Presentability of Solvable Groups. VI, 178 pages. 1987.

Vol. 1262: E. Hlawka (Hrsg.), Zahlentheoretische Analysis II. Seminar, 1984–86. V, 158 Seiten. 1987.

Vol. 1263: V.L. Hansen (Ed.), Differential Geometry. Proceedings, 1985. XI, 288 pages. 1987.

Vol. 1264: Wu Wen-tsün, Rational Homotopy Type. VIII, 219 pages. 1987.

Vol. 1265: W. Van Assche, Asymptotics for Orthogonal Polynomials. VI, 201 pages. 1987.

Vol. 1266: F. Ghione, C. Peskine, E. Sernesi (Eds.), Space Curves. Proceedings, 1985. VI, 272 pages. 1987.

Vol. 1267: J. Lindenstrauss, V.D. Milman (Eds.), Geometrical Aspects of Functional Analysis. Seminar. VII, 212 pages. 1987.

Vol. 1268: S.G. Krantz (Ed.), Complex Analysis. Seminar, 1986. VII, 195 pages. 1987.

Vol. 1269: M. Shiota, Nash Manifolds. VI, 223 pages. 1987.

Vol. 1270: C. Carasso, P.-A. Raviart, D. Serre (Eds.), Nonlinear Hyperbolic Problems. Proceedings, 1986. XV, 341 pages. 1987.

Vol. 1271: A.M. Cohen, W.H. Hesselink, W.L.J. van der Kallen, J.R. Strooker (Eds.), Algebraic Groups Utrecht 1986. Proceedings. XII, 284 pages. 1987.

Vol. 1272: M.S. Livšic, L.L. Waksman, Commuting Nonselfadjoint Operators in Hilbert Space. III, 115 pages. 1987.

Vol. 1273: G.-M. Greuel, G. Trautmann (Eds.), Singularities, Representation of Algebras, and Vector Bundles. Proceedings, 1985. XIV, 383 pages. 1987.

Vol. 1274: N. C. Phillips, Equivariant K-Theory and Freeness of Group Actions on C*-Algebras. VIII, 371 pages. 1987.

Vol. 1275: C.A. Berenstein (Ed.), Complex Analysis I. Proceedings, 1985–86. XV, 331 pages. 1987.

Vol. 1276: C.A. Berenstein (Ed.), Complex Analysis II. Proceedings, 1985–86. IX, 320 pages. 1987.

Vol. 1277: C.A. Berenstein (Ed.), Complex Analysis III. Proceedings, 1985–86. X, 350 pages. 1987.

Vol. 1278: S.S. Koh (Ed.), Invariant Theory. Proceedings, 1985. V, 102 pages. 1987.

Vol. 1279: D. Ieşan, Saint-Venant's Problem. VIII, 162 Seiten. 1987.

Vol. 1280: E. Neher, Jordan Triple Systems by the Grid Approach. XII, 193 pages. 1987.

Vol. 1281: O.H. Kegel, F. Menegazzo, G. Zacher (Eds.), Group Theory. Proceedings, 1986. VII, 179 pages. 1987.

Vol. 1282: D.E. Handelman, Positive Polynomials, Convex Integral Polytopes, and a Random Walk Problem. XI, 136 pages. 1987.

Vol. 1283: S. Mardešić, J. Segal (Eds.), Geometric Topology and Shape Theory. Proceedings, 1986. V, 261 pages. 1987.

Vol. 1284: B.H. Matzat, Konstruktive Galoistheorie. X, 286 pages. 1987.

Vol. 1285: I.W. Knowles, Y. Saitō (Eds.), Differential Equations and Mathematical Physics. Proceedings, 1986. XVI, 499 pages. 1987.

Vol. 1286: H.R. Miller, D.C. Ravenel (Eds.), Algebraic Topology. Proceedings, 1986. VII, 341 pages. 1987.

Vol. 1287: E.B. Saff (Ed.), Approximation Theory, Tampa. Proceedings, 1985–1986. V, 228 pages. 1987.

Vol. 1288: Yu. L. Rodin, Generalized Analytic Functions on Riemann Surfaces. V, 128 pages, 1987.

Vol. 1289: Yu. I. Manin (Ed.), K-Theory, Arithmetic and Geometry. Seminar, 1984–1986. V, 399 pages. 1987.

Vol. 1290: G. Wüstholz (Ed.), Diophantine Approximation and Transcendence Theory. Seminar, 1985. V, 243 pages. 1987.

Vol. 1291: C. Mœglin, M.-F. Vignéras, J.-L. Waldspurger, Correspondances de Howe sur un Corps p-adique. VII, 163 pages. 1987

Vol. 1292: J.T. Baldwin (Ed.), Classification Theory. Proceedings, 1985. VI, 500 pages. 1987.

Vol. 1293: W. Ebeling, The Monodromy Groups of Isolated Singularities of Complete Intersections. XIV, 153 pages. 1987.

Vol. 1294: M. Queffélec, Substitution Dynamical Systems – Spectral Analysis. XIII, 240 pages. 1987.

Vol. 1295: P. Lelong, P. Dolbeault, H. Skoda (Réd.), Séminaire d'Analyse P. Lelong – P. Dolbeault – H. Skoda. Seminar, 1985/1986. VII, 283 pages. 1987.

Vol. 1296: M.-P. Malliavin (Ed.), Séminaire d'Algèbre Paul Dubreil et Marie-Paule Malliavin. Proceedings, 1986. IV, 324 pages. 1987.

Vol. 1297: Zhu Y.-l., Guo B.-y. (Eds.), Numerical Methods for Partial Differential Equations. Proceedings. XI, 244 pages. 1987.

Vol. 1298: J. Aguadé, R. Kane (Eds.), Algebraic Topology, Barcelona 1986. Proceedings. X, 255 pages. 1987.

Vol. 1299: S. Watanabe, Yu.V. Prokhorov (Eds.), Probability Theory and Mathematical Statistics. Proceedings, 1986. VIII, 589 pages. 1988.

Vol. 1300: G.B. Seligman, Constructions of Lie Algebras and their Modules. VI, 190 pages. 1988.

Vol. 1301: N. Schappacher, Periods of Hecke Characters. XV, 160 pages. 1988.

Vol. 1302: M. Cwikel, J. Peetre, Y. Sagher, H. Wallin (Eds.), Function Spaces and Applications. Proceedings, 1986. VI, 445 pages. 1988.

Vol. 1303: L. Accardi, W. von Waldenfels (Eds.), Quantum Probability and Applications III. Proceedings, 1987. VI, 373 pages. 1988.

Vol. 1304: F.Q. Gouvêa, Arithmetic of p-adic Modular Forms. VIII, 121 pages. 1988.

Vol. 1305: D.S. Lubinsky, E.B. Saff, Strong Asymptotics for Extremal Polynomials Associated with Weights on $\mathbb{R}$. VII, 153 pages. 1988.

Vol. 1306: S.S. Chern (Ed.), Partial Differential Equations. Proceedings, 1986. VI, 294 pages. 1988.

Vol. 1307: T. Murai, A Real Variable Method for the Cauchy Transform, and Analytic Capacity. VIII, 133 pages. 1988.

Vol. 1308: P. Imkeller, Two-Parameter Martingales and Their Quadratic Variation. IV, 177 pages. 1988.

Vol. 1309: B. Fiedler, Global Bifurcation of Periodic Solutions with Symmetry. VIII, 144 pages. 1988.

Vol. 1310: O.A. Laudal, G. Pfister, Local Moduli and Singularities. V, 117 pages. 1988.

Vol. 1311: A. Holme, R. Speiser (Eds.), Algebraic Geometry, Sundance 1986. Proceedings. VI, 320 pages. 1988.

Vol. 1312: N.A. Shirokov, Analytic Functions Smooth up to the Boundary. III, 213 pages. 1988.

Vol. 1313: F. Colonius, Optimal Periodic Control. VI, 177 pages. 1988.

Vol. 1314: A. Futaki, Kähler-Einstein Metrics and Integral Invariants. IV, 140 pages. 1988.

Vol. 1315: R.A. McCoy, I. Ntantu, Topological Properties of Spaces of Continuous Functions. IV, 124 pages. 1988.

Vol. 1316: H. Korezlioglu, A.S. Ustunel (Eds.), Stochastic Analysis and Related Topics. Proceedings, 1986. V, 371 pages. 1988.

Vol. 1317: J. Lindenstrauss, V.D. Milman (Eds.), Geometric Aspects of Functional Analysis. Seminar, 1986–87. VII, 289 pages. 1988.

Vol. 1318: Y. Felix (Ed.), Algebraic Topology – Rational Homotopy. Proceedings, 1986. VIII, 245 pages. 1988

Vol. 1319: M. Vuorinen, Conformal Geometry and Quasiregular Mappings. XIX, 209 pages. 1988.

Vol. 1320: H. Jürgensen, G. Lallement, H.J. Weinert (Eds.), Semigroups, Theory and Applications. Proceedings, 1986. X, 416 pages. 1988.

Vol. 1321: J. Azéma, P.A. Meyer, M. Yor (Eds.), Séminaire de Probabilités XXII. Proceedings. IV, 600 pages. 1988.

Vol. 1322: M. Métivier, S. Watanabe (Eds.), Stochastic Analysis. Proceedings, 1987. VII, 197 pages. 1988.

Vol. 1323: D.R. Anderson, H.J. Munkholm, Boundedly Controlled Topology. XII, 309 pages. 1988.

Vol. 1324: F. Cardoso, D.G. de Figueiredo, R. Iório, O. Lopes (Eds.), Partial Differential Equations. Proceedings, 1986. VIII, 433 pages. 1988.

Vol. 1325: A. Truman, I.M. Davies (Eds.), Stochastic Mechanics and Stochastic Processes. Proceedings, 1986. V, 220 pages. 1988.

Vol. 1326: P.S. Landweber (Ed.), Elliptic Curves and Modular Forms in Algebraic Topology. Proceedings, 1986. V, 224 pages. 1988.

Vol. 1327: W. Bruns, U. Vetter, Determinantal Rings. VII,236 pages. 1988.

Vol. 1328: J.L. Bueso, P. Jara, B. Torrecillas (Eds.), Ring Theory. Proceedings, 1986. IX, 331 pages. 1988.

Vol. 1329: M. Alfaro, J.S. Dehesa, F.J. Marcellan, J.L. Rubio de Francia, J. Vinuesa (Eds.): Orthogonal Polynomials and their Applications. Proceedings, 1986. XV, 334 pages. 1988.

Vol. 1330: A. Ambrosetti, F. Gori, R. Lucchetti (Eds.), Mathematical Economics. Montecatini Terme 1986. Seminar. VII, 137 pages. 1988.

Vol. 1331: R. Bamón, R. Labarca, J. Palis Jr. (Eds.), Dynamical Systems, Valparaiso 1986. Proceedings. VI, 250 pages. 1988.

Vol. 1332: E. Odell, H. Rosenthal (Eds.), Functional Analysis. Proceedings, 1986–87. V, 202 pages. 1988.

Vol. 1333: A.S. Kechris, D.A. Martin, J.R. Steel (Eds.), Cabal Seminar 81–85. Proceedings, 1981–85. V, 224 pages. 1988.

Vol. 1334: Yu.G. Borisovich, Yu. E. Gliklikh (Eds.), Global Analysis – Studies and Applications III. V, 331 pages. 1988.

Vol. 1335: F. Guillén, V. Navarro Aznar, P. Pascual-Gainza, F. Puerta, Hyperrésolutions cubiques et descente cohomologique. XII, 192 pages. 1988.

Vol. 1336: B. Helffer, Semi-Classical Analysis for the Schrödinger Operator and Applications. V, 107 pages. 1988.

Vol. 1337: E. Sernesi (Ed.), Theory of Moduli. Seminar, 1985. VIII, 232 pages. 1988.

Vol. 1338: A.B. Mingarelli, S.G. Halvorsen, Non-Oscillation Domains of Differential Equations with Two Parameters. XI, 109 pages. 1988.

Vol. 1339: T. Sunada (Ed.), Geometry and Analysis of Manifolds. Procedings, 1987. IX, 277 pages. 1988.

Vol. 1340: S. Hildebrandt, D.S. Kinderlehrer, M. Miranda (Eds.), Calculus of Variations and Partial Differential Equations. Proceedings, 1986. IX, 301 pages. 1988.

Vol. 1341: M. Dauge, Elliptic Boundary Value Problems on Corner Domains. VIII, 259 pages. 1988.

Vol. 1342: J.C. Alexander (Ed.), Dynamical Systems. Proceedings, 1986–87. VIII, 726 pages. 1988.

Vol. 1343: H. Ulrich, Fixed Point Theory of Parametrized Equivariant Maps. VII, 147 pages. 1988.

Vol. 1344: J. Král, J. Lukeš, J. Netuka, J. Veselý (Eds.), Potential Theory – Surveys and Problems. Proceedings, 1987. VIII, 271 pages. 1988.

Vol. 1345: X. Gomez-Mont, J. Seade, A. Verjovski (Eds.), Holomorphic Dynamics. Proceedings, 1986. VII, 321 pages. 1988.

Vol. 1346: O. Ya. Viro (Ed.), Topology and Geometry – Rohlin Seminar. XI, 581 pages. 1988.

Vol. 1347: C. Preston, Iterates of Piecewise Monotone Mappings on an Interval. V, 166 pages. 1988.

Vol. 1348: F. Borceux (Ed.), Categorical Algebra and its Applications. Proceedings, 1987. VIII, 375 pages. 1988.

Vol. 1349: E. Novak, Deterministic and Stochastic Error Bounds in Numerical Analysis. V, 113 pages. 1988.